Einstein's Space-Time

An Introduction to Special and General Relativity

Einstein's Space-Time

An Introduction to Special and General Relativity

Rafael Ferraro

Universidad de Buenos Aires

Consejo Nacional de Investigaciones Científicas y Técnicas (República Argentina)

Rafael Ferraro
Universidad de Buenos Aires
Consejo Nacional de Investigaciones Científicas y Técnicas (República Argentina)
Email: ferraro@iafe.uba.ar

ISBN 978-1-4419-2419-3 e-ISBN 978-0-387-69947-9

9 8 7 6 5 4 3 2 1

springer.com

Preface

Between 1994 and 1999, I had the pleasure of lecturing Special and General Relativity in the Facultad de Ciencias Exactas y Naturales of the Universidad de Buenos Aires. These lectures were targeted to undergraduate and graduate students of Physics. However, it is increasingly apparent that interest in Relativity extends beyond these academic circles. Because of this reason, this book intends to become useful to students of related disciplines and to other readers interested in Einstein's work, who will be able to incorporate entirely the fundamental ideas of Relativity starting from the very basic concepts of Physics.

To understand the Theory of Relativity it is necessary to give up our intuitive notions of space and time, i.e., the notions used in our daily relation with the world. These *classical* notions of space and time are also the foundations of Newtonian mechanics, which dominated Physics for over two centuries until they clashed with Maxwell's electromagnetism. Classical physics assumed that space is immutable and its geometry obeys the Euclidean postulates. Furthermore, distances and time intervals are believed *invariant*, i.e., independent of the state of motion. Both preconceptions about the nature of space and time rely firmly on our daily experience, in such a way that the classical notions are imprinted in our thought with the status of "true." Therefore, we tend to resist abandoning the classical notions of space and time, and to replace them with other notions that are not evident in the phenomena observed in our daily life. This state of affairs can take the student of Relativity to a mode of thinking vitiated by the coexistence of old and new conceptions of space and time, which would lead to several perplexities and paradoxes. The historical approach of this text attempts to guide the reader along the same road followed by Physics between the seventeenth and twentieth centuries, thus reproducing the intellectual process that led to the relativistic way of conceiving the space-time. After this journey, the reader will become more receptive to the consequences of the change of paradigm.

Chapters 1–7 contain a complete course on Special Relativity that can be adapted to an introductory level by means of a suitable choice of issues. In this sense, the text is conceived to allow its reading in several levels, since some subjects that could hinder the reading by beginners have been located as *Complements* inside boxes. The historical road toward Special Relativity is covered in the first two chapters. They begin with the dispute between Leibniz and Newton about the relational or absolute character of space—an issue that paves the way for introducing General Relativity in Chapter 8—and the analysis of the role played by absolute distances and times in the foundations of Newton's Dynamics. The two mechanical theories of light—corpuscular model and wave theory of luminiferous ether—are developed in detail in Chapter 2, thus arriving at the tensions arisen between Maxwell's electromagnetic ether theory and the Principle of relativity in the second half of the nineteenth century. The challenge posed by experimental results (Arago, Hoek, Airy, Michelson-Morley, etc.) and the theoretical efforts to solve it (Fresnel, Lorentz, Poincaré, etc.) prepare the reader for the arrival of Einstein's theory. Chapter 3 introduces the postulates of Special Relativity and the notions of space and time derived from them. Here mathematics is entirely basic, because no differential calculus is needed. The understanding of length contraction, time dilatation, Lorentz transformations, relativistic addition of velocities, etc. only entails elementary concepts of kinematics. This Chapter deals with replacing classical notions of space and time with those notions that emerge from postulating that Maxwell's laws are valid in any inertial frame. Therefore, the central goal here is to enable a full understanding of the interrelation between length contraction and time dilatation and its consequences. Chapter 3 ends with the transformation of a plane wave, where its connection with the foundations of Quantum Mechanics is emphasized. Chapter 4 develops the geometric properties of Minkowski space-time (invariance of the interval, causal structure, light cone, etc.) together with some advanced topics like Wigner rotation. Chapter 5 teaches the transformations of the electromagnetic field, charge and current densities, potentials, etc. in the context of the ordinary vector language, starting from the preservation of the polar and axial characters of electric and magnetic fields, and the essential features of a plane electromagnetic wave. It also contains the fields of moving charges, dipoles, and continuous media. Chapter 6 is devoted to relativistic Dynamics, and explains the changes suffered by Newton's Dynamics in order to bring their laws into agreement with the Principle of relativity under Lorentz transformations. The reformulation of the Dynamics leads to the "mass-energy equivalence," so the rich phenomenology in atomic and nuclear physics concerned with this topic is covered. Chapter 7 introduces the four-tensor formulation of Special Relativity, and applies it to several subjects that enter into play also in General Relativity: volume and hypersurfaces, energy-momentum tensor of a fluid, electromagnetism, Fermi–Walker transport, etc. Chapter 8 displays General Relativity—the relativistic theory for the gravitational-inertial field—starting from Mach's criticism of Mechanics as a trigger of Einstein's thought. Once again, the explanation transits the historical way by focusing on the beginning of the new concepts. This chapter introduces the mathematical tools of the geometric language—covariant derivative, Riemann tensor, etc.—avoiding excesses of mathematical complexity. The Einstein equations for the geometry of space-time are formulated, and their consequences in the weak field approximation are studied to connect with Newtonian gravity. The issue of the number of degrees of freedom and constraint equations is tackled, and exemplified in the context of gravitational waves. Chapter 9 explains the main features of the Schwarzschild black hole, and applies General Relativity to Cosmology. The isotropic and homogeneous cosmological models are displayed within the context of the recent progress of the observational cosmology. The chapter ends with an updated account of the set of experimental results confirming General Relativity. Some special topics, not essential for a first reading of the book but meaningful for advanced students, are contained in the Appendix. In short, the final chapters are adequate for an introductory course on General Relativity and Cosmology.

I wish to thank the *Instituto de Astronomía y Física del Espacio* (CONICET -UBA), where most of the lectures took place, for its hospitality. IAFE is the institute where a significant number of relativistic physicists from Buenos Aires got our start under the wise guidance of Mario Castagnino. I am indebted to Gerardo Milesi, Daniel Sforza, Claudio Simeone, and Marc Thibeault, who have contributed in several ways to the realization of this book: by adding bibliographic cites, correcting proofs, or enriching the contents with their comments. A special gratitude for my wife Mónica Landau—who has also collaborated in preparing this work—and my son Damián and my daughter Sofía, for their encouragement and patience during this long period of writing.

Rafael Ferraro
Buenos Aires, January 2005

Preface to the English Edition

This English edition is basically the translation of the first Spanish edition published last year. A few changes and additions were made, and errors were corrected.

I wish to express my gratitude to Luis Landau and Claudio Simeone for their cooperation in the realization of this edition. I am especially grateful to Alicia Semino for her thorough supervision of the English version of the manuscript.

Rafael Ferraro
Buenos Aires, October 2006

Contents

List of Complements

List of Biographies

1

Space and Time Before Einstein

1.1. ABSOLUTE SPACE AND TIME

The notions of space and time that have dominated the Physics until the beginning of the twentieth century are strongly bound to the thought of Isaac Newton (1642–1727). However, in the seventeenth century two philosophical currents confronted about the nature of space and time. While Newton defended the idea of *absolute* space and time, whose existence does not depend on physical phenomena, the *relationists* thought that space and time are not things in themselves, but emerge from relations among material objects. The main exponent of this idea was Gottfried W. Leibniz (1646–1716). In his correspondence with Samuel Clarke (a friend and disciple of Newton), Leibniz stated that

relationism vs. absolutism

> . . . [space] *is that order which renders bodies capable of being situated, and by which they have a situation among themselves when they exist together, as time is that order with respect to their successive position. But if there were no creatures, space and time would only be in the ideas of God.*
>
> *. . . there is no real space out of the material universe . . .*
>
> *. . . 'tis unreasonable it should have motion any otherwise, than as its parts* [of the universe] *change the situation among themselves; because such a motion would produce no change that could be observed, and would be without design. . . There is no motion, when there is no change that can be observed.*

The Newtonian notion of absolute space was found unacceptable by Leibniz, because two universes whose bodies occupied different "absolute" positions but identical relative positions would be indiscernible.

For his part, Newton considered that the space has its own reality, independently of the bodies residing in it; so, even the empty space would be conceivable. In Newton's Physics space and time are the seat of the physical phenomena; but the phenomena produce no effects on them, as the space always remains equal to itself and the time passes uniformly. In *Philosophiae Naturalis Principia Mathematica* Newton says,

Newtonian absolute space and time

> *Absolute, true, and mathematical time, of itself, and from its own nature flows equably without regard to anything external, and by another name is*

called duration: relative, apparent, and common time, is some sensible and external (whether accurate or unequable) measure of duration by the means of motion, which is commonly used instead of true time. . .

Absolute space, in its own nature, without regard to anything external, remains always similar and immovable. Relative space is some movable dimension or measure of the absolute spaces; which our senses determine by its position to bodies; and which is vulgarly taken for immovable space. . .

Absolute motion is the translation of a body from one absolute place into another; and relative motion, the translation from one relative place into another.

I. Newton, *Principia* (London, 1687),
Definitions (Scholium)

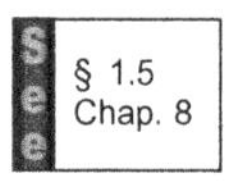
§ 1.5
Chap. 8

The controversy between relationists and absolutists quieted down in the following centuries, due to the success of the Newtonian science. Actually it remained in a latent state, because it would resurge at the end of the nineteenth century.

1.2. GEOMETRIC PROPERTIES OF THE SPACE

Euclidean geometry

Euclides' 5th postulate

The only space geometry known at the time of Newton was Euclides' geometry. Although little is known about Euclides' life—not even there is certainty that he really existed—it is believed that he taught in Alexandria around 300 BC. Euclidean geometry—the same geometry that is nowadays studied at school—is just a logical system derived from a set of definitions, postulates, and axioms. Even so, it was considered that it suitably described the metric properties of the space. Their axioms and postulates are completely intuitive and natural. Only the one known as 5th postulate—whose content is equivalent to state that one and only one parallel passes through any point outer to a given straight line—seemed to be little natural for many mathematicians, who wondered whether it would be possible to build a different geometry by replacing that postulate with some other. It is said that Karl Friedrich Gauss (1777–1855), intrigued by this question, decided to make an experiment to find out whether or not the 5th postulated was "true." One of the consequences of this postulate is that the inner angles of any triangle add two right angles. Then Gauss decided to measure the angles of the triangle formed by the tops of the mounts Brocken, Hohenhagen, and Inselberg, distant each other several tens of kilometers. The sum of the experimental determinations of the angles showed a good agreement with the result predicted by Euclidean geometry, within the margin of error of the experiment. In this way, Gauss verified that the space is approximately Euclidean, at least in regions similar to the one involved in his experiment.

Cartesian coordinates

Accepting that the geometric properties of the space are those of Euclidean geometry, we can use them to build *Cartesian coordinate systems* to locate a point with respect to a reference body. We can represent the reference body

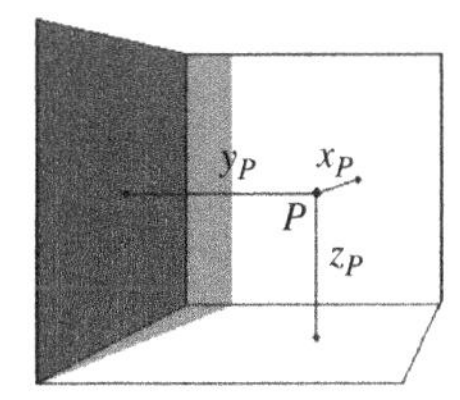

Figure 1.1. Cartesian coordinates (x, y, z) of a point P.

by means of three mutually orthogonal planes. The Cartesian coordinates of a point P are obtained by measuring the distances between P and each one of the planes. This procedure implies to draw by P the (sole) perpendicular line to each plane (Figure 1.1). The straight lines where the planes intersect are the Cartesian axes. We can calibrate these axes to measure the Cartesian coordinates directly on them, the coordinate origin being the point O where the axes intersect (Figure 1.2).

If the Cartesian coordinates of two points P and Q are known in a given *reference system*, then the distance d_{PQ} between both points can be computed by means of Pythagoras theorem (Figure 1.3):

distance

$$d_{PQ}{}^2 = (x_P - x_Q)^2 + (y_P - y_Q)^2 + (z_P - z_Q)^2 \tag{1.1}$$

If the points P and Q move with respect to the reference body, then their Cartesian coordinates change with time. In that case the calculation of the distance requires a *simultaneous* determination of the coordinates of P and Q, and the result is a function of time.

Physics before Einstein regarded distance as an *invariant* quantity (independent of the reference system). Distance was considered as a property of the pair of absolute positions occupied by the points P and Q at the time when d_{PQ} is evaluated. This assumption plays a fundamental role in the coordinate transformation between two reference systems that will be displayed in §1.4.

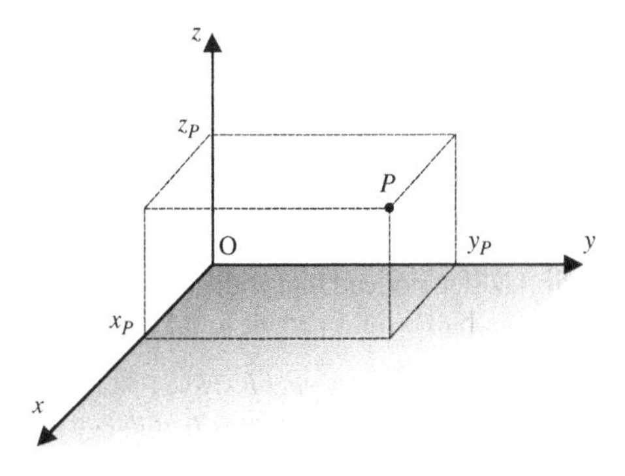

Figure 1.2. Cartesian coordinates of a point P measured on the Cartesian axes.

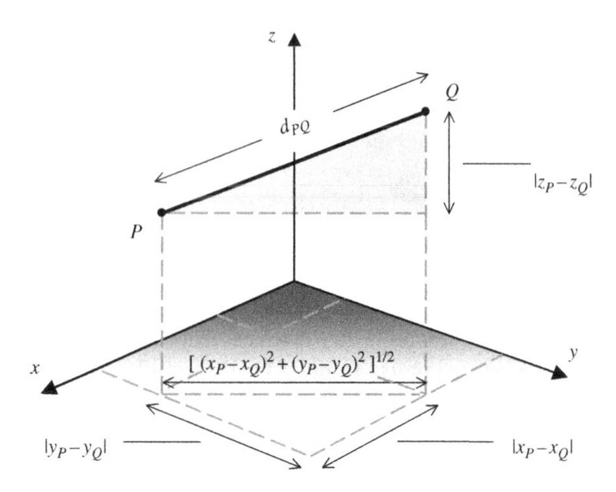

Figure 1.3. Distance d_{PQ} between two points P and Q.

1.3. GALILEO AND THE LAWS OF MOTION

During the centuries of Aristotelian tradition that preceded Galileo Galilei (1564–1642), the motion of a body was associated with the action of a force on the body. In words of Aristotle himself: *the moving body stops when the pushing force gives up acting*. Let us imagine a stone that drops from the top of the mast of a sailing boat. According to the Aristotelian thinking, the stone will not fall at the foot of the mast because, when it leaves the mast, the force needed to conserve its state of motion (communicated by the boat) will disappear. Therefore the stone will not follow the advance of the boat but fall backwards of the mast.

Galileo looks at the inertia

Galileo defied the Aristotelian thinking by asserting that the stone does conserve the state of motion it had before dropping, so it will follow the boat and fall at the foot of the mast. Although Galileo discussed this question under the form of a "thought experiment" (the real experiment was made later by Pierre Gassendi in 1641), he had examined the persistence of the motion in the laboratory by combining the procedures that characterize the scientific method: the control of the theory through the experiment and the extrapolation of the experimental results to ideal conditions where secondary effects do not take part. In this way, Galileo studied the persistence of the motion (*inertia*) by throwing small balls on tables. Although the ball stops because of the friction with the table surface and the air, Galileo noticed that if the conditions were ideal—i.e., in the absence of friction—the ball would persist in its state of motion.

> *...any velocity once imparted to a moving body will be rigidly maintained as long as the external causes of acceleration or retardation are removed, a condition which is found only on horizontal planes; for in the case of planes which slopes downwards there is already present a cause of acceleration,*

while on planes sloping upward there is retardation; from this follows that motion along a horizontal plane is perpetual; for, if the velocity be uniform, it cannot be diminished or slackened, much less destroyed.

Galileo, *Dialogues Concerning Two New Sciences* (Leiden, 1638)

Although the crude observation would indicate that the motion always stops (by cause of frictions), Galileo realized that the essential thing is the tendency of the motion to persist.

In the thought experiment of the stone and the boat, Galileo reveals that the persistence of the motion implies that the experiment has the same result either the boat is moored or sailing: in both cases the stone falls at the foot of the mast. If in a real experiment it were observed that the stone falls toward the stern when the boat sails, as the Aristotelian thinking would sustain, this would not mean that the movement does not persist but that the friction with air restrains the movement of the stone. An equal result will be obtained in a moored boat if wind blows from prow (i.e., whenever the same relative motion of air with respect to the boat occurs). In both cases this effect could be eliminated by performing the experiment in a closed cabin fixed to the boat.

the state of motion is undetectable

Hence, Galileo broke with the idea that the motion needs a force. On the other hand, he showed that it is not possible to detect the motion of the boat by means of an experiment made in its interior: the laws of Physics on the

Galileo Galilei (1564–1642). From 1581 to 1585 he took lessons in medicine, philosophy, and mathematics at the University of Pisa, close to the place where he was born. There he had a chair of mathematics in 1592, when he would have performed the experiment in the inclined tower to prove that the free fall motion does not depend on the properties of the bodies. In 1609 he constructed a telescope at the University of Padua, which allowed him to discover the craters of the Moon, the satellites of Jupiter, and the structure of the Milky Way. In 1613, working in the court of Toscana, he announced that the succession of the phases of Venus, and its relation with the apparent size of the planet, favored the Copernicus' system. Because of this, Galileo was accused of heresy by the Inquisition in 1633, forced to renounce the idea that the Earth moves, and confined to house arrest at his residence of Arcetri (Florence). His last works were clandestinely taken abroad and published in Leiden (the Netherlands), exerting great influence. Galileo countered the old naive empiricism with the "arranged" experiment of modern science, which employs *ad hoc* equipment to interrogate Nature about the validity of a mathematically formulated physical theory. His research on inertia and the laws of motion, by means of inclined planes and pendulums, prepared the way to the work of Newton. Galileo was not the first to attain this knowledge: around 575 I. Philoponus had correctly criticized Aristotle's thought about the fall of the bodies, the law of uniformly accelerated movement was well known from 14th century, and J. of Groot and S. Stevin had already published their own experimental verification of the fall law in 1586. However, these facts in no way mar the figure of Galileo as the father of the modern science.

boat sailing do not differ from the laws on the moored boat. We can verify this conclusion daily in any means of transport: the velocity of a train is not detectable in its interior (by watching through the window we just detect that the vehicle has a motion *relative* to the earth). Instead, we do observe the *variations* of the velocity, either in direction or magnitude: when the train takes a curve or jumps at the union of two stretches of railroad, when brakes or accelerates.

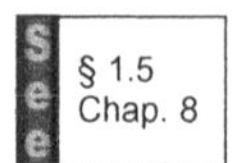

1.4. CHANGE OF COORDINATES BETWEEN FRAMES IN RELATIVE MOTION

The possibility of using the laws of Physics in different reference systems—like the earth frame or the boat frame in the previous section—leads us to find out how the Cartesian coordinates of a moving body transform when the reference system is changed. Let us consider two different reference systems S and S' whose relative movement is a *translation* (the relative orientation of the Cartesian axes does not change as time goes) with a constant velocity $\mathbf{V}$. For simplicity, the frames S and S' will be chosen in such a way that their axes are coincident at $t = 0$, and their x–x' axes have the direction of the relative movement (see Figure 1.4). At an arbitrary time t the distance between both coordinate origins, O and O′, is $d_{OO'} = V\,t$.

The transformation of the Cartesian coordinate x of a point P can be obtained from Figure 1.5, where it can be seen that

$$d_{OP} = d_{OO'} + d_{O'P} \tag{1.2}$$

together with the relation between distances and Cartesian coordinates: $d_{OP} \equiv x$, $d_{O'P} \equiv x'$. Then

$$x = V\,t + x' \tag{1.3}$$

Galileo transformation assumes the invariance of the distances

Although the coordinate transformation (1.3) seems to us completely evident and natural, it should be emphasized that result (1.3) is closely linked to the invariant character ascribed to distances in classical physics. In principle,

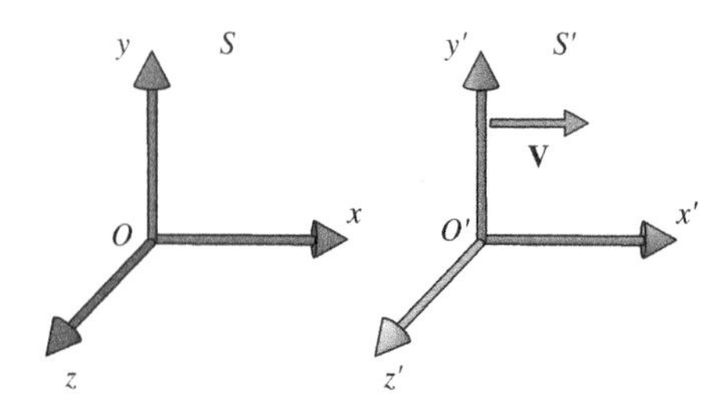

Figure 1.4. Reference systems S and S′ in relative motion.

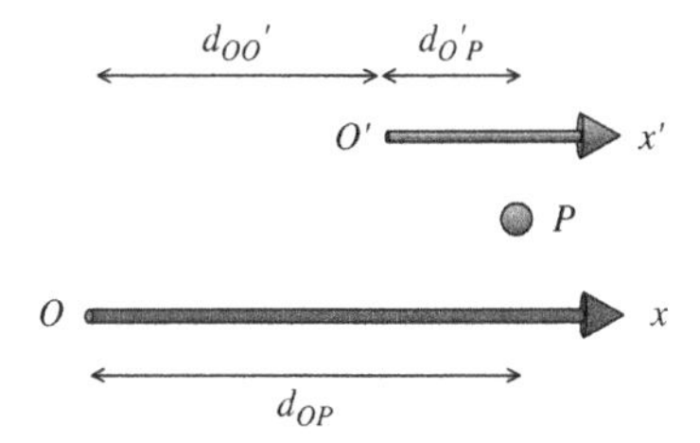

Figure 1.5. The distance d_{OP} is the sum of $d_{OO'}$ and $d_{O'P}$.

Eq. (1.2) has only a meaning when all the distances are measured in the same frame, either S or S'. But coordinate x is the distance d_{OP} *measured in* S, whereas coordinate x' is the distance $d_{O'P}$ *measured in* S'. Therefore, if Eq. (1.2) is read in the frame S then d_{OP} can be replaced by x, but $d_{O'P}$ should not be substituted with x'. However, owing to the invariant character ascribed to distances (the distances have the same value in any reference system), it becomes correct to replace the value of $d_{O'P}$ measured in S by the coordinate x' of the point P.

The coordinate transformation (1.3) would not be consistent unless an additional assumption were accepted: the invariance of time. In fact, if the time needed for the displacement $d_{OO'}$ were t' when measured in S', then Eq. (1.2) regarded from S' would lead to $x = V\,t' + x'$. This result is compatible with Eq. (1.3) only if $t' = t$. Therefore both assumptions about the nature of space and time—i.e., that distances and times are invariant—are strongly intertwined and mutually necessary.

invariant distances require invariant times

Although the classical conjectures about the nature of distances and times seem to be unquestionably confirmed by daily experience, it should be remarked that daily experience occurs in a narrow range of relative velocities. So, it would be more prudent to state that those assumptions are suitable for that range of velocities.

The coordinates y, z are the distances between point P and two coordinate planes mutually orthogonal that intersect on the $x$axis (see Figures 1.1 and 1.2). The frames S and S' in Figure 1.4 are equally oriented, and their x–x' axes are superimposed. Therefore those coordinate planes are shared by S and S'. This property together with the invariance of the distances imply that the coordinates y, z of a point are the same in both reference systems. In sum, the Cartesian coordinates of a point at a given instant t transform as

Galileo transformations

$$\begin{aligned} x' &= x - V\,t \\ y' &= y \\ z' &= z \end{aligned} \tag{1.4}$$

which are called *Galileo transformations*, in honor to whom began the modern study of the laws of motion.[1] These three equations can be joined in one vector equation for the transformation of the *position vector* of the point P, $\mathbf{r} = (x, y, z)$:

$$\mathbf{r}' = \mathbf{r} - \mathbf{V}\, t \tag{1.5}$$

Due to its vector character, Eq. (1.5) is valid even when the direction of $\mathbf{V}$ does not coincide with that of the x–x' axes. In fact, vector relations are not affected by rotations of Cartesian axes. The transformations (1.4) are completed with the invariance of time:

$$t' = t \tag{1.6}$$

By differentiating Eq. (1.5) with respect to t $(= t')$, and taking into account that $\mathbf{V}$ does not depend on t, the transformation of the velocity $\mathbf{u} = \mathrm{d}\mathbf{r}/\mathrm{d}t$ is obtained, which is known as *Galileo addition theorem of velocities*:

addition theorem of velocities

$$\mathbf{u}' = \mathbf{u} - \mathbf{V} \tag{1.7}$$

See Chap. 3 § 4.1

Clearly, the value of the distance (1.1) between simultaneous positions of two points P and Q is not modified by Galileo transformations. This invariance cannot surprise us, since it has already been assumed in the building of the coordinate transformation.

1.5. PRINCIPLE OF INERTIA

As referred in §1.3, Galileo considered that the essential feature of the movement is the tendency to persist. This concept was raised by Newton to the rank of First Law of Dynamics, also known as *Principle of inertia*:

> *Every body perseveres in its state of rest, or of uniform motion in a right line, unless it is compelled to change that state by forces impressed thereon.*
>
> I. Newton, *Principia* (London, 1687), Axioms

We could hardly sustain that this principle is a strict experimental result. On the one hand it is not evident how to recognize whether a body is free of forces or not. Even if a unique body in the universe were thought, it is undoubted that its movement could not be rectilinear and uniform in every reference system. But on the other hand, if a body has a rectilinear and uniform movement in a given reference system (i.e., its velocity $\mathbf{u}$ is constant), then its motion will also be rectilinear and uniform in any reference system translating with constant velocity

[1] The name of the transformations was given by P. Frank in 1909.

V relative to the former frame; this is a consequence of the addition theorem of velocities (1.7). Therefore if there exists a reference system for which the Principle of inertia is valid, then it will exist a family of reference systems where the Principle of inertia is satisfied. Which is this family? Why such a privilege is conferred to this family, so selecting it from the rest of the reference systems?

The absolute space is a necessity in Newton's theoretical scheme, otherwise these questions would not find an answer. According to Newton, the Principle of inertia is valid in a reference system that is at absolute rest (at rest in absolute space), and also in any other which translates with constant velocity relative to the first one. Such reference systems are called *inertial frames*; their privilege is conferred by the way they move in the absolute space. Therefore Newton's absolute space determines the inertial trajectories of the bodies without receiving any consequence, since it remains itself immutable.

absolute space selects the family of inertial frames

Newton refers to the inertial frames, and the determinations of positions and velocities relative to such frames, when he speaks about *relative spaces* and *sensible measures* (see the citation at the end of §1.1):

> *...because the parts of space cannot be seen, or distinguished from one another by our senses, therefore in their stead we use sensible measures of them...*
>
> *And so, instead of absolute places and motions, we use relative ones; and that without any inconvenience in common affairs; but in philosophical disquisitions, we ought to abstract from our senses, and consider things themselves, distinct from what are only sensible measures of them.*
>
> I. Newton, *Principia* (London, 1687), Definitions (Scholium)

In practice, an inertial frame is recognized not because of its state of motion relative to the absolute space but for the degree of verification of the Principle of inertia. Actually this attitude leads us to a vicious circle, since we need a body free of forces to carry out such a verification. But, what does it mean and how could it be guaranteed that a body is free of forces? Then, we have to content ourselves by adopting as inertial frames those reference systems where the Principle of inertia and the rest of the fundamental laws of Physics are fulfilled to a "satisfactory" degree.

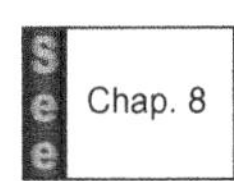
See Chap. 8

1.6. PRINCIPLE OF RELATIVITY

As mentioned in §1.3, the persistence of the motion (or inertia) discovered by Galileo leads to the impossibility of discerning the state of motion of the boat in the thought experiment of the boat and the stone. In fact, the result of the experiment is the same whether the boat is sailing or moored. This conclusion is formalized in the *Principle of relativity*:

> *The same fundamental laws of Physics are fulfilled in all inertial frames.*

Principle of relativity

While the absolute space selects a set of privileged reference systems—the inertial frames—the Principle of relativity tells us that all the inertial frames are equivalent. As a consequence, an experimenter cannot detect the (absolute) state of motion of his inertial laboratory by means of experiments made therein, since the laws of Physics in his laboratory do not differ from the laws fulfilled in any other laboratory that translates rectilinearly and uniformly in the absolute space.

The equivalence of the inertial frames requires that the laws of Physics behave properly under coordinate transformations. In our first approach to the Principle of relativity, we combined the Principle of inertia with the Galileo addition theorem of velocities in order to state that if the Principle of inertia is valid in a given reference system, then it will be valid in any other reference system translating with constant velocity relative to the first one. Here it is opportune to emphasize that we would have reached an identical conclusion if the transformation of spatial and temporal coordinates had been any other linear transformation. In fact, in a rectilinear uniform movement the Cartesian coordinates are linear functions of the time; this kind of dependence is preserved by any linear transformation.

Instead, the other fundamental laws of Newtonian physics are in a more close relation with Galileo transformations. In Newtonian mechanics the behavior of a physical system is described by two complementary types of law: on the one hand, the Second Law of Dynamics states that if a force **F** acts on a particle of mass m, then the particle will acquire an acceleration **a** such that

2nd Law of Dyanmics

$$\mathbf{F} = m\,\mathbf{a} \tag{1.8}$$

On the other hand, there are laws to describe the interactions between particles, that tell us which is the value of the force **F** in Eq. (1.8). This kind of laws includes, for instance, Newton's law of universal gravitational attraction. In order to verify that these laws satisfy the Principle of relativity under Galileo transformations we must take into account that the acceleration $\mathbf{a} = \mathrm{d}\mathbf{u}/\mathrm{d}t$ is a Galilean invariant, as it results from differentiating with respect to t $(= t')$ in the addition theorem of velocities (1.7) (remind that velocity **V** in (1.7) is constant):

$$\mathbf{a}' = \mathbf{a} \tag{1.9}$$

Galilean relativity requires forces depending on distances

Therefore, in order that the Second Law of Dynamics (1.8) be valid in any inertial frame, the force must also be invariant (the mass is assumed invariant). This will be the case if the laws for the interactions assert that the forces depend only on the distances between interacting particles (not on the velocities, etc.)—as it happens with the gravitational interaction. Thus the invariance of the distances will imply the invariance of the forces, and the Galilean Principle of relativity will be so fulfilled.

1.7. PHYSICAL PHENOMENA HAVING A PRIVILEGED REFERENCE SYSTEM

forces depending on the velocity

The conclusions of the previous section do not exclude forces depending on velocities from Newtonian physics. When a body moves within a viscous fluid, the friction force exerted by the fluid on the body increases with the speed of the body relative to the fluid. In particular, if the body is at rest relative to the fluid, then the friction force will be null. Although the velocity is not a Galilean invariant (see Eq. (1.7)), the velocity playing a role in this example is not the velocity relative to an arbitrary inertial frame, but the velocity relative to the fluid. Thus, the fluid becomes a privileged reference system. It is clear that this privilege cannot be taken as a violation of the Principle of relativity because the type of phenomenon under consideration *naturally* privileges the fluid frame. In other words, although the speed of a point P is not a Galilean invariant, what is coming into play here is the relative speed between two bodies, which is in fact a magnitude independent of the reference system:

$$\mathbf{u}_P{}' - \mathbf{u}_Q{}' = \mathbf{u}_P - \mathbf{u}_Q \tag{1.10}$$

Isaac Newton (1642–1727). He was born an orphan in Woolsthorpe (Lincolnshire). His mother remarried when Isaac was 3 years old, leaving the boy under the tutelage of his grandmother until she became a widow again seven years later. By suggestion of his uncle, Isaac completed the studies that allowed him to enter the Trinity College of Cambridge in 1661, where he studied philosophy and mathematics. The plague of 1665 forced him to return to the family farm. There he began the research in mathematics, mechanics, astronomy, and optics that revolutionized human knowledge. Graduating in 1668, he obtained the Lucasian chair of mathematics of the University of Cambridge in 1669. Newton developed differential and integral calculus, which he called *method of fluxions*. In *Principia* (1687)—recognized as the most important scientific work ever written—he enunciated the laws of mechanics and gravitation, and used them to predict the orbits of planets and comets and explain the motion of the tides. Through mathematics and axioms, Newton gave mechanics a rational and deductive character, making possible precise experimental verification of its laws. In *Opticks* (1704) he promoted the corpuscular model of light, although he also resorted to the wave concept, expressing his puzzlement about the nature of the light. He discovered that white light decomposes into colors of the rainbow. He constructed a reflector telescope to avoid the chromatic aberration of the lenses. Susceptible to the criticisms and reticent to publish his works, Newton was a passionate alchemist and mystic. After undergoing his second nervous breakdown, he retired from research in 1693. Then he became Master of the Mint and member of the Parliament. From 1703 to his death, he presided over the Royal Society, ordering a report to a committee of experts to settle his controversy with Leibniz about the paternity of differential and integral calculus; Newton himself took care of writing of the apocryphal report. He did not marry nor have children. In spite of his differences with the Church, he was buried in the Westminster Abbey.

Complement 1A: *Mechanical waves in a perfect fluid*

In the absence of an external field, the only force on an element of a perfect fluid comes from the pressure gradient. Then Newton's Second Law acquires the form

$$-\nabla p = \rho \frac{d\mathbf{u}}{dt} \tag{1A.1}$$

where ρ is the mass density, p is the pressure, and $\mathbf{u}$ is the velocity in each point of the fluid at each time. This law must be joined to the continuity equation which expresses the mass conservation,

$$\frac{\partial \rho}{\partial t} + \nabla \cdot (\rho\, \mathbf{u}) = 0 \tag{1A.2}$$

and the state equation which relates pressure and density:

$$p = p(\rho) \tag{1A.3}$$

In order to obtain the wave equation the following requirements must be fulfilled:

1. The density and the pressure are close to their equilibrium values ρ_0 and $p_0 = p(\rho_0)$; then a perturbation $\psi(\mathbf{r},t)$ can be introduced such that

$$\rho = \rho_o(1+\psi) \quad |\psi| \ll 1 \tag{1A.4}$$

$$p = p_o + \left.\frac{\partial p}{\partial \rho}\right|_{\rho_o} (\rho - \rho_o) + \ldots \cong p_o + {c_S}^2 \rho_o \psi \tag{1A.5}$$

where ${c_s}^2 \equiv (\partial p/\partial \rho)|\rho_0$ is a property of the fluid having units of squared velocity.

2. The velocities are small (this supposes to adopt the reference system fixed to the medium at equilibrium), and the velocity gradients are also small. The acceleration of a fluid element is approximated by

$$\frac{d\mathbf{u}}{dt} = \frac{\partial \mathbf{u}}{\partial t} + (\mathbf{u} \cdot \nabla)\, \mathbf{u} \cong \frac{\partial \mathbf{u}}{\partial t} \tag{1A.6}$$

These approximations can be replaced in (1A.1) and (1A.2) to obtain

$${c_s}^2 \nabla \psi \cong -\frac{\partial \mathbf{u}}{\partial t}, \qquad \frac{\partial \psi}{\partial t} \cong -\nabla \cdot \mathbf{u} \tag{1A.7a–b}$$

Taking the divergence in (1A.7a) and differentiating with respect to t in (1A.7b), the right sides become equal. Thus the wave equation is obtained

$$\frac{1}{{c_s}^2} \frac{\partial^2}{\partial t^2} \psi(\mathbf{r}, t) - \nabla^2 \psi(\mathbf{r}, t) \cong 0 \tag{1A.8}$$

where $\nabla^2 \equiv \partial^2/\partial x^2 + \partial^2/\partial y^2 + \partial^2/\partial z^2$ is the Laplacian. Solving the wave equation it results that the perturbation—a compression and rarefaction wave—propagates in the medium with speed c_s.

Another example of physical phenomenon having a privileged reference system is the propagation of mechanical waves. Once again, this fact must not be regarded as a violation of the Principle of relativity because the phenomenon requires a *material medium* for its realization. The sound waves, for instance, propagate in air, water, or a solid, but they cannot propagate in vacuum. In fact, a mechanical wave is nothing but the disturbance of the material medium where the wave propagates. Therefore, the reference system fixed at the propagation medium is *naturally* privileged. The usual form of the *wave equation* describing the propagation of the disturbance is only valid in the frame fixed to the medium. In particular, the velocity of propagation of the wave is written in the wave equation, appearing as a quantity determined by the properties of the medium; this feature clearly shows that the wave equation cannot be valid in any inertial frame because the velocity is not a Galilean invariant.[2]

mechanical waves

The existence of this naturally privileged frame does not violate the Principle of relativity because the wave equation could be used in *any* inertial frame where the material medium were at rest. Actually, the equation for mechanical waves is nothing but the result of performing approximations that are valid in the frame fixed to the medium, starting from the fundamental laws of Mechanics which positively satisfy the Galilean Principle of relativity (see an example in Complement 1A).

1.8. MAXWELL'S ELECTROMAGNETISM

The laws governing the electromagnetic field were conceived by James Clerk Maxwell (1831–1879), after the discovering of the electromagnetic induction made by M. Faraday in 1831. In the electromagnetic theory the magnetic forces depend on the velocities of the interacting charges. Besides, Maxwell's laws (1873) can be rewritten as wave equations in those cases where there are not sources (see Complement 1.B). Since the theory predicted a speed of propagation for the electromagnetic waves that matched the measured value of the speed of light, Maxwell was led to conclude that light is an electromagnetic phenomenon. In accordance with §1.7, the presence of wave equations and forces depending on velocities in Maxwell's theory means that the sole interpretation consistent with the notions of space and time that prevailed in the nineteenth century is that electromagnetism is a mechanical phenomenon which happens in a material medium. If this interpretation is correct then Maxwell's laws will be valid only in the inertial frame fixed to that medium. The electromagnetic

[2] The wave equation is a partial differential equation (see Complement 1A). Its change of form under Galileo transformations can be verified by replacing:

$$\frac{\partial}{\partial t}=\frac{\partial t'}{\partial t}\frac{\partial}{\partial t'}+\frac{\partial x'}{\partial t}\frac{\partial}{\partial x'}=\frac{\partial}{\partial t'}-V\frac{\partial}{\partial x'},\quad \frac{\partial}{\partial x}=\frac{\partial t'}{\partial x}\frac{\partial}{\partial t'}+\frac{\partial x'}{\partial x}\frac{\partial}{\partial x'}=\frac{\partial}{\partial x'},\quad \frac{\partial}{\partial y}=\frac{\partial}{\partial y'},\quad \frac{\partial}{\partial z}=\frac{\partial}{\partial z'}$$

material medium was identified with the *ether* of the wave model of light, in view of the electromagnetic character of light phenomena. We shall dedicate Chapter 2 to the efforts spent in detecting the ether or, at least, the state of motion of a laboratory relative to the ether. Instead, in this section we shall show that the use of Maxwell's laws in different reference systems, joined to the invariance of distances and times accepted in classical physics, lead to absurdities.

Let us consider two parallel infinite line charge distributions, as shown in Figure 1.6. The upper line is electrically neutral because it contains both types of charges in equal quantities; the lower line is charged. Besides, both lines carry electric currents with equal directions that result from the motion of one type of charge. In this stationary charge configuration there is no electric interaction between the lines, since the electric interaction requires that both lines be charged. However, the lines do magnetically interact because electric currents of equal direction magnetically attract each other. This is what electrostatics and magnetostatics say (i.e., Maxwell's theory restricted to the case where fields do not depend on time). Now we shall try a change of reference system, and we shall pretend to apply Maxwell's laws also in the new frame. In changing frames we shall use the assumptions about the invariance of distances and times which support Galileo transformations. In particular, if distances are invariant then we can assure that the upper line will be electrically neutral in any reference system.

incompatibility of Maxwell's laws, the Principle of relativity, and Galileo transformations

In the frame that moves with the positive charges, we see the charge configuration shown in Figure 1.7. In this case, the same Maxwell's laws say that there is no interaction at all. There is no electric interaction, since it would be necessary that both lines had net electric charge. There is no magnetic charge, since it would be required that both lines carried electric current. The results obtained in each frame are incompatible, since the existence or not of an interaction must be an absolute fact, independent of the chosen reference system. The absurdity is a consequence of the application of Maxwell's laws in both frames, together with Galileo transformations. As long as the notions of space and time of classical physics remain valid, the use of Maxwell's laws in different reference systems will be unacceptable. In such a case, it is imperative to know which is the frame where Maxwell's laws are applicable. In the next chapter

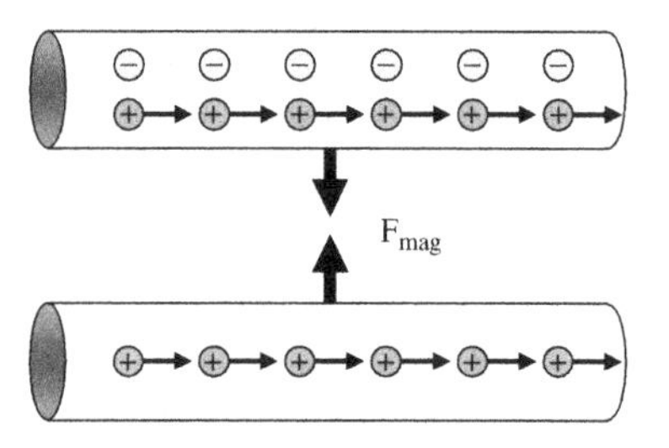

Figure 1.6. Magnetic force between electric currents of equal direction.

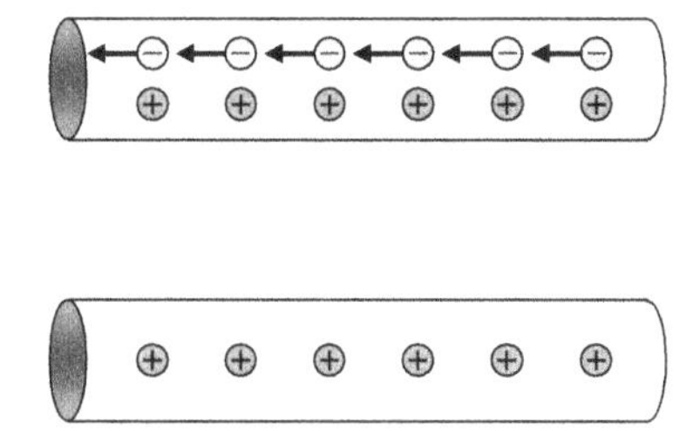

Figure 1.7. There is neither electric interaction (one of the lines is neutral), nor magnetic interaction (one of the lines does not carry electric current).

we will follow the history of the searching for this privileged reference system: the ether frame. The failure of this enterprise will lead to understand that the classical notions of space and time will have to be modified.

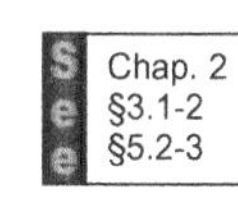
See Chap. 2 §3.1-2 §5.2-3

James Clerk Maxwell (1831–1879). Born in Edinburgh, James Clerk Maxwell grew up in a southwest Scottish countryside, where he was educated by his parents. At the age of 10, he returned to his home town to enter the Academy of Edinburgh. At the age of 14, he presented a work on ovals in the Royal Society of Edinburgh. In 1854, after a 4-year degree course, he graduated in mathematics at Trinity College. He remained in Cambridge until 1856, when he presented his work *On Faraday's Lines of Force*, explaining the interdependence between electric and magnetic fields. He won the Adams Prize for demonstrating that the stability of the rings of Saturn is possible only if they are formed by small solid particles. In 1860, he obtained the chair of Natural Philosophy in the King's College of London. There he deduced the speed of propagation of electromagnetic waves; the value closely approximated the speed of the light, which induced Maxwell to propose that "light consists in the transverse undulations of the same medium which is the cause of electric and magnetic phenomena" (1862). The definitive form of Maxwell's electromagnetic theory came out in 1873, developed after he left London to return to Scotland. In his last years, Maxwell supervised the edition of Cavendish's works and the building of the Cavendish Laboratory.

Thanks to Maxwell, the idea of a *field* as a magnitude having its own identity appeared in Physics. Maxwell realized that the electric current that is induced in a circuit in the presence of a variable magnetic field—which was observed by Faraday in 1831—is produced by an electric field that exists even in the absence of the circuit. A variable magnetic field is the source of an electric field, and a variable electric field is the source of a magnetic field. In addition, he made remarkable contributions to the kinetic theory of gases (1866); using the hypothesis that the behavior of a gas is the result of the movement of its constituent molecules, he deduced—independently of L. Boltzmann—the statistical distribution of velocities of the molecules as a function of the thermodynamic notion of temperature.

Complement 1B: *Maxwell's equations*

Maxwell's equations are first-order differential equations describing the local behavior of the electric and magnetic fields $\mathbf{E}(\mathbf{r}, t)$ and $\mathbf{B}(\mathbf{r}, t)$. The sources of the fields are the electric charge density $\rho(\mathbf{r}, t)$ and the electric current density $\mathbf{j}(\mathbf{r}, t)$. The values of the sources ρ and $\mathbf{j}$, together with proper boundary conditions for $\mathbf{E}$ and $\mathbf{B}$, determine the configuration of the fields:

$$\nabla \cdot \mathbf{B} = 0 \qquad \nabla \times \mathbf{E} + \frac{\partial \mathbf{B}}{\partial t} = 0 \tag{1B. 1a–b}$$

$$\nabla \cdot \mathbf{E} = \frac{\rho}{\varepsilon_0} \qquad \nabla \times \mathbf{B} - \mu_0\, \varepsilon_0\, \frac{\partial \mathbf{E}}{\partial t} = \mu_0\, \mathbf{j} \tag{1B. 2a–b}$$

Taking divergence in Eq. (1B.2-b) and using (1B.2-a) it results the continuity equation which expresses the local conservation of the electric charge:

$$\frac{\partial \rho}{\partial t} + \nabla \cdot \mathbf{j} = 0 \tag{1B.3}$$

The force exerted by the electromagnetic field on a charge q moving with velocity **u** is called *Lorentz force*:

$$\mathbf{F} = q\,(\mathbf{E} + \mathbf{u} \times \mathbf{B}) \tag{1B.4}$$

In the places where the sources are null the Cartesian components of the fields satisfy wave equations like (1A.8). These can be obtained from (1B.1-b) and (1B.2-b) differentiating with respect to t and using the identity $\nabla \times (\nabla \times \mathbf{C}) \equiv \nabla\,(\nabla \cdot \mathbf{C}) - \nabla^2\, \mathbf{C}$:

$$\mu_0\, \varepsilon_0\, \frac{\partial^2 \mathbf{E}}{\partial t^2} - \nabla^2 \mathbf{E} = 0 \qquad \mu_0\, \varepsilon_0\, \frac{\partial^2 \mathbf{B}}{\partial t^2} - \nabla^2 \mathbf{B} = 0 \tag{1B.5}$$

In (1B.5) $c \equiv (\mu_0 \varepsilon_0)^{-1/2}$ is a constant with dimensions of velocity (it is the speed of propagation of electromagnetic waves) whose value should be experimentally determined. As can be seen in (1B.2), μ_0 and ε_0 separately involve units of current and charge respectively, which are not independent units as it is evident in (1B.3). Thus the adoption of a unit of current (or, alternatively, of charge) is equivalent to choose a value for μ_0(or ε_0). The International System (SI) unit of electric current—the Ampère (A)–is defined by choosing $\mu_0 = 4\pi \times 10^{-7}\,\mathrm{NA}^{-2}$. The value of ε_0 is then established by means of an electrostatic experiment, resulting $\varepsilon_0 = 8.854187817 \times 10^{-12}\,\mathrm{N}^{-1}\,\mathrm{A}^2\,\mathrm{m}^{-2}\,\mathrm{s}^2$. Therefore $c = 2.997924580 \times 10^8$ m/s. Since 1983 this value of c was adopted by definition, which amounts to stop considering the units of length and time as independent.

Note: In Gaussian System the magnetic field is redefined by means of the substitution $\mathbf{B} \rightarrow \mathbf{B}/c$, which gives equal units to both electric and magnetic fields. The unit of charge—the stat-coulomb (u.e.e.)—results from the choice $\varepsilon_0 = (4\pi)^{-1}$.

2

In Search of the Ether

2.1. TWO MODELS FOR THE LIGHT

In 1704, Isaac Newton published *Opticks*, where he affirmed that light is something that propagates like particles ejected from the light source. The statement was based on the apparent rectilinear propagation of the light rays. Within the context of this *corpuscular model* of the light, Newton tried to explain the reflection and refraction (the change of the ray direction when light goes through the surface separating two transparent media), as a consequence of the action of repulsive and attracting short-range forces acting on the corpuscles in the vicinity of the surface. In the case of refraction, the ray bends toward the *normal* direction (the direction perpendicular to the surface separating both media) when the light corpuscles penetrate an optically "denser" medium, as happens with water or glass with respect to the air. Newton ascribed this phenomenon to the existence of a force attracting the light particles toward the denser medium, so that the normal component of the corpuscles velocity increases when they penetrate that medium. In refraction the light decomposes in colors with different degrees of deviation (*dispersion*), what forced to assume that there exist light particles of different types (colors), the attraction toward the denser medium being different for each type of particle.

corpuscular model

Nevertheless, the corpuscular model was not able to tackle all the known optical phenomena. While Newton experimented with the decomposition of light, he was aware of F.M. Grimaldi's discovery that light does not produce sharp shadows—as it should happen if light traveled in straight lines—but the border of the shadows displays a set of fringes alternately bright and dark (*diffraction*, 1665). From this period also comes the discovery of the birefringence in calcite (E. Bartholin, 1669). Newton himself observed the interference phenomenon of fringes appearing when a plane-convex lens is placed over a glass plate (*Newton's rings*), which he called *inflection*. Actually Newton frequently alternated concepts from the wave model with others from the corpuscular model in order to interpret the optical phenomena.

In 1678 the Dutch physicist Christiaan Huygens (1629–1695) presented his *Traité de la lumière* at the Academie Royale des Sciences of Paris. There Huygens defended the *wave model* for the propagation of light with these words:

wave model

> *...when one considers the extreme speed with which light spreads on every side, and how, when it comes from different regions, even from those directly opposite, the rays traverse one another without hindrance, one may well understand that when we see a luminous object, it cannot be by any transport of matter coming to us from this object, in the way in which a shot or an arrow traverses the air; for assuredly that would too greatly impugn these two properties of light, especially the second of them. It is then in some other way that light spreads; and that which can lead us to comprehend it is the knowledge which we have of the spreading of Sound in the air.*
>
> *...there is no doubt at all that light also comes from the luminous body to our eyes by some movement impressed on the matter which is between the two; ... this movement, impressed on the intervening matter, is successive; and consequently it spreads, as Sound does, by spherical surfaces and waves.*
>
> Chr. Huygens, *Traité de la lumière* (Leiden, 1690)

Huygens' Principle

According to Huygens, these surfaces or *wave fronts* propagate in such a way that each point of a wave front behaves as a source of a *secondary* wave front; the successive wave fronts result to be the envelope of the secondary wave fronts (Huygens' Principle). The *ray* is the direction from the secondary source to the point where the secondary wave is tangent to the envelope. Figure 2.1 shows Huygens' construction for a wave emitted by a point-like source, propagating in an isotropic and homogeneous medium. The isotropy and homogeneity of the medium imply that the wave propagates with the same velocity in any direction at any point of the medium. From this it results that both the wave front and the secondary wave are spherical. Figure 2.2a shows that a plane wave front bends in a non-homogeneous (but isotropic) medium; the secondary waves are still spherical, but they do not have the same radius because the speed of propagation is different at each point. The rays go from the center of each secondary wave to the point touching the envelope; since the secondary wave is spherical the rays result to be perpendicular to the wave front (in the frame fixed to the medium). This property is lost when the medium is not isotropic. Figure 2.2b shows a plane wave propagating in homogeneous but anisotropic medium. In this case

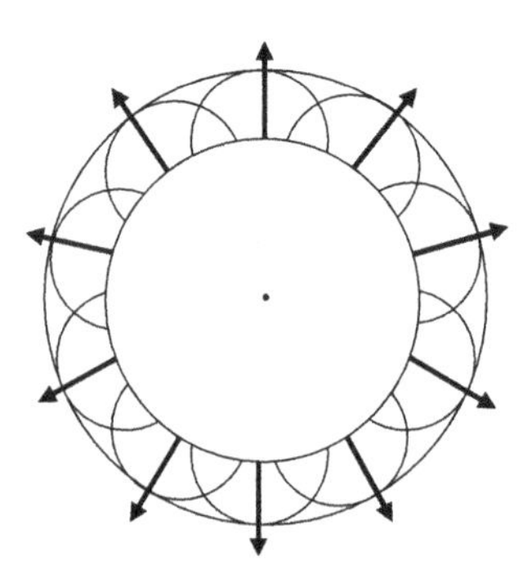

Figure 2.1. Huygens' construction for a wave emitted by a point-like source, propagating in an isotropic and homogeneous medium.

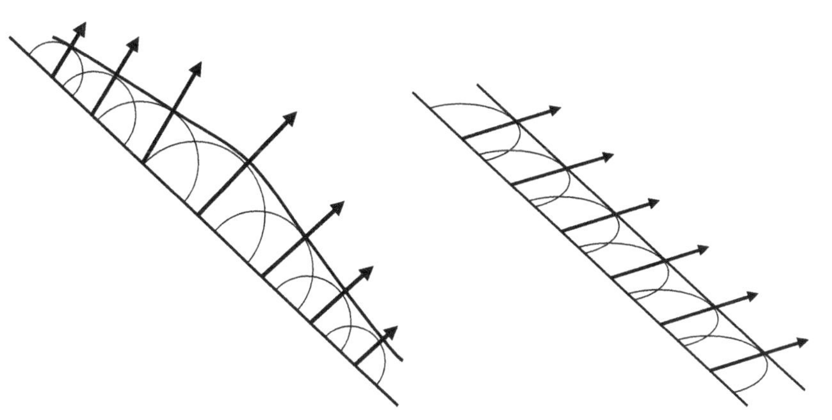

Figure 2.2a. Huygens' construction in a non-homogeneous but isotropic medium.

Figure 2.2b. Huygens' construction in an anisotropic but homogeneous medium.

the secondary waves are not spherical, but they are all equal as a consequence of the homogeneity. As a result, the wave front remains plane, but the rays are not perpendicular to the wave front.

Let us now consider how the wave model explains reflection and refraction. In Figure 2.3 a plane wave front AB strikes the surface AD separating two transparent isotropic and homogeneous media. The angle of incidence i is the angle between the rays and the direction normal to the interface AD. In order to apply Huygens' construction to this case, we need the additional hypothesis that points on the surface AD become secondary sources as long as they are reached by the wave front. Except for normal incidence, the points belonging to the same front do not reach the interface AD simultaneously: the first point to become a secondary source is point A. Point D is struck by the incident front after the time $t = d_{BD}/c_1$ has elapsed (c_1 is the speed of propagation in

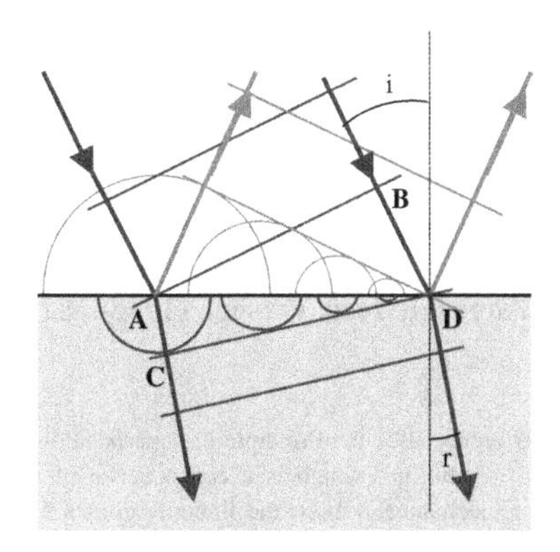

Figure 2.3. Description of the reflection and refraction of light, according to the wave model.

the first medium). In the meantime the secondary wave emitted in A traveled radii $c_1 t$ in medium 1 and $c_2 t$ in medium 2 (Figure 2.3 assumes $c_2 < c_1$). The secondary waves emitted from the intermediate points of segment AD developed radii proportional to the distance between the secondary source and the point D. The envelopes of secondary waves are again plane fronts. From the construction it results that the angle between the reflected rays and the normal is equal to the angle of incidence i (law of reflection). Instead, the wave transmitted to medium 2 will bend toward the normal if $c_2 < c_1$; this means that the wave model requires that the speed of propagation be smaller in the optically denser medium. The dispersion comes from the different speeds of propagation of different colors. Since the angle BAD is equal to i, and the angle ADC is equal to r, it results

Snell's law

$$\frac{\sin i}{\sin r} = \frac{d_{BD}}{d_{AC}} = \frac{c_1\, t}{c_2\, t} = \frac{c_1}{c_2} \equiv n_{21} \tag{2.1}$$

where n_{21} is the refractive index of medium 2 relative to medium 1. This is how wave model explains Snell's law (1618) for refraction.

In refraction we find a first discrepancy between both models. As above mentioned, the corpuscular model asserted that the ray bending toward the normal happens because a force attracts the corpuscles toward the denser medium. If this were the case, the speed of propagation would increase when the corpuscles enter the denser medium, contrarily to the wave model implications.

Until the beginning of the nineteenth century the corpuscular model enjoyed a better acceptation than the wave model, due in part to the influence of Newton's thought. Although Newton was aware of the difficulty of explaining interference fringes within the corpuscular model framework, the wave model was relegated because of its apparent inability to explain the well-defined shadows projected by the objects (except for the weak effect of diffraction observed by Grimaldi). Furthermore, Huygens' construction by itself is insufficient to describe the interference, since it does not incorporate the concepts of amplitude and phase. In order that these concepts can emerge it is necessary to have the wave equation and its solutions. The wave equation was written for the first time in 1746, when J. d'Alembert studied the propagation of waves in a string. In 1802 the English physicist and physician Thomas Young (1773–1829) correctly explained the interference of light within the wave model framework.[1] In 1808, E. Malus discovered the effects of the polarization by reflection while he observed the reflection in a window through a crystal of calcite. Young suggested that Malus' observation could be understood if light were a transversal wave.[2] The French physicist Augustin Jean Fresnel (1788–1827) took into account

the wave model consolidates in the 19th century

[1] The explanation of the interference uses the Principle of superposition (the sum of two solutions of the wave equation is also a solution), which is a consequence of the linear character of the wave equation. In the case of mechanical waves, the linearity comes from an approximation (see Complement 1.A). Instead, Maxwell's equations are directly linear.

[2] The transversal nature of electromagnetic waves is displayed in Chapter 5.

this idea, and joined D.F.J. Arago to perform experiments confirming Young's hypothesis. Fresnel gave a polished mathematical treatment to the wave model, which allowed him to explain the well-defined shadow projected by the objects once he computed the diffraction patterns produced by several types of obstacles. By this time, the wave model was able to explain all the known phenomenology; it will not run into difficulties until the end of the nineteenth century when the photoelectric effect and the behavior of black body radiation were discovered. Even so, there existed defenders of the corpuscular model after Fresnel's death. In 1850, J.B.L. Foucault measured the speed of light propagation in water, concluding that it was lower than the speed in air, in complete agreement with the wave model.

2.2. FIRST DETERMINATION OF THE SPEED OF LIGHT

Galileo was the first to attempt to elucidate whether the speed of light is finite or not. He had thought that the issue could be solved by two distant persons that exchange light signals by covering and uncovering lanterns. If they agree in answering each signal immediately after its reception, then they could establish the time elapsed in the round-trip of the light, and so the speed of light could be computed. Galileo (1638) tells, "I have tried the experiment only at a short distance, less than a mile, from which I have not been able to ascertain with certainty whether the appearance of the opposite light was instantaneous or not."

In 1676 the Danish astronomer Olaf Römer was engaged in a project of the Academie Royale des Sciences of Paris researching celestial events that were suitable for an accurate determination of the geographical longitude, aimed to improve the production of maps and facilitating the navigation. The idea—proposed by Galileo—consisted of elaborating a table with Paris times for a periodic celestial event. Thus, the solar time for the same event observed in other point of the globe, would allow to determine the longitude of the place (with respect to Paris). The celestial event under consideration was the succession of eclipses of Io, one of the four Jupiter's big moons that Galileo discovered in 1610 by using his rudimentary telescope, and the one closest to the planet. Owing to the characteristic of its orbit, Io is eclipsed each time it passes behind Jupiter. This happens each 42.5 hours on average. However, it was known that the time between consecutive eclipses underwent a delay when Jupiter and the Earth moved away; instead the eclipses succeeded each other faster when the Earth and Jupiter approached. When Römer announced in Paris that the anomalous behavior could be the consequence of a finite speed of light, he had to face a strong opposition. Many scientists, under the influence of R. Descartes, still believed that light propagated instantaneously. Römer thought that the delay was produced by the longer distance the light traveled from Jupiter. The cumulative delay from the shortest distance between the Earth and Jupiter to the longest one totalized around 22 minutes. This process takes a few more than half a year (since the orbital period of Jupiter is almost 12 years, then the direction

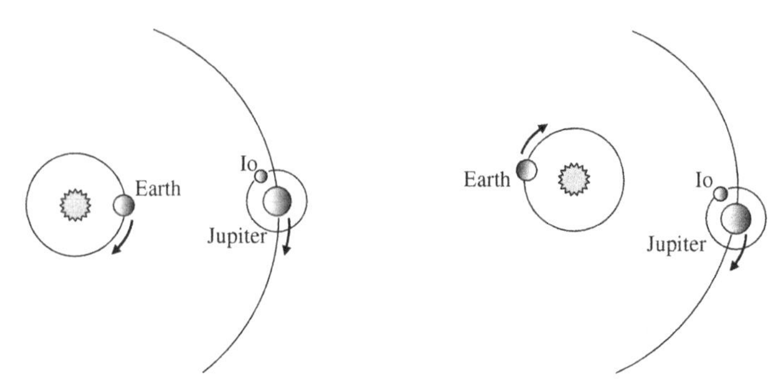

Figure 2.4a. The Earth and Jupiter when the approach is maximum.

Figure 2.4b. The Earth and Jupiter six and a half months later.

Sun–Jupiter has only a little change in half a year) (Figures 2.4a,b). According to Römer, 22 minutes should be the time elapsed when the light travels the diameter of the terrestrial orbit. Using the estimation of 293,000,000 km for the diameter of the terrestrial orbit, Römer obtained 222,000 km/s for the speed of light. The order of magnitude was correct, but the result was affected by important errors in the measurements of times (the real delay is 16 minutes and 36 seconds).

2.3. THE ABERRATION OF LIGHT

Although Römer's interpretation was immediately supported by Huygens and Newton, the scientific community did not completely accept that light propagates with a finite speed until 1729, when the English astronomer J. Bradley communicated the observation of the phenomenon of *aberration* of light. Bradley discovered the aberration while he was trying to measure the *parallax* of the star Gamma Draconis, with the aim of demonstrating the movement of the Earth around the Sun, hence dealing a definitive blow to the anti-Copernicans.

stellar parallax

The stellar parallax is the change of the direction of observation of a star when it is watched from different points of the terrestrial orbit. As a result of the changes of position of the Earth, the successive directions of observation span the contour of an ellipse after a year (see Figure 2.5). The detection of this ellipse would be the evidence of the movement of the Earth. The shorter the distance to the star, the larger the angle made by the major axis of the ellipse. The minor axis also depends on the *ecliptic latitude* of the star—the angle between the direction of observation and the ecliptic plane. So the minor axis makes a null angle when the star is on the ecliptic plane, but is equal to the major axis when the direction Sun–star is perpendicular to the ecliptic plane (the ellipse becomes a circle that copies the almost circular shape of the terrestrial orbit) Figure 2.5

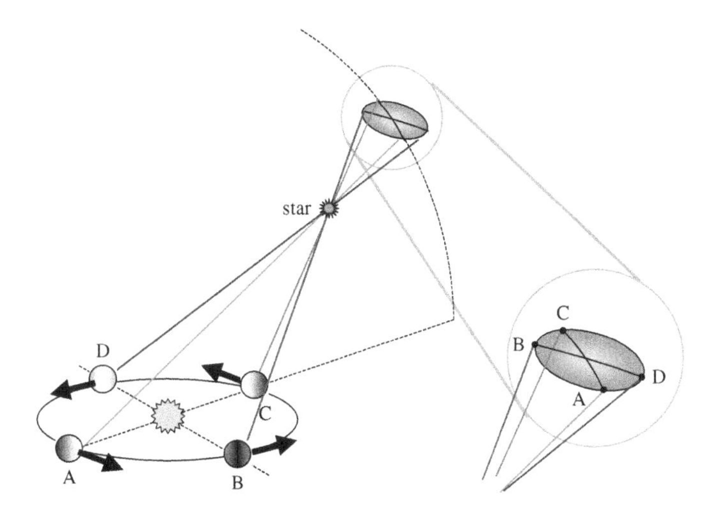

Figure 2.5. Annual parallax of a star.

shows that the ecliptic latitude of the star is maximum when the Earth is in position C (highest point of the ellipse) and is minimum when the Earth is in position A (lowest point of the ellipse).

Bradley set out to determine the ellipse developed by a same star throughout a year. For this purpose he had to register the annual variation of the direction of observation of his telescope (ecliptic latitude and longitude). He selected Gamma Draconis, a star that crosses the zenith of the observation place (a vicinity of London) whose brightness allows to observe it even by day, and whose almost zenithal position diminishes the refraction effects occurring when the light enters the atmosphere. Bradley found two facts that surprised him: the ellipse major semi-axis was about 20 arcseconds (much greater than the expected one), and the ellipse developed in a way that did not fit the behavior associated with the parallax. In fact, the minimum ecliptic latitude of the star did not happen at the expected point of the terrestrial orbit (position A in Figure 2.5) but it occurred when the movement of the Earth was directed toward the star (position B). Neither the A–B–C–D development of the ellipse resulted as expected (Figure 2.5), but it happened as shown in Figure 2.6. Bradley observed other stars and found, to his great surprise, that all the stars developed ellipses with equal major axes, without any evidence of dependence on the distance to the star.

aberration of light

Bradley had detected an effect that was not related to the Earth orbital position but to its state of motion: the star "moved" in the direction of the movement of the Earth, and its displacement did not depend on the distance to the star. This unforeseen effect resulted to be larger than that coming from parallax,

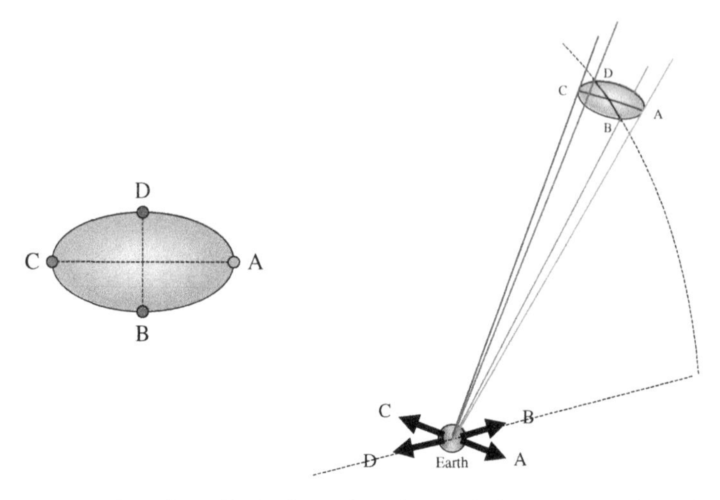

Figure 2.6. Ellipse observed by Bradley, and its relation with the movement of the Earth.

what means that this last remained hidden.[3] To understand the phenomenon observed by Bradley, let us consider the simpler case of a star placed in a direction perpendicular to the Earth movement (Figure 2.7). In the absence of motion, the telescope should be exactly oriented toward the star so that the star

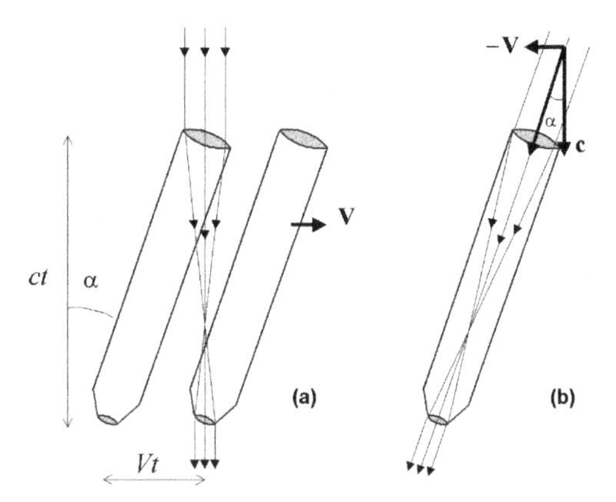

Figure 2.7. Explanation of the aberration of light: (a) frame where the telescope moves with velocity V; (b) frame fixed to the telescope. In order that a star placed in the direction perpendicular to the movement of the telescope can be seen at the center of the field of view, the telescope must be oriented at the angle $\alpha = \text{arctg } V/c$ (angle of aberration).

[3] The parallax of a star will be measured in the next century by F.W. Bessel (1838), who determined the value of 0.3 arcseconds for the near star *61 Cygni*. The nearest star—*Alfa Centauri*, distant 4.3 light year—has a value of 1.524 arcseconds.

occupies the center of the field of view. However, if the telescope moves, then it will be necessary to lean the telescope in the direction of its motion in order that the star remains in the center of the field of view of the observer accompanying the telescope. Figure 2.7 shows the phenomenon as seen in the frame where the telescope moves, and in the frame fixed to the telescope. In this last case we used the Galileo addition theorem of velocities (1.7) for transforming the speed of propagation of light to the frame fixed to the telescope. If the Earth moves with velocity c, then the *angle of aberration* α is such that

$$\text{tg } \alpha = \frac{V}{c}$$

In the general case, let θ be the angle between the velocity $\mathbf{V}$ and the ray direction in the frame where light travels with speed c, and θ' the corresponding angle in the frame moving with velocity $\mathbf{V}$ relative to the former one. The relation between θ' and θ results from Galileo addition theorem of velocities (1.7):

c u′ θ θ′ V

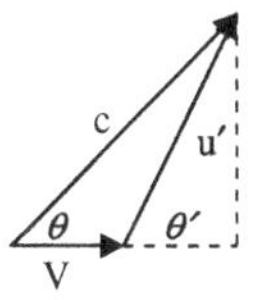

$$\text{tg } \theta' = \frac{\sin \theta}{\cos \theta - \frac{V}{c}} \tag{2.2}$$

The angle of aberration $\alpha \equiv \theta' - \theta$ is obtained by using the identity $\text{tg}(\theta' - \theta) = (\text{tg } \theta' - \text{tg } \theta)/(1 + \text{tg } \theta' \text{ tg } \theta)$:

$$\text{tg}\, \alpha = \frac{V}{c} \frac{\sin \theta}{1 - \frac{V}{c} \cos \theta} \tag{2.3}$$

If $V << c$ the angle of aberration is small, being approximated by

$$\alpha \cong \frac{V}{c} \sin \theta \tag{2.4}$$

resulting to be proportional to the component of $\mathbf{V}$ that is perpendicular to the ray direction.[4]

Clearly, if the Earth always moved with the same velocity $\mathbf{V}$ then the phenomenon would not be detectable because the entire sky would be displaced in a permanent way, hence it would be impossible to notice such displacement. However, vector $\mathbf{V}$ changes in the course of the year: the Earth moves around the Sun at about 30 km/s, with a consequent change of the direction of $\mathbf{V}$. After 6

[4] Here it could seem that the use of Galileo addition theorem of velocities suppose to consider the light as composed by particles; in such case the aberration would appear for the same reason that one leans the umbrella forward when running under a vertical rain (as Bradley himself believed). However, in the wave model the ray direction transforms in the same way as the velocity of a particle (see §2.5 and Figure 2.9).

months this change implies that $|\Delta \mathbf{V}|$ is about 60 km/s (considering two opposite points of the orbit, and assuming that the state of motion of the Sun does not significantly vary during that time). Therefore the angle of aberration changes, and the star displaces toward one and the other side drawing an ellipse. The positions B and D in Figure 2.6 define the ends of the ellipse minor axis because the component of the terrestrial velocity that is perpendicular to the direction to the star becomes minimal at B and D.

If light propagated instantaneously ($c = \infty$) there would be no aberration. Thus the effect discovered by Bradley—whose interpretation took him more than a year of reflection—confirmed that light travels with a finite speed. Bradley's measurement of 20.25 arcseconds for the ellipse major semi-axis gives the value of c with an error of 1% (the value of the aberration accepted at present is 20.49552 arcseconds).

See § 3.14

2.4. FIRST TERRESTRIAL METHOD TO MEASURE *C*

Almost two centuries went by from Römer's work until the speed of light could be measured by means of a terrestrial method. In 1849, A.H.L. Fizeau utilized a toothed wheel to divide a light beam in pulses. After traveling a distance l, each pulse is reflected in a mirror and returns to the wheel (Figure 2.8). If the speed of revolution of the wheel is such that the next tooth gets in the way of the returning pulse, then the observer field of view turns out to be dark. By increasing the velocity of the wheel, the reflected light begins to pass through the next gap between teeth. By gradually increasing the velocity, the intensity of light reaches a maximum. Fizeau obtained this maximum by revolving the wheel at 25 revolutions per second. Since the wheel had 720 teeth, the time between consecutive gaps was $1/(25 \times 720)\text{s} = 1/18,000\,\text{s}$. This was the time elapsed

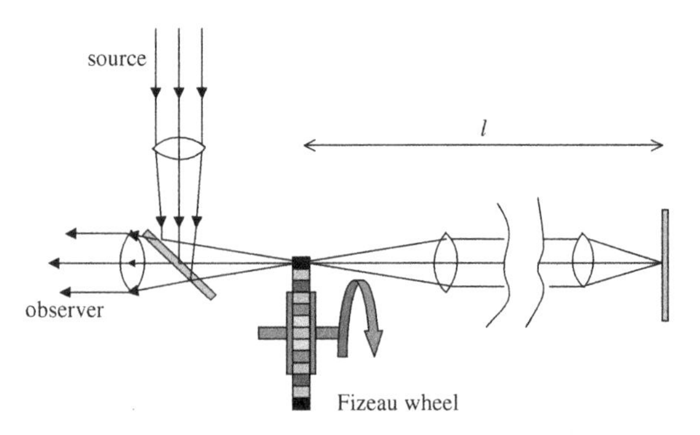

Figure 2.8. Fizeau's device to determine the time of the round-trip of light. The toothed wheel divides the light beam in pulses passing between the teeth. After the pulses are reflected on the mirror, they can be recovered through the following space between teeth, provided that the wheel revolves fast enough.

during the round-trip of the light. Being the mirror at the distance of 8.5 km from the wheel, the light then traveled 17 km in 1/18,000 s, which means a speed of 306,000 km/s. In 1850, Foucault improved Fizeau's device by substituting the wheel with a rotating mirror; in this way he measured the speed of light in air and water, confirming that the light propagates slower in water than air, as predicted by the wave model.

Foucault confirms the wave model

2.5. THE LUMINIFEROUS ETHER

The use of a model based on mechanical waves to describe the light phenomena (see §1.8) requires a material medium to be the seat of the light perturbation; this medium should be present in all the places where light propagates. This propagation medium was called *luminiferous ether*, and would be identified with the electromagnetic ether once Maxwell realized that light is an electromagnetic wave. The speed c of light propagation in ether must be regarded as a property of the medium. The discovering of polarization at the beginning of the nineteenth century showed the transversal character of light waves (differing from sound, which is a compression and rarefaction longitudinal wave), and livened up the discussion about the physical nature of ether. In order to be the seat of transversal waves (where the perturbation of the medium is perpendicular to the direction of propagation), the ether must be elastic. This elastic material should be endowed with enigmatic properties. On one hand it should be tenuous enough to become practically intangible and to allow the course of the planets without apparent loss of energy. On the other hand the ether should be extremely rigid, in order to produce an enormous speed of propagation. G.G. Stokes (1845) disputed that this was a trouble, by remarking that rigidity is a concept that depends on the scale of time involved in a phenomenon: the asphalt behaves like a rigid body when is cut by a hit, but flows and it is penetrable in much larger scales of time. The absence of longitudinal waves in this elastic medium was a source of more controversies. In the interior of transparent substances, the properties of the ether should change in order to explain the change of propagation speed: according to Snell's law (2.1), light propagates inside a substance with a speed c/n relative to the ether, where n is the *absolute refractive index* of the substance, i.e., the index relative to the ether.

An important issue was to establish how the ether interacts with the rest of the bodies. In the beginning it was thought that the ether was not disturbed by the passing of a body, since it penetrated all the bodies remaining itself immutable in a state that could be called, with reason, absolute rest. Its universality and immutability seemed to endow ether with the rank of absolute reference system, in the Newtonian sense. The aberration of light was regarded as a confirmation of this presumption. In fact, in the ether frame the rays are perpendicular to the wave front, as a consequence of the isotropic propagation. Instead, in the Earth frame, although the wave fronts keep their orientation, the rays (i.e., the direction of propagation of the energy, which goes from the position of the secondary

transformation of the ray direction

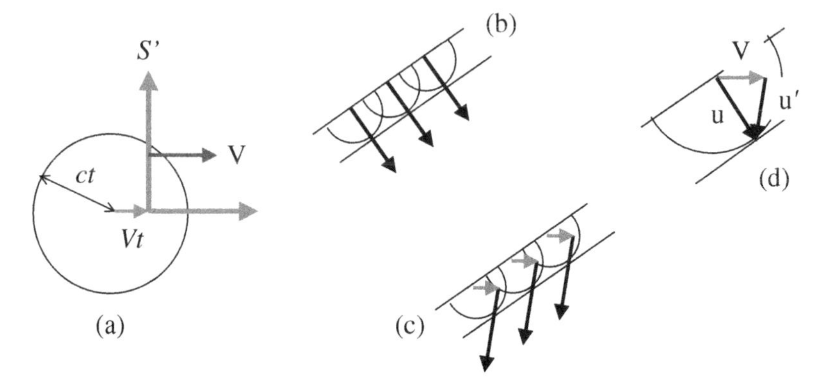

Figure 2.9. (a) Spherical wave front emitted at $t=0$ from the origin of a frame S' moving with velocity V relative to the ether. The sphere is centered at the position occupied by the source in the frame fixed to the ether at the time of emission. In system S' the emitter point (the coordinate origin) does not coincide with the center of the spherical wave, thus the energy does not propagate along the direction normal to the wave front. (b) A plane wave front and its rays as they are seen in the frame fixed to the ether. (c) The same plane wave front and its rays as they are seen in a frame S' in motion relative to the ether. (d) The velocity **u** corresponding to the propagation of the energy, which is called *ray velocity*, transforms according to the Galileo addition theorem of velocities; its change of direction causes the aberration.

source to the point where the secondary wave is tangent to the envelope) are not perpendicular to the front, as it also happens in anisotropic media (Figure 2.9). It could be said that the "ether wind" generates an anisotropy in the Earth frame that is responsible for the change of the ray direction (aberration). The velocity of the energy traveling from the emitter position to the wave front is called *ray velocity*. As shown in Figure 2.9d, the ray velocity (which has the direction of the ray) transforms according to the Galileo addition theorem of velocities (1.7).

2.6. SEARCHING FOR THE ABSOLUTE TERRESTRIAL MOTION: DRAGGING OF ETHER

Since a direct detection of ether seemed to be impossible, the experiments were directed to reveal, at least, the movement of the Earth relative to the ether (which amounted to detecting the absolute motion). These experiments involved the speed of light relative to the Earth (i.e., the Galilean composition between c and the terrestrial velocity). The first of them was performed by Arago in 1810 (at that time Arago still advocated the corpuscular model; nevertheless this posture does not affect the conception of his experiment but, at most, the interpretation of the results). Since the speed of propagation takes part in the explanation of Snell's law (2.1) (in a different way for each model of light), Arago thought the refraction of a ray would depend on the movement of the refractive substance. Arago covered a half of the objective lens of a telescope with an achromatic prism; so he could collect starlight directly through the objective lens or previously deviated by the prism. Thus the observation of these two different images of a star required two different orientations of the telescope. The angle between the two directions of the telescope is equal to the deviation of the rays produced by the prism. Arago thought that the motion of the prism (i.e., the movement of the Earth) would leave its mark on this angle. This effect would become more evident if the compared angles correspond to a star observed in the direction of the movement of the Earth and another star observed in the opposite direction (the speeds of light relative to the Earth would be $c + V$ and $c - V$ respectively). Assuming that the motion of the Earth relative to the ether is mainly due to its motion around the Sun at a rate of 30 km/s, then the observations should be performed on the directions tangent to the orbit (a pair of stars close to the ecliptic and passing the meridian at 6 a.m. and 6 p.m. respectively should be chosen). It was expected a difference of some tens of arcseconds. However, Arago did not obtain the expected result: he could only observe little changes of a few arcseconds, without any defined behavior, which could be ascribed to experimental errors. The experiment was repeated 6 months later with similar result. The motion of the Earth relative to the ether was not evidenced.

For Stokes, the null result of Arago's experiment was consistent with his way of regarding the interaction between ether and other bodies. Stokes (1845) thought that it was not correct to state that the Earth passes through the ether without disturbing it. According to Stokes, the Earth drags the ether contained in its interior, and carries a layer of ether adhered on its surface which gradually becomes detached until it remains at rest in the absolute space (universal ether). In this way the Arago's experiment was performed in a frame fixed to a local ether, so the Snell's law was fulfilled, and no additional deviation could be expected.

Stokes: the moving bodies carry a layer of ether adhered on their surfaces

However, the idea of a local ether fixed to the Earth surface made difficult the understanding of the aberration of light. In fact, if this idea were true the change of the ray direction would require a change of the orientation of the wave front, since the rays are perpendicular to the wave front in the ether frame. Stokes made a model to describe the modification of the wave fronts; in this explanation he considered the ether as an incompressible and irrotational fluid. The model was objected by H.A. Lorentz in 1886, who remarked that such fluid could not be attached to a moving Earth (with respect to the universal ether) because that would be an inadmissible boundary condition for its velocity field.

Fresnel proposed other explanation for the null result of Arago's experiment, also based on a dynamical interaction between moving bodies and ether. According to Fresnel (1818) the moving substances communicate only a fraction $f\mathbf{V}(f < 1)$ of their (absolute) motion $\mathbf{V}$ to the ether contained inside them. Thus the interior of any transparent substance houses a *local ether* moving with velocity $f\,\mathbf{V}$ relative to the universal ether. Taking into account that the ray velocity transforms like the velocity of a particle (Figure 2.9), then it results from (1.7) that

Fresnel: moving substances partially drag the ether contained inside them

Complement 2A: *The partial dragging of ether*

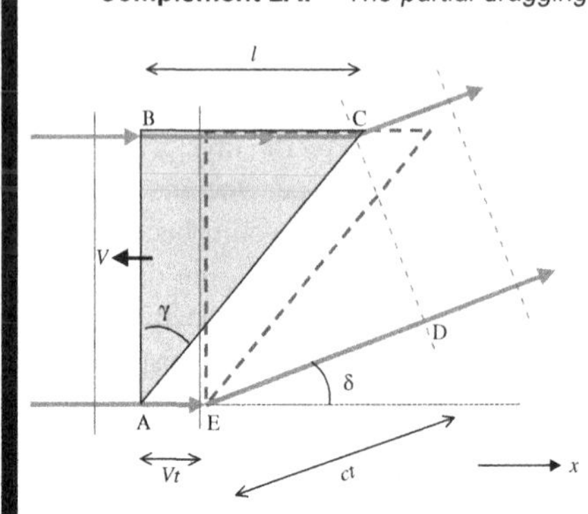

The Figure shows the rays in the frame fixed to the universal ether, when a prism moves with absolute velocity V opposite to the incident ray. According to (2.5) and (1.7), the speed of light relative to the prism is $c/n - fV + V$, f being the dragging coefficient. Thus, light travels the length l along the line BC in the time

$$t = \frac{l}{c/n + (1-f)\,V} \cong \frac{n\,l}{c} - \frac{n\,l\,V}{c}\,(1-f)$$

(if $V << c$). During this time the prism moves a distance $V\,t$, and the ray passing by E travels a distance $c\,t$ completely outside the prism.

In the frame fixed to the ether the ray is perpendicular to the wave front. This property and the length ct of segment ED determine the angle of deflection $\delta(V)$. Instead, in the frame fixed to the prism the emergent ray travels the line AD with speed c'. In the figure on the right α is the angle of aberration ($\alpha \cong V\,c^{-1}\sin\delta$), and speed c' results from applying the cosine theorem in AED: $c'^2 = c^2 + V^2 + 2\,V\,c\cos\delta$. Then

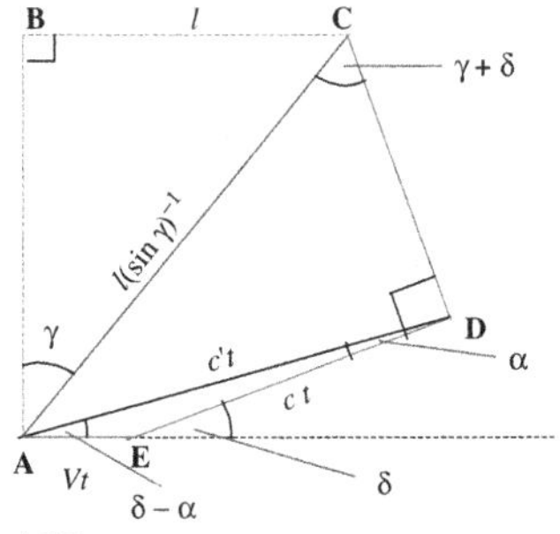

$$c' = c\,(1 + V\,c^{-1}\cos\delta) + O(V^2/c^2) \qquad (2A.1)$$

On the other hand, applying the sine theorem in ACD:

$$l(\sin\gamma)^{-1}\sin(\gamma+\delta) = c'\,t\sin(\pi/2-\alpha) = c'\,t\cos\alpha = c'\,t + O(V^2/c^2)$$

and replacing the time t:

$$(\sin\,\gamma)^{-1}\,\sin(\gamma+\delta) = n - V\,c^{-1}\,n^2\,(1-f) + V\,c^{-1}\,n\,\cos\,\delta + O(V^2/c^2) \qquad (2A.2)$$

which is the equation to obtain $\delta(V)$. If $V = 0$ (prism at rest with respect to the universal ether), then the deflection δ_o in (2A.2) is the one coming from Snell's law:

$$\sin(\gamma+\delta_o) = n\,\sin\,\gamma \qquad (2A.3)$$

The null result of Arago's experiment (§2.6) could be explained if the value of f prevented, at the lower order in V/c, the detection of the absolute motion of the prism. This requires that the outgoing ray direction in the frame fixed to the prism, $\delta - \alpha$, turns out to be coincident with the value δ_o corresponding to a prism at absolute rest; i.e., $\delta(V) = \delta_o + \alpha \cong \delta_o + Vc^{-1}\sin\,\delta_o$. This form of $\delta(V)$ must be used in (2A.2) to obtain the searched value of f. Replacing (2A.3) in (2A.2) and expanding $\sin(\gamma + \delta(V))$ at the first order in V/c it is

$$\sin\,\delta_o\,\cos(\gamma+\delta_o) = -n^2\,(1-f)\,\sin\,\gamma + \sin(\gamma+\delta_o)\,\cos\,\delta_o + O(V/c) \qquad (2A.4)$$

This equation will be fulfilled if the value of the dragging coefficient is

$$f = 1 - n^{-2} \qquad (2A.5)$$

$$\mathbf{u}\big|_{\text{universal ether}} = \mathbf{u}\big|_{\text{local ether}} + f\,\mathbf{V} \tag{2.5a}$$

On the other hand the *phase velocity*—which measures the displacement per unit of time of the wave front along the direction of propagation $\hat{\mathbf{n}}$ normal to it—transforms like the projection of a velocity vector on the direction $\hat{\mathbf{n}}$:

$$\text{phase velocity}\big|_{\text{universal ether}} = \frac{c}{n} + f\,\mathbf{V}\cdot\hat{\mathbf{n}} \tag{2.5b}$$

where c/n is the phase velocity relative to the local ether (the transparent substance is supposed isotropic).[5]

The dragging coefficient f could have a suitable value in order that the alteration of the refraction of the rays caused by the absolute motion of the prism and the aberration of rays in the prism frame *compensate* each other. If this were the case, the direction of the outgoing rays in the prism frame would not differ from the one corresponding to a prism at absolute rest; no tracks of the terrestrial motion would be left in the image through the prism, what would explain the null result of Arago's experiment. Complement 2A shows that the value $f = 1 - n^{-2}$ proposed by Fresnel [6] allows such a compensation at the lowest order in V/c. In this way a curious cancellation between two different effects, both of them owing to the motion of the Earth relative to the universal ether—the partial dragging of the ether contained in the prism and the aberration of light—leads to the result that the directions of incident and outgoing rays fulfill Snell's law *in the prism frame* (at least, at the lower order in V/c). It could be concluded that the privilege of the ether frame had been damaged.

Fresnel's dragging coefficient

The calculus of the dragging coefficient (Complement 2A) repeatedly resorted to the invariance of distances and times, both explicitly (a time or a distance is computed in a reference system and then used in another reference system) and implicitly (by using Galileo addition theorem of velocities (1.7)). For this reason Fresnel's partial dragging could be regarded as a way to make compatible an experimental result and the (at that time) indisputable notions of space and time. But the only truly indisputable thing was that the experiment did not reveal the movement of the Earth relative to the ether.

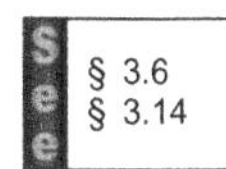

[5] If the substance is non-dispersive then the phase velocity and the modulus of the ray velocity are equal in the frame fixed to the local ether.

[6] Fresnel considered the ether as an elastic medium. Therefore the squared speed of propagation of the light waves was inversely proportional to the ether density. In order that the speed of light in a transparent substance were $c/n < c$, the ether density inside the substance had to exceed the universal ether density by a factor n^2. Fresnel postulated that the *excess* of ether density (with respect of the universal ether) was completely dragged by the substance while the rest of the interior ether remained at rest. According to Fresnel, the motion of the center of mass of the ether contained in the substance was communicated to the light. Further details on Arago and Fresnel works can be found in R. F. and D. M. Sforza, European Journal of Physics **26** (2005), 195–204.

2.7. FIZEAU'S EXPERIMENT

In 1851, Fizeau performed an experiment with the aim of measuring the dragging of the ether caused by a current of water moving with a velocity v relative to the laboratory. This experiment would make possible to decide between the disagreeing positions of Stokes and Fresnel. Figure 2.10 shows a scheme of the experimental device used by Fizeau. A light beam is divided into two parts—rays 1 and 2—by a half-silvered glass plate. An arrangement of mirrors forces the rays to cover the circuit of water in opposite directions. The idea is to compare the travel time of both rays to determine if a possible dragging of ether, either partial or total, favors ray 1, which is the one traveling in the direction of circulation of the water. A difference between ray travel times produces a phase shift that can be measured by means of interferometric techniques.

In order to evaluate the dragging of the ether due to the current of water, we have to notice that the dragging is caused by the motion of the water relative to the universal ether. This means that the dragging involves the composition of the movement v of the water relative to the laboratory with the motion of the laboratory (i.e., the Earth) relative to the universal ether at an unknown rate $\mathbf{V}$.

Even if the water did not flow, the last contribution would be present. However, since both rays travel closed paths which decompose into *physically identical* outward and return stretches, then it can be stated that the phase shift remains invariant if $\mathbf{V}$ changes to $-\mathbf{V}$. On the contrary, if water flowed in the opposite direction ($v \to -v$), then the difference $\Delta t = t_2 - t_1$ between the travel times of both rays would change sign because changing the direction of the flow is equivalent to an exchange of the roles of the rays. From these remarks

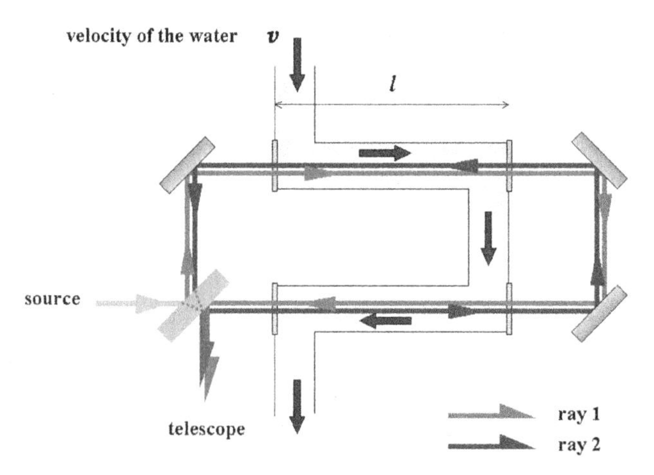

Figure 2.10. Fizeau's experiment is aimed to detect the dragging of ether by a current of water. The dragging shortens the travel time of ray 1 (which travels in the direction of the current of water), and increases the one of ray 2. The difference between travel times produces a phase shift that can be measured in the interference of both rays.

it becomes clear that the movement of the Earth can only contribute to the expansion of Δt as a power series with terms of even order in V/c, while the flow of water adds odd powers of v/c. If Δt is a magnitude of first order in v/c, then we will neglect terms of second order in V/c (anyway the experimental device was not sensitive enough to detect them). Therefore Δt can be calculated as if the Earth were at rest: the velocities of rays 1 and 2 relative to the laboratory will be taken as $c/n + fv$ and $c/n - fv$ respectively. Thus the difference between travel times is approximately equal to

$$\Delta t = t_2 - t_1 \cong \frac{2l}{\frac{c}{n} - fv} - \frac{2l}{\frac{c}{n} + fv} = \frac{nl}{c} \frac{4nfv/c}{1 - \left(nf\frac{v}{c}\right)^2} \cong \frac{4nl}{c} \frac{v}{c} nf \qquad (2.6)$$

where we took into account that the sole contributions to Δt come from both stretches of length l that light covers inside the water.

In order to observe the travel time difference Δt, the rays must reach the telescope in slightly divergent paths (this requirement can be achieved, for instance, if the half-silvered glass plate is not exactly oriented at 45°). In these conditions, and working with an extensive source, interference fringes are produced—*Fizeau fringes*. These fringes are localized at the plane where the divergent rays intersect, and can be observed by focusing them with a telescope (see Complement 2B). If the phase shift of the rays changes, then the position of the fringes also changes. Fizeau compared the position of the fringes for water at rest and flowing water. In the last case the delay of ray 2 with respect to ray 1 is the time Δt given in (2.6); this delay amounts to $N = \Delta t/T = c\Delta t/\lambda$ wavelengths (T is the period and λ is the wavelength). Each wavelength entering the delay means that the fringes will displace a distance equal to that existing between two consecutive interference maxima. The values of v and l utilized by Fizeau (7 m/s and 1.49 m respectively) give $N = 0.46 f$. Stokes' hypothesis corresponds to $f = 1$ (total dragging), while Fresnel's partial dragging implies the value $f = 1 - n^{-2} = 0.43$ ($n_{water} = 1.33$). The displacement of fringes measured by Fizeau was compatible with Fresnel's dragging coefficient within an error of 15%.

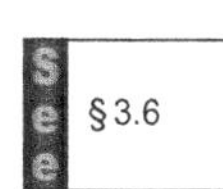

2.8. HOEK'S EXPERIMENT

In 1868, M. Hoek performed an experiment that involved the motion of the Earth relative to the ether at the first order in V/c. The experimental arrangement is similar to that used by Fizeau, since a light beam is divided into two parts that are forced to travel closed paths until they meet again with a probable phase shift (Figure 2.11). However, in this case the closed paths do not decompose into physically equivalent stretches: the rays go through a recipient filled with water, at rest in the laboratory, *in a sole stretch of their trips*. Since the rays pass through the water in opposite directions, it results that changing $\mathbf{V}$ for $-\mathbf{V}$ amounts to exchange the roles of the rays. Therefore the difference Δt of travel

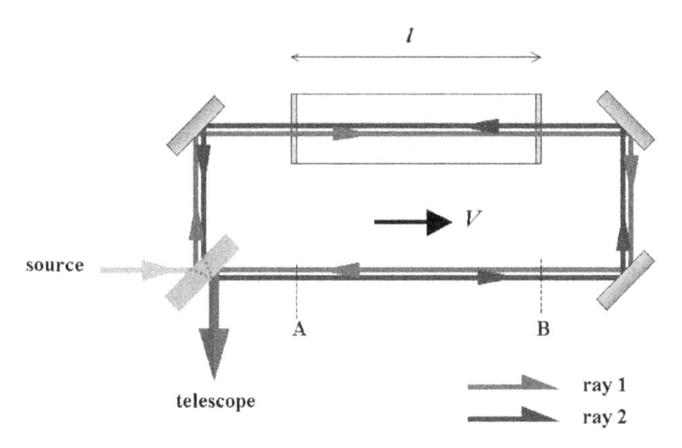

Figure 2.11. Hoek's experiment is aimed to establish the velocity V of the Earth relative to the ether, by observing the phase shift between rays 1 and 2. The recipient filled with water would allow for a result of first order in V/c.

times of the rays will receive contributions of odd order in V/c. The observation of the phase shift produced by the different times traveled by the rays could demonstrate the movement of the earth relative to the universal ether. Although V/c is unknown, the experiment should be sensitive enough to detect a value of order 10^{-4}. In fact, if worst comes to worst, the Earth could be at rest relative to the universal ether when the experiment is performed. In such a case, 6 months later the Earth should be moving at a rate of 60 km/s, as a consequence of its movement around the Sun (assuming that the state of motion of the Sun relative to the universal ether does not undergo appreciable changes in that period of time). The experimental device was ready for detecting contributions of first order in $V/c \sim 10^{-4}$.

Once again, the detection of the phase shift relies on the examination of the interference fringes. However, in Fizeau's experiment it was possible to stop the flow of water to compare the position of the fringes with the water at rest and moving with respect to the laboratory. On the contrary, if the Earth is moving with respect to the universal ether, we cannot stop it to observe a change in the position of the fringes. Nevertheless, we can turn on 180° the table where the arrangement is mounted; this procedure is equivalent to changing $\mathbf{V}$ for $-\mathbf{V}$. As above mentioned, the difference Δt will change its sign: therefore the fringes will displace when the table turns.

Let us compute the differences of travel times Δt. Clearly, we only have to evaluate the times traveled by each ray inside the water and along the segment AB outside the water; in fact the rest of the paths do not contribute to Δt because they cancel out. Although the direction of $\mathbf{V}$ is not known, we shall compute the travel times as if the table were oriented along $\mathbf{V}$ (Figure 2.11). The ray 1 goes through the water with speed $c/n + f\,V - V$ relative to the laboratory (we used Eqs. (2.3) and (1.7)); while in the stretch AB the ray 1 travels with speed

$c+V$ relative to the laboratory. Therefore the time employed in traveling both stretches is, at the lower order in V/c,[7]

$$\begin{aligned} t_1 &= \frac{l}{\frac{c}{n}+fV-V}+\frac{l}{c+V} \cong \frac{nl}{c}\left(1+(f-1)\frac{nV}{c}\right)^{-1}+\frac{l}{c}\left(1+\frac{V}{c}\right)^{-1} \\ &= \frac{l}{c}\,(n+1)+\frac{l}{c}[(1-f)\,n^2-1]\frac{V}{c} \end{aligned} \tag{2.7}$$

The travel time of ray 2 is obtained from (2.7) by replacing V with $-V$. Equation (2.7) shows that the term linear in V/c, which would be responsible for the contribution of the lowest order to Δt, vanishes when the dragging coefficient has the value given by Fresnel. In such a case, the effect of the motion of the Earth relative to the ether could only be revealed at the third order in V/c.

Hoek's experiment did not exhibit any displacement of fringes when the table turned. As already happened in Arago's experiment, the movement of the Earth relative to the ether was not perceptible. Once again the partial dragging of the ether proposed by Fresnel was able to give an explanation for such a failure, at least for the first order in V/c, within the context of the notions of space and time that were current in the nineteenth century.

2.9. AIRY'S EXPERIMENT

In 1871, G.B. Airy performed an experiment whose conception was similar to Arago's conception. As already explained in §2.3, owing to the aberration of light the direction of a telescope focusing a star in the center of its field of view does not coincide with the direction of the starlight ray in the ether frame. The angle between both directions is the angle of absolute aberration, and depends on the velocity of the Earth relative to the ether (Eq. (2.3)). Only the changes of this angle over time are observable; so this effect is not useful to demonstrate the motion of the Earth relative to the ether. Nevertheless, if the telescope were filled with water, then the rays would undergo an additional refraction, what would compel to give an extra angle to the telescope in order to keep the star in the center of the field of view. Since the additional refraction would be the consequence of the lack of alignment between telescope and ray in the ether frame (Figure 2.7a), the observation of the extra angle would be the evidence of the movement of the Earth relative to the ether.

In the former argument it was not considered a possible role of Fresnel's partial dragging of the ether. As shown in §2.6, the effect of the dragging on the direction of the rays is equivalent to the fulfillment of Snell's law—at first order

---

[7] The result (2.7) is obtained by using the approximation $(1+x)^\alpha \cong 1+\alpha x$, valid if $x << 1$.

in V/c—in the frame fixed to the transparent medium (in this case, the frame fixed to the telescope). Figure 2.7b displays the ray paths in the frame fixed to the telescope; it results that if Snell's law were valid in this frame, then the filling of the telescope with water would not force any change of the orientation of the telescope in order to keep the star at the center of the field of view; it would only be necessary to correct the focus, what would have happened anyway whether the Earth were or not at rest in the universal ether.

Airy verified that the filling of the telescope with water did not displace the star from the center of the field of view. This new test was consistent with those by Arago and Hoek: the three of them failed in detecting the motion of the Earth relative to the universal ether, and for all of them Fresnel's partial dragging of the ether offered a satisfactory explanation (at first order in V/c) within the context of the notions of space and time that prevailed in the nineteenth century.

2.10. MICHELSON-MORLEY EXPERIMENT

Maxwell's proposal

In 1879, Maxwell remarked that an experiment of astronomical type was able to reveal the motion of the Earth at the first order in V/c, without involving Fresnel's dragging of ether. The delay time of the eclipses of Io—which were used to demonstrate that the speed of light is finite (see §2.2), and is nothing but the time passed when the light travels the diameter of the terrestrial orbit—should diminish when Jupiter were placed in the direction of the movement of the Solar System with respect to the ether, and should increase when Jupiter were placed in the opposite direction. If the Solar System moves with velocity V relative to the ether, then in the first case the light will cover the diameter d of the terrestrial orbit in the time $d/(c+V)$, while in the second case the travel time will be $d/(c-V)$. Through the almost 12 years that Jupiter takes to complete its orbit, the delay time of the eclipses of Io would continuously vary between these two extreme values, which could be determined. The velocity V would result from the difference between them, being a quantity of first order in V/c :

$$\Delta t = \frac{d}{c-V} - \frac{d}{c+V} = \frac{2\,d\,V}{c^2 - V^2} \cong \frac{2d}{c}\,\frac{V}{c} \tag{2.8}$$

The astronomical data received by Maxwell shortly before his death were not precise enough to solve the issue. Maxwell emphasized as well that, in those experiments where the light travels a round-trip in the same medium, the movement of the laboratory would produce corrections to the travel time of second order in V/c. As noticed in §2.8, the experiment has to be prepared to detect at least the value $V/c \sim 10^{-4}$; therefore, an effect of second order would mean that the travel time is corrected in its ninth significant digit!

The challenge of working with such level of precision was taken by the American physicist Albert Abraham Michelson (1852–1931), who had already excelled in improving Foucault's method of measuring the speed of light. Based

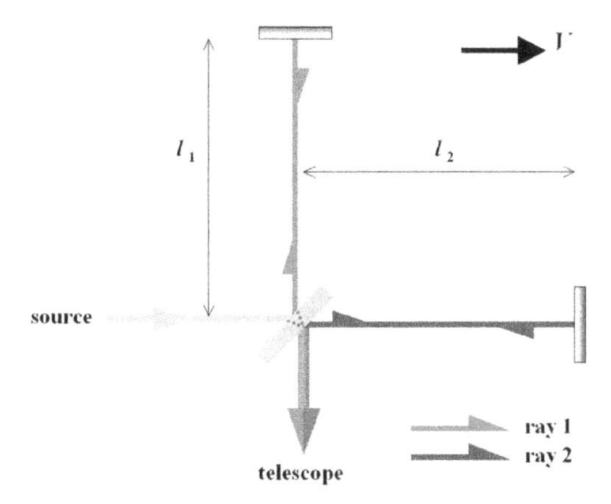

Figure 2.12. In Michelson's interferometer a beam is divided into two parts by a half-silvered glass plate. After they travel different round-trip paths, they meet again and interfere forming interference fringes that are observed with a telescope.

Michelson's interferometer

on the techniques used by Fizeau, Michelson designed the interferometer schematized in Figure 2.12. A light beam is divided by a half-silvered glass plate. After the separation, each part travels different round-trips, and eventually meet again at the dividing glass plate. If the laboratory (the Earth) moved with velocity V relative to the ether, then the travel time of each ray would depend on V in a different way, since they travel different paths. If, for instance, we assume that the (unknown) direction of motion of the Earth is coincident with that of ray 2 (Figure 2.12), then ray 2 will go with velocity $c - V$, and return with velocity $c + V$, both of them referred to the laboratory frame (Figure 2.13). Thus the round-trip travel time of ray 2 is

$$t_2 = \frac{l_2}{c - V} + \frac{l_2}{c + V} = \frac{2l_2/c}{1 - \dfrac{V^2}{c^2}} \tag{2.9}$$

The round-trip travel time of ray 1 can be more easily calculated in the ether frame. We remark that times and distances are considered invariant; thus the so computed time t_1 will be valid in any reference system. Figure 2.14 shows ray 1 in the way it is seen in the ether frame. The direction of the ray differs from the one in the laboratory frame (aberration); in fact, when ray 1 reaches the mirror, and then the glass plate, it finds them not at the initial positions but at displaced positions, owing to the movement of the laboratory with velocity V. In the frame fixed to the ether, light travels with speed c in all directions. Both the outgoing and returning stretches are covered in the same time $t_1/2$; this time is obtained by noticing that each stretch is the hypotenuse of a right triangle whose short sides are l_1 (the same distance than in the frame fixed to

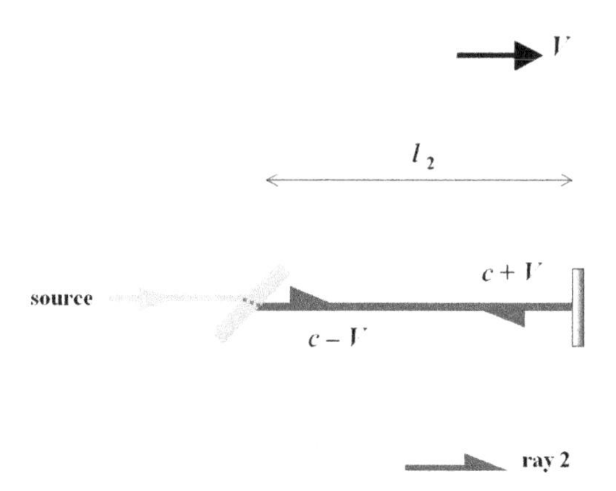

Figure 2.13. Path of ray 2 in the frame fixed to the laboratory.

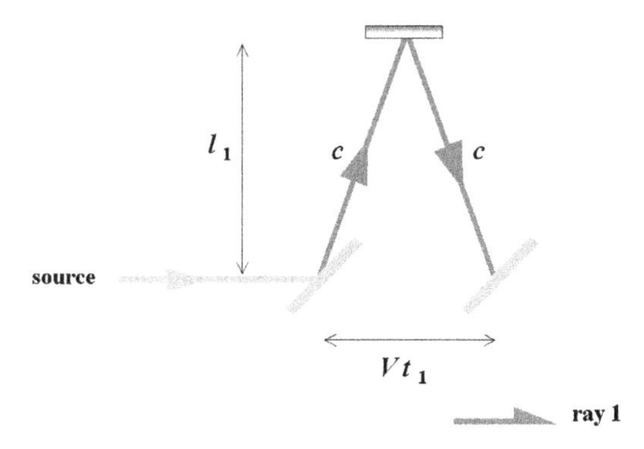

Figure 2.14. Path of ray 1 in the frame fixed to the ether.

the laboratory, because it is supposed invariant) and $V\ t_1/2$. Then by applying Pythagoras theorem,

$$\left(\frac{c\,t_1}{2}\right)^2 = l_1^2 + \left(\frac{V\,t_1}{2}\right)^2$$

and solving for t_1:

$$t_1 = \frac{2l_1/c}{\sqrt{1-\frac{V^2}{c^2}}} \tag{2.10}$$

Contrarily to Hoek's experiment, in this case the travel times depend on the squared V/c, so nothing is changed by turning the interferometer on 180°. Instead, if the interferometer is turned on 90° then rays 1 and 2 will exchange roles. In this way, while the difference between travel times at the original position is

$$\Delta t_{0^\circ} = t_2 - t_1 = \frac{2l_2/c}{1 - \frac{V^2}{c^2}} - \frac{2l_1/c}{\sqrt{1 - \frac{V^2}{c^2}}} \tag{2.11}$$

the same difference at the final position becomes

$$\Delta t_{90^\circ} = t_2 - t_1 = \frac{2l_2/c}{\sqrt{1 - \frac{V^2}{c^2}}} - \frac{2l_1/c}{1 - \frac{V^2}{c^2}} \tag{2.12}$$

In practice, the direction of the laboratory movement relative to the ether is not known. However, by gradually turning the interferometer it should be observed that the phase shift coming from the difference of travel times continuously varies between two extreme values distant themselves 90°. The phenomenon should be demonstrated by the displacement of the interference fringes. Since the difference between the two extreme values of Δt is (see Note 7)

$$\begin{aligned} \Delta t_{90^\circ} - \Delta t_{0^\circ} &= \frac{2}{c}(l_2 + l_1)\left[\frac{1}{\sqrt{1 - \frac{V^2}{c^2}}} - \frac{1}{1 - \frac{V^2}{c^2}}\right] \\ &\cong \frac{2}{c}(l_2 + l_1)\left(1 + \frac{1}{2}\frac{V^2}{c^2} - \left(1 + \frac{V^2}{c^2}\right)\right) = -\frac{l_2 + l_1}{c}\frac{V^2}{c^2} \end{aligned} \tag{2.13}$$

then the rotation of the interferometer would change the phase shift of rays in a quantity equivalent to $N = c|\Delta t_{90^\circ} - \Delta t_{0^\circ}|/\lambda = (l_2 + l_1)/\lambda \times V^2/c^2$ wavelengths (λ is the wavelength). Each wavelength implies that the fringes will displace a distance equal to the one existing between two consecutive interference maxima. The experiment was performed for the first time at Postdam Observatory in 1881. There Michelson did not observe any shift of fringes, as if the Earth were at rest relative to the ether. Michelson thought that the result proved that the Earth carries a layer of ether adhered to its surface; in this way the laboratory was actually at rest in a local ether.

Even if the Earth did not drag ether, the null result of this experiment could be the consequence of a circumstantial Earth's state of motion close to rest, at the time the experiment was performed. In order to rule out this possibility, the experiment should be repeated 6 months later to detect a velocity of the same order than the speed of the Earth relative to the Sun (it was already remarked that this velocity was the minimum expected effect; for this reason it determines the sensitivity of the experimental device). However, Lorentz realized that Michelson had omitted in his calculations the square root appearing in Eq. (2.10) (he attributed a speed c relative to the laboratory to ray 1). The rectification of this

Complement 2B: ***Fringes localized in an air wedge***

The rays coming from an extensive source are reflected in two plates separated by an air wedge of angle α, and then gathered by a telescope. Only those rays emerging almost perpendicularly to the plates can reach the telescope. Each ray outgoing from the source originates two phase-shifted rays—those resulting from the reflection on each plate—with a difference between optical path lengths depending on the distance x. Thus, it produces an interference pattern with the form of fringes. These "Fizeau fringes" are localized on the plane close to the plates where each pair of rays intersects.

Since $\alpha << 1$, the difference between optical path lengths can be approximated by $2\, x\, \alpha$. This quantity must be smaller than the coherence length of the source in order that the interference can exist. The distance Δx between two consecutive interference maxima is such that $2\, \Delta x\, \alpha = \lambda$. For a wedge of angle equal to $1''$, this means a distance between fringes of around 1 mm.

In the experiments by Fizeau, Hoek, and Michelson the pairs of phase-shifted and lightly divergent rays are the result of dividing a beam emitted by an extensive source in a half-silvered glass plate, and forcing each part to travel different optical paths. The effect of wedge, responsible for the divergence, is reached by properly deviating the separating plate or the mirrors.

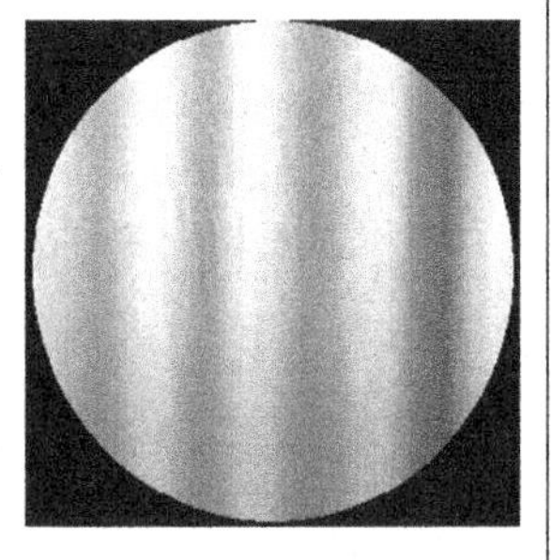

error diminished the shift (2.13) by half, and led the minimum expected effect to the limit of the experimental error. Therefore, it was necessary to improve the design in order to reduce the experimental error.

1887 experiment

Michelson joined Edward William Morley (1838–1923) in Cleveland. In 1886 they repeated Fizeau's experiment, confirming to a higher precision the result favorable to the partial dragging of the ether suggested by Fresnel. Then Michelson and Morley began to work on an interferometer more sensitive than the one used in 1881. The optical path length was considerably extended with the aim to increase the displacement of the fringes. The lengths l_1 and l_2 were augmented by a factor of ten by resorting to successive reflections in a set of mirrors (Figure 2.15). In order to absorb vibrations and facilitate the rotation of the arrangement, the table was a granite block floating on the mercury contained in a ring-shaped iron basin (Figure 2.16). The compensating plate shown in Figure 2.15 has the same thickness than the half-silvered plate; it guarantees

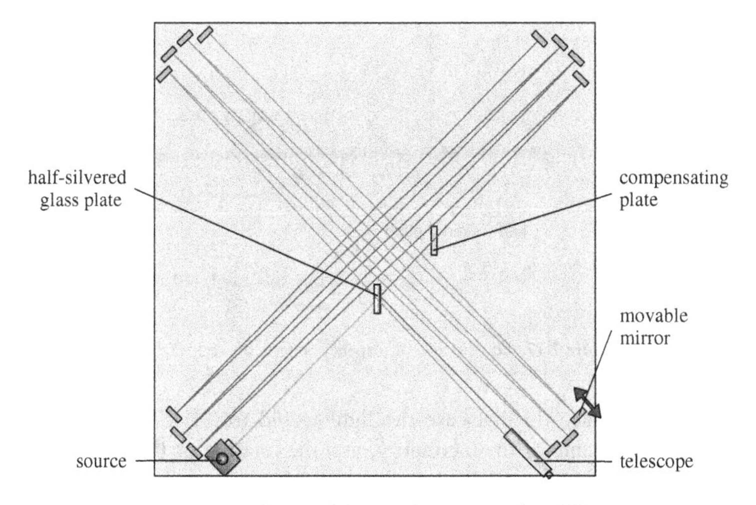

Figure 2.15. Design of the interferometer used in 1887.

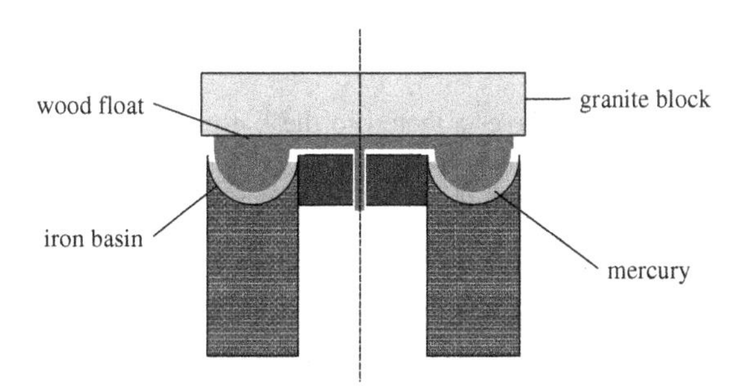

Figure 2.16. Vertical section of the mounting of the interferometer.

that both rays undergo equal dispersive effects, allowing the use of white light (Figure 2.17). Due to its highly non-monochromatic character (small coherence length), the white light requires that both rays have very close optical path lengths to preserve the interference pattern. Because of this reason the fringes are obtained by moving one of the mirrors with a millimetric screw. The light source was an Argand lamp (a type of oil lamp), where salt crystals could be burned in case that a nearly monochromatic light were needed. Since the lengths of the arms were $l_1 \cong l_2 \approx 11\,\mathrm{m}$ ($l/\lambda \approx 2 \times 10^7$), a shift of at least 0.4 fringes was expected (corresponding to $V^2/c^2 \sim 10^{-8}$). The experiment was performed during 4 days in July 1887, and the result was again null: the observed shifts were, at most, 20 times smaller than the minimum expected shift. Although Michelson and Morley promised to repeat the measurements every 3 months, they would not return to look for the "ether wind" but after 10 years, and separately. They were convinced that the null result was produced by the layer of ether adhered

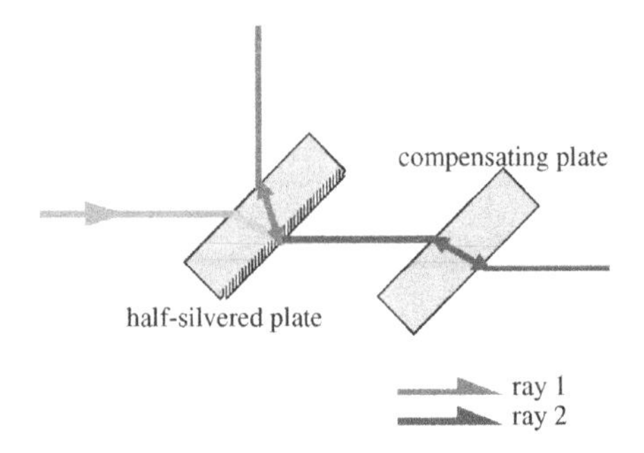

Figure 2.17. Equalization of dispersive effects on each ray.

to the Earth's surface. In that case the light would travel with speed c in all directions in the frame of the laboratory, and the rotation of the interferometer would have no effect at all. Michelson and Morley proposed that the experiment were repeated to a certain altitude above sea level—where, according to their conviction, the ether layer would become detached—using glass walls to isolate the device (with the intention of moderating the debate about whether the cellar they utilized as laboratory transported or not the ether contained in its interior).

The idea of an ether layer adhered to the Earth was the same hypothesis already tried by Stokes, which could not offer a proper explanation of the stellar aberration (see §2.6). On the other hand, it was in disagreement with the result obtained by Fizeau in 1851—and confirmed by Michelson and Morley in 1886—which indicated that inside a substance in movement just a partial dragging takes place (of the magnitude suggested by Fresnel). In 1892, O. Lodge tried to determine if a light ray passing tangent to a wheel that turns at great speed receives some effect from the ether layer that the wheel would possibly carry adhered: he did not observe any effect.

2.11. FITZGERALD-LORENTZ LENGTH CONTRACTION

In 1892 the Dutch physicist Hendrik Antoon Lorentz (1853–1928) showed that the null result of Michelson–Morley experiment could be understood by assuming that the ether exerts an action on moving bodies which contracts their lengths. The contraction should take place along the direction of motion. Instead, the dimensions that are transversal to the direction of motion would not undergo changes. If l_0 is the length of one of the dimensions of a body at rest relative to the ether, then this length would decrease to

FitzGerald–Lorentz length contraction

$$l = l_0\sqrt{1 - \frac{V^2}{c^2}} \tag{2.14}$$

when the body moves with respect to the ether at velocity V along that dimension. In such a case, the length l_2 in Eq. (2.9) should be replaced by $l_2\ (1 - V^2/c^2)^{1/2}$, but l_1 would not be affected. Then, the travel times t_2 and t_1 in Eqs. (2.9–2.10) would have the same functional dependence on V/c. Therefore, the rotation of the interferometer (i.e., the interchange of roles between rays 1 and 2) would not introduce any change in the phase difference of the rays, and the null result of the experiment would be justified.[8]

The Irish mathematician and physicist G.F. FitzGerald had already proposed the contraction hypothesis, in a qualitative way, in a scarcely divulged article written in 1889 that was not known by Lorentz. Since Eq. (2.14) will reappear within the context of relativistic ideas, it is necessary to emphasize that both Lorentz and FitzGerald regarded the contraction as a real and objective dynamical effect. The contraction was thought as a consequence of the interaction between the moving body and the ether. Therefore, such effect should be perceived in any frame, so meaning that the contraction was an absolute fact coming from the absolute motion of the body. The body had the same contracted length in all reference systems, in agreement with the prejudice about the invariance of distances.

Actually, Lorentz did not write Eq. (2.14) but its lowest order approximation (anyway, he only tried to explain a null effect experimentally verified up to the second order in V/c). Lorentz included the contraction hypothesis in the first of his three articles (1892, 1895, and 1899), which were devoted to explain the electromagnetic properties of matter by assuming that matter would be composed of elemental electric charges. Lorentz successively called them charged particles, ions, and electrons (at that time the atomic structure was just to be discovered). In *Theory of Electrons* (Lorentz, 1909) the elemental charges move inside a *stationary* ether—identified with the absolute space—creating electric currents. Concerning the propagation of electromagnetic waves in transparent substances, the lowest approximation of Lorentz's theory leads to results that are comparable with Fresnel's partial dragging hypothesis. Moreover, it avoids the problems of Fresnel's dragging in dispersive media (in these media the refractive index depends on the frequency and the dragging coefficient should be different for each frequency). In his theory of electrons, Lorentz dynamically explains all the effects depending on the first order in V/c. However, the length contraction appears as an *ad hoc* hypothesis.[9]

---

[8] Even so, Δt still depends on V. Therefore, it could be expected that the change of V month-by-month—and daily variations as well—modifies the value of Δt. In 1932, R.J. Kennedy and E.M. Thorndike examined a *fixed* interferometer during several months, without observing apparent changes in Δt.

[9] In his theory of electrons, Lorentz obtains macroscopic properties, such as conductivity, electric permitivity, and magnetic permeability, from Maxwell's laws for the electromagnetic field in the presence of just elemental charges moving in stationary ether (which acquires the status of sole continuous medium). In the same way he also describes optical effects like dispersion of light and rotation of the polarization plane by a magnetic field (Faraday rotation). In 1895, Lorentz proposes the force (1B.4) exerted by an electromagnetic field on an "ion."

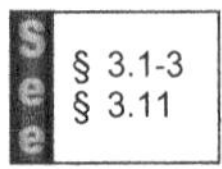
§ 3.1-3
§ 3.11

Length contraction acquired the form of Eq. (2.14) when J. Larmor related the contraction hypothesis to the coordinate transformation leaving invariant the Maxwell's laws. These had been independently obtained by Lorentz (1899, 1904) and Larmor (1900) and were called *Lorentz transformations* by Poincaré.[10]

2.12. THE TWILIGHT OF THE ETHER

The null result obtained by Michelson and Morley was confirmed in several repetitions of the experiment (Michelson, 1897; Morley-Miller, 1898, 1905, 1906; Tomaschek, 1924). D.C. Miller worked in open-air at Mt. Wilson (1742 m) by using a glass box to protect the device, being the only one to report non-null results (1922, 1925). Later Miller's results were attributed to a bad control of temperature (which was crucial for the stability of the dimensions of the device). Subsequent experiments reiterated the null result (Kennedy (1926), Illingworth (1927), Piccard and Stahel in a hot-air balloon (1926) and at Mt. Rigi (1927), Joos (1930), etc.).

At the beginning of the twentieth century the hope of revealing the state of absolute motion by means of electromagnetic phenomena started to fade. Not only the interferometric experiments but other negative results went in the same direction: the absence of birefringence caused by the contraction (Rayleigh, 1902; Brace, 1904) and the absence of torque between the plates of a charged condenser (Trouton and Noble, 1903).

In view of the disconcert produced by the null results, there appeared attempts of explaining them outside the context of the wave equation, but keeping in force the belief in the invariance of distances and times that lead to the Galilean addition of velocities. The *emission theories* proposed that light travels with speed c in *the reference system fixed to the source* (Ritz, 1908). It is clear that this hypothesis cannot explain by itself all experimental results. In the case of Michelson–Morley experiment, the light source and the mirrors are at rest in the laboratory. Thus emission theories indicate that the speed of light is c relative to the laboratory, so it should not be expected that the experiment reveals the absolute motion of the Earth. However, Tomaschek (1924) used stellar light, and Miller used solar light in some occasions, without obtaining different results.[11]

emission theories

While the scientific community was debating about *dynamical* explanations for the results of the experiments (partial dragging of ether, length contraction caused by the interaction with the ether), Albert Einstein (1879–1955)—a physicist circumstantially working at the patent office in Bern—published in

Einstein: Maxwell's laws are valid in any inertial frame

[10] In 1887, W. Voigt obtained Lorentz transformations apart from a global scale factor $(1 - V^2/c^2)^{1/2}$.

[11] It could be argued that, even if the source of light is located outside of the Earth, the light is then re-emitted by the half-silvered mirror of the interferometer. Therefore the emission theories could not be ruled out. However, in 1964, G.C. Babacock and T.G. Bergman proved in the laboratory that the speed of light does not change after it passes through a moving transparent slide.

1905 an article where he exposed a completely different vision of the problem (1905b). Einstein thought that Maxwell's laws are fundamental laws of Nature that should then integrate the set of laws satisfying the Principle of relativity, so remaining valid in any inertial frame. In Einstein's opinion, the electromagnetic field has its own entity and does not need a "materialization" by means of the idea of ether. Electromagnetism is a theory by itself and does not need Mechanics to support it. According to Einstein the inertial frames are not provided with a property "**V**." The absolute motion and the "ether wind" are notions deprived of physical meaning. Only *relative* motion has a physical meaning.

The validity of Maxwell's laws in all inertial reference systems implies the elimination of the privileged frame, so meaning that the notion of absolute motion does not have physical content at all. Besides it implies that *light propagates with the same velocity* c in all inertial frames. Since the invariance of a finite velocity is not compatible with Galilean addition of velocities (1.7), then the path followed by Einstein demands the rejection of the prejudices about the invariance of distances and times that supported Galilean transformations (1.5–1.6). These classical notions of space and time must be replaced by other notions of space and time being compatible with the validity of Maxwell's laws in all inertial frames.

classical notions of space and time must be abandoned

It is evident that Einstein's point of view easily explains null results like those of Michelson–Morley and Trouton–Noble. If the speed of light in vacuum is c in all laboratories, for any direction of propagation, then there is not change in the travel times when the interferometer is rotated. If Maxwell's laws are valid in all laboratories, then there is no torque on the plates of a condenser at rest relative to any laboratory. So, it can be supposed that these experiments influenced Einstein's thought. However, Einstein pointed out in several occasions that "partial dragging of ether" was significant enough to suggest that our notions of space and time should be changed. In Einstein's perspective, Eq. (2.5) should not be interpreted as the result of a dynamical interaction between a transparent substance and ether. Instead, Eq. (2.5) should be regarded as a composition of velocities—at the lower order in V/c—expressing the speed of light in a material medium that moves with velocity V *relative* to the laboratory. But such kind of composition of motions demands to abandon our prejudices about the nature of space and time.

Although Einstein's position could appear as the "simplest" one, the scientific thought at the end of the nineteenth century was structured on the *classical* concepts of space and time, which are those that seem to us *a priori* natural in the context of daily life. The coordinate transformations leaving invariant Maxwell's laws—the Lorentz transformations (see §3.11)—were known before Einstein's work. However, they did not receive the interpretation that Einstein's theory will confer them. Length contraction was regarded as an absolute fact coming from the interaction between matter and ether, and "time dilatation" did not affect, according to Lorentz, the absolute time but involved a "mathematically auxiliary time."

J.H. Poincaré glimpsed what was about to come. In 1899, Poincaré considered "very probable that optical phenomena depend only on the relative motions of material bodies, luminous sources and optical apparatus concerned, and that this is true not merely as far as quantities of the order of the square of aberration, but rigorously." In spite of this conceptual approach to Einstein's theory, Poincaré would not reach a full understanding of it. At least this is the conclusion if one considers that still in 1909 Poincaré thought that the length contraction had to be added as an additional hypothesis to Einstein's postulates.

By proposing a new way to regard space and time, Einstein produced in addition a deep change in the laws of Dynamics, and the allowed form of the laws for interactions (a form that Maxwell electromagnetic theory certainly has). The set of fundamental laws of classical mechanics was built on the observance of the Principle of relativity under Galileo transformations. When these laws were adjusted to satisfy the Principle of relativity under Lorentz transformations, an unsuspected consequence appeared: the possibility of turning mass into other types of energy.

3

Space and Time in Special Relativity

3.1. POSTULATES OF SPECIAL RELATIVITY

In 1905, Einstein published the article entitled "On the electrodynamics of moving bodies" (1905b), where he reformulated the notions of space and time starting from two postulates:

> *...the same laws of electrodynamics and optics will be valid for all frames of reference for which the equations of mechanics hold good.* (Principle of relativity)
>
> *...light is always propagated in empty space with a definite velocity c which is independent of the state of motion of the emitting body.*

The postulates refer to the validity of Maxwell's laws in all inertial frames. In this way, Einstein includes Maxwell's laws in the set of fundamental laws satisfying the Principle of relativity and eliminates any possibility of detecting the state of (absolute) motion of an inertial frame by means of electromagnetic experiments:

> *...the phenomena of electrodynamics as well as of mechanics possess no properties corresponding to the idea of absolute rest.*

The invariance of a finite velocity, in this case the speed of light, is not compatible with Galileo transformations (1.6), in particular with Galilean addition of velocities (1.7). Therefore Einstein's postulates imply the abandonment of Galileo transformations and, along with them, the assumptions about the nature of space and time that supported them: the invariance of distances and times. As already pointed out in §1.4, these two assumptions mutually imply; the abandonment of one of them implies the abandonment of the other one. In fact, let us call *proper length* L_0 the length of a bar in a reference system where the bar is at rest, and let us suppose that the same bar has a different length in a reference system moving relative to the bar. This supposition does not privilege any reference system, because the length will be equal to L_0 in *any* reference system where the bar is at rest, and

proper length

the change of length will be the same in *any* reference system where the bar exhibits the same relative movement. Let us consider a particle that moves along the bar with speed V relative to the bar (Figure 3.1). In the bar frame the particle takes a time $\Delta t = L_0/V$ to travel between the ends of the bar (Figure 3.1a). Instead, in the particle frame the length of the bar is different—let us say L—and the time elapsed between the same pair of *events* (i.e., the passages of the particle in front of each end of the bar) is $\Delta\tau = L/V$. In these results we take advantage of the fact that *the same velocity V characterizes the relative motion in both frames* (although the directions are opposite; see Figure 3.1b). The quotient between both times results

relative distances require relative times

$$\frac{\Delta t}{\Delta\tau} = \frac{L_0}{L} \tag{3.1}$$

This means that relative lengths imply relative times, and absolute lengths ($L = L_0$) imply absolute times ($\Delta\tau = \Delta t$).

proper time

Let us call *proper time* $\Delta\tau$ the time interval between two events when measured in the reference system where the events happen at the same position (whenever such a reference system exists). In the former example, the *proper frame* of the pair of events is the particle frame (Figure 3.1b).

The Newtonian notions of absolute space and time suppose that each side in Eq. (3.1) is, separately, equal to 1. However, in Special Relativity the notions of space and time will be subordinated to the postulate of invariance of the speed of light. Let us accept then that the length and time ratios, L_0/L and $\Delta t/\Delta\tau$, can differ from 1 as a consequence of this postulate. In that case, the only variable that can determine the value of each quotient is the relative velocity between the frames where the terms of the quotient are measured. In Figure 3.1, L_0/L could only depend on the modulus $V \equiv |\mathbf{V}|$ of the relative velocity. In fact, it cannot depend on the position of the body or its orientation because it is accepted that

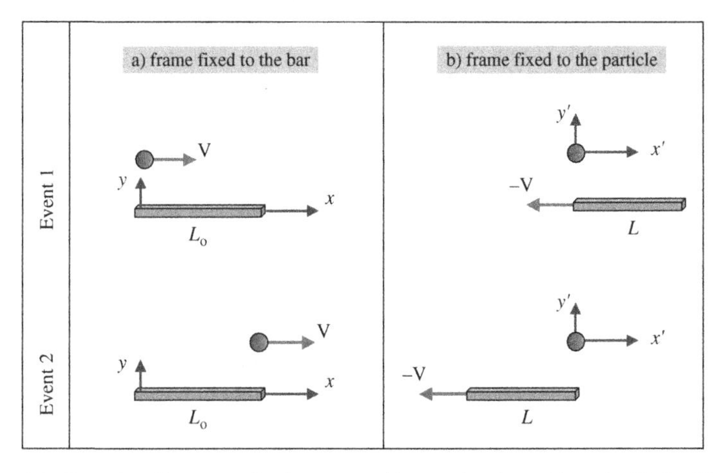

Figure 3.1. Events 1 and 2 correspond to the passages of the particle in front of the ends of the bar. The pair of events is represented in the frame fixed to the bar (left) and the frame fixed to the particle (right).

space is homogeneous and isotropic and has an Euclidean geometrical structure. It cannot depend on time either because time is believed to be homogeneous.

An equal analysis is valid for the ratio between proper time $\Delta\tau$ passed in the proper frame of a pair of events, and the corresponding time interval Δt in a reference system moving with respect to the proper frame. These considerations are entirely general, and could be applied to the case of Figure 3.1 and others. However, in Figure 3.1 the *same* relative velocity V determines both quotients L_o/L and $\Delta t/\Delta\tau$, i.e., the speed of the relative motion bar-particle. Therefore Eq. (3.1) means that the function of V determining the ratio L_o/L coincides with the one fixing the ratio $\Delta t/\Delta\tau$. Let us call $\gamma(V)$ such function:

$$\frac{L_o}{L} = \gamma(V) \qquad\qquad \frac{\Delta t}{\Delta\tau} = \gamma(V) \tag{3.2}$$

3.2. LENGTH CONTRACTIONS AND TIME DILATATIONS

Function $\gamma(V)$ in Eq. (3.2) will result from the postulate of invariance of the speed of light. Let us consider the case of Figure 3.2, where a source of light is placed at the end of a bar of a proper length L_o. The source emits a light pulse that is reflected in a mirror located at the other end of the bar, then returning to the position of the source. In the bar frame, where the length of the bar is L_o, the emission and the reception of the pulse are a pair of events happening at the same position (the location of the source). Thus the elapsed time in the bar frame is, in this case, the proper time; its value is $\Delta\tau = 2L_o/c$ (the speed of light is c in any inertial reference system). In another frame S the time Δt elapsed between both events reflects that the bar moves with speed V and its length is L instead of L_o. By decomposing Δt as $\Delta t = \Delta t_{\text{going}} + \Delta t_{\text{return}}$, the time of each stage can be obtained by calculating the distance covered by the light. Since the speed of light is c *as well* in frame S, the paths measure (Figure 3.2):

$$c\,\Delta t_{\text{going}} = V\,\Delta t_{\text{going}} + L \qquad\qquad c\,\Delta t_{\text{returning}} = L - V\,\Delta t_{\text{returning}} \tag{3.3}$$

Then

$$\Delta t = \Delta t_{\text{going}} + \Delta t_{\text{returning}} = \frac{L}{c-V} + \frac{L}{c+V} = \frac{2L}{c}\frac{1}{1-\dfrac{V^2}{c^2}} \tag{3.4}$$

In this case the relation between Δt and the proper time elapsed between the pair of events is

$$\frac{\Delta\tau}{\Delta t} = \frac{L_o}{L}\left(1 - \frac{V^2}{c^2}\right) \tag{3.5}$$

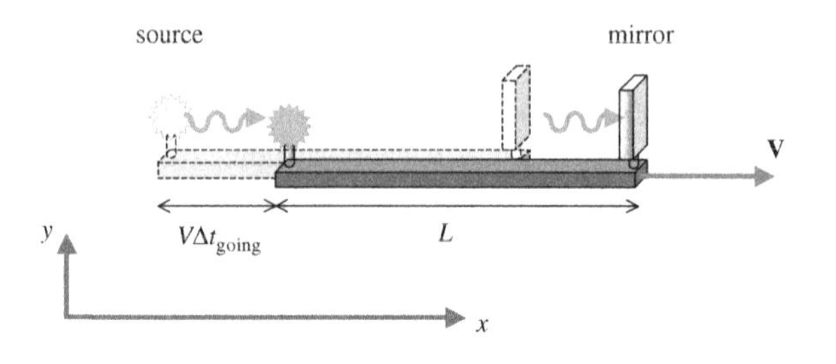

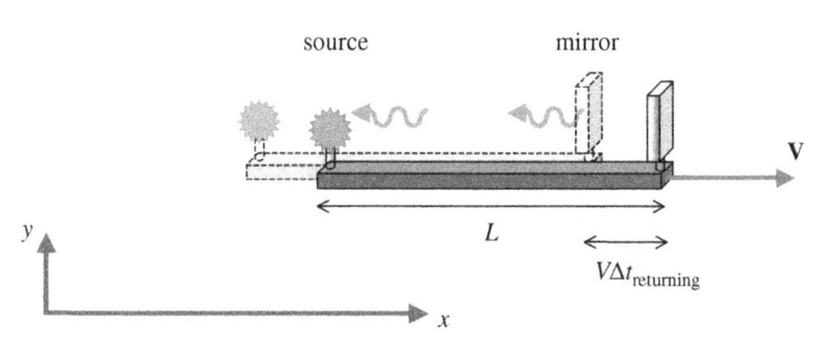

Figure 3.2. A light pulse performs a roundtrip between the ends of a bar that moves with speed V relative to the frame S.

Combining Eqs. (3.5) and (3.2) it is obtained

$$\gamma(V) = \frac{1}{\sqrt{1-\dfrac{V^2}{c^2}}} \tag{3.6}$$

Replacing in Eq. (3.2):

length contraction

$$L = L_o \sqrt{1-\frac{V^2}{c^2}} \tag{3.7}$$

time dilatation

$$\Delta t = \frac{\Delta\tau}{\sqrt{1-\dfrac{V^2}{c^2}}} \tag{3.8}$$

Equation (3.7) expresses the contraction of the dimension of a body along the direction of its movement relative to the reference system. Unlike FitzGerald–Lorentz contraction, which was an absolute contraction produced by the interaction with ether, Eq. (3.7) means that the length of a same bar is different in different frames, being maximal in the bar frame. Whereas V in FitzGerald–Lorentz contraction was the absolute speed of the body, V in Eq. (3.7) is its velocity relative to the reference system where its length is L_o.

Equation (3.8) refers to the time passed between a pair of events accepting a reference system where both of them occur in the same position (proper frame of the pair of events). The time interval between the events is minimal in the proper frame, and is indicated with $\Delta\tau$ (proper time). Equation (3.8) expresses the time interval Δt passed between the same pair of events in another frame moving with speed V relative to the proper frame (time dilatation).

In the range of relative velocities much smaller than the speed of light—the range of daily life—the effects of length contraction and time dilatation are too small to be perceived by our senses. Even for a relative speed V equal to the tenth part of the speed of light, lengths and times would be modified only by 0.5%. This is the reason why our prejudice about the absolute nature of distances

Albert Einstein (1879–1955). Born in Ulm (Germany), Albert Einstein spent his childhood in Munich where his family had an electromechanical industry. The crisis of the family company forced his parents and sister to move to Milan in 1894, and then to Pavía. Albert remained in a school in Munich, but soon fled to be with his family. At that time he decided to resign his German citizenship, and all religious dogma. He finished high school in Switzerland, and enrolled at the Polytechnic of Zurich. There he graduated in 1900 as a schoolteacher in mathematics and physics. He obtained Swiss citizenship in 1901. In 1903, Albert married Mileva Marič, a student of the Polytechnic, with whom he had three children: a daughter born before the marriage, whose destiny is not known, and two sons. Albert did not get a permanent teaching position, so the father of his companion Marcel Grossmann recommended him for a job at the patent office in Bern. There he found time to devote himself to research, although in a place far from the academic circles. While working in this office, in 1905, Einstein wrote his articles on photoelectric effect, Special Relativity, and Brownian motion. The last one led him to earn his Ph.D. at Zurich. In 1908 he obtained a position at the University of Bern. Subsequently, Einstein was designated as professor at Zurich, Prague, Berlin, and Princeton. Between 1907 and 1916 he developed the relativistic formulation of gravitational interaction (General Relativity), with later applications to cosmology. Other works treated statistical physics and quantum theory. In 1919 he divorced Mileva and married his cousin Elsa. In 1921, Einstein received the Nobel Prize in Physics for his discovery of the law of the photoelectric effect.

In politics, Einstein considered himself an internationalist liberal Jew. In religion, he felt imbued with what he called cosmic religiosity, which he considered the highest stimulus for the scientific research: "Only those who realize the immense efforts and, above all, the devotion which pioneer work in theoretical science demands, can grasp the strength of the emotion out of which alone such work, remote as it is from the immediate realities of life, can issue" (1934). With Einstein away, in 1933, Nazism confiscated his goods in Berlin; Einstein never returned to Germany. The increasing anti-Semitism moved him to support the creation of a Jewish state in Palestine. Once settled in Princeton, Einstein decided to write to President Roosevelt, a month before the beginning of World War II, urging him to go ahead with the project of a nuclear bomb, in view of the risk that Hitler might obtain it first. Einstein was always an anti-war activist, and felt great repentance when he knew that the bomb would be really used. In 1952, Einstein declined the offer of presiding over the State of Israel. He died in Princeton, and his body was cremated.

and times seems to agree with daily experience. These prejudices are nothing but good approximations when $V << c$.

the speed of light as a limiting velocity

See § 3.11 §, 4.1

Length contraction (3.7) and time dilatation (3.8) become singular when $V = c$. Actually, if the bar moved with speed $V \geq c$ relative to frame S in Figure 3.2, then the pulse of light would not reflect in the mirror because the light propagating with speed c would be unable to reach the mirror. In order that the events can have absolute character (i.e., the occurrence of an event is a fact verifiable in any reference system), it should not be feasible that a body reaches the speed of light. The invariant speed would be a limiting velocity as well. As it will be seen in Chapter 6, relativistic Dynamics prevents that an accelerated body reaches the speed of light.

3.3. THE MUON JOURNEY

One of the first experiments that proved the phenomenon of time dilatation and length contraction (Rossi and Hall, 1941; Frisch and Smith, 1963) consisted in the observation of populations of high-energy cosmic-ray muons, to different altitudes above sea level. The muons are unstable particles that disintegrate to give an electron, a neutrino, and an antineutrino. The muon half-life is $T = 1.53 \times 10^{-6}$ s. This means that an initial population of muons will be reduced to half after that time (in statistical terms). This period has been measured for low-energy muons, i.e., muons approximately at rest; so T is a proper time because it is measured in the proper frame of muons. Frisch and Smith examined populations of high-energy muons that are produced by cosmic rays entering the atmosphere (the muons' speeds were 99.52% of the speed of light). While the muons travel toward the Earth surface, they disintegrate; in this way the minor the altitude above sea level, the lower the population of muons. The experimental results indicated that the population diminishes $(27.5 \pm 3.0)\%$ for an altitude difference of 1908 m. Notice that a muon traveling at the speed of $0.9952\ c$ covers 1908 m in a time $\Delta t = 1908\,m/(0.9952\ c) = 6.39 \times 10^{-6}$ s, i.e., around four times its half-life. According to Newtonian notions of space and time, the population after that time would be reduced to $(1/2)^4 = 1/16$ of the original population, a reduction much greater than that really observed.[1] However, since the mentioned half-life was measured in the proper frame of muon, we should transform Δt to the proper frame before making statements about the reduction of the population. In the proper frame of muon, the trip demands a time

muons disintegrate slower in the frame where they move

$$\Delta\tau = \sqrt{1 - \frac{V^2}{c^2}}\Delta t = \sqrt{1 - 0.9952^2}\ 6.39 \times 10^{-6}\ \text{s} = 6.25 \times 10^{-7}\ \text{s}$$

[1] Although in real experiment the muons counted at different altitudes do not belong to a same set of muons, the conclusions are not altered because the populations of muons at different altitudes are stationary.

Then the population would be really reduced by a factor $(1/2)^{\Delta\tau/T} = (1/2)^{0.41} = 0.75$; i.e., it will diminish 25%, in excellent agreement with the experimental result.

The experiment here examined is nothing but a realization of the situation described in Figure 3.1, where the particle (the muon) covers the proper length $L_o = 1908\,\mathrm{m}$ in $6.39 \times 10^{-6}\,\mathrm{s}$, at the speed $V = 0.9952\,c$. In the proper frame of the muon the length is contracted to $L = (1 - 0.9952^2)^{1/2} 1908\,\mathrm{m} = 186.72\,\mathrm{m}$, being then covered in a time of $6.25 \times 10^{-7}\,\mathrm{s}$ (at the same relative speed). In this way, time dilatation and length contraction are two complementary visions of a same physical fact.

3.4. LENGTHS TRANSVERSAL TO MOTION

The former results on length contractions and time dilatations are direct consequences of the existence of an invariant finite velocity. On the contrary, the postulate of invariance of the speed of light does not affect the dimensions that are transversal to the direction of movement. Figure 3.3 describes a situation involving a dimension perpendicular to the relative motion between two reference systems. In frame S' the bar length D' is covered by a light pulse in a round-trip. Since the departure and return of the pulse occur at the same position in frame S', then the time elapsed between both events is a proper time in S', and its value is $\Delta\tau = 2D'/c$. In frame S the pulse is not perpendicular to the relative motion $S' - S$, as a consequence of the aberration of light. In S, both the emitted and reflected rays have the same angle of aberration, because all directions being perpendicular to the relative motion $S' - S$ are equivalent (this equivalence is due to the isotropy of space). The trajectory of the pulse is then symmetrical, as described in Figure 3.3. In spite of the aberration, the speed of light is c in S as well, in agreement with the postulate of invariance of

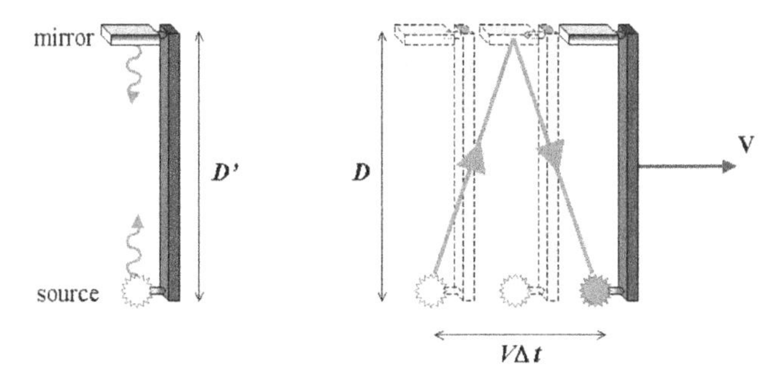

Figure 3.3. Trajectory of a light ray involving a dimension transversal to the relative movement between two frames, as viewed in the frame where the source and the mirror are fixed (left), and in a frame where they are in movement (right).

the speed of light. The time elapsed in the round-trip is calculated in frame S by means of Pythagoras theorem of Euclidean geometry, which leads to the result (2.10):

$$\Delta t = \frac{2\, D/c}{\sqrt{1 - \frac{V^2}{c^2}}} \tag{3.9}$$

In Eq. (3.9) D is the length of the transversal dimension in frame S. The quotient of time intervals corresponding to each frame is

$$\frac{\Delta t}{\Delta \tau} = \frac{D/D'}{\sqrt{1 - \frac{V^2}{c^2}}}$$

invariance of the lengths that are tranversal to the relative movement

Comparing with Eq. (3.8) one concludes that

$$D = D' \tag{3.10}$$

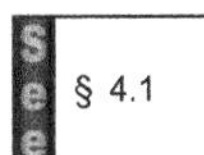
§ 4.1

i.e., dimensions that are transversal to the relative motion $S' - S$ do not undergo changes.

3.5. COMPOSITION OF MOTIONS

Figure 3.4 shows a particle moving in the direction of the relative movement between two frames S and S', with velocities $\mathbf{u}$ and $\mathbf{u}'$ relative to S and S' respectively. The pair of events indicated in Figure 3.4 corresponds to the passage of the particle by both ends of a fixed bar in S' whose proper length is L_o; so that $u'_x \Delta t' = L_o$. In frame S, the displacement $u_x \Delta t$ between these two events can be decomposed into the sum of the bar length L and its displacement $V\Delta t$:

$$u_x\, \Delta t = L + V\, \Delta t = \gamma(V)^{-1}\, L_o + V\, \Delta t = \gamma(V)^{-1}\, u'_x\, \Delta t' + V\, \Delta t \tag{3.11}$$

In Eq. (3.11) the intervals Δt and $\Delta t'$ can be substituted by using their relation with the proper time $\Delta \tau$. The proper time $\Delta \tau$ is, in this case, the time elapsed in the frame where the particle is at rest. Since the proper frame of the particle moves with velocities $\mathbf{u}$ and $\mathbf{u}'$ relative to S and S', then Eq. (3.8) shows that the intervals Δt, $\Delta t'$ can be written as

$$\Delta\, t = \frac{\Delta \tau}{\sqrt{1 - \frac{u^2}{c^2}}} = \gamma(u)\, \Delta \tau \tag{3.12a}$$

$$\Delta\, t' = \frac{\Delta \tau}{\sqrt{1 - \frac{u'^2}{c^2}}} = \gamma(u')\Delta \tau \tag{3.12b}$$

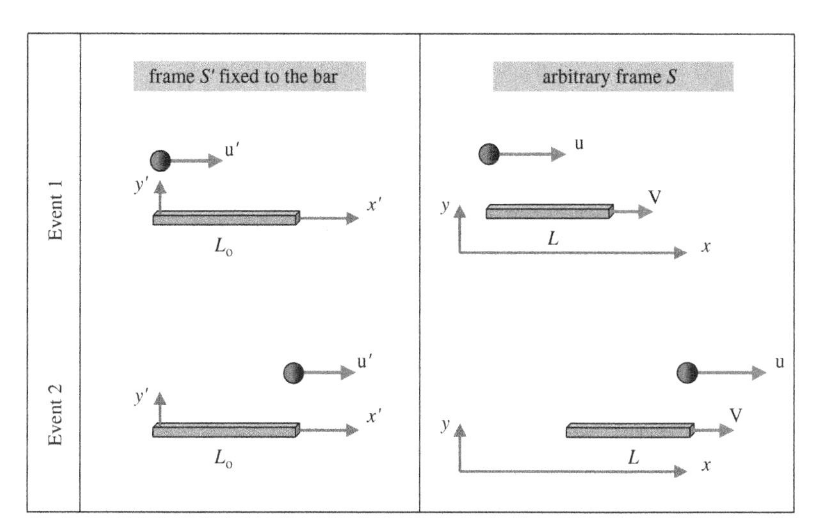

Figure 3.4. A pair of events defined by the passages of a particle by the ends of a bar, as they are viewed in the frame S' fixed to the bar (left) and an arbitrary frame S (right).

Therefore

$$\gamma(u')^{-1}\,\Delta\,t' = \gamma(u)^{-1}\,\Delta\,t \tag{3.13}$$

By replacing (3.12a,b) in Eq. (3.11) it results

$$\gamma(u')\;u'_x = \gamma(u)\;\gamma(V)\;(u_x - V) \tag{3.14}$$

Equation (3.14) is a relativistic transformation of velocities. For velocities much smaller than c, factors γ approximate to 1, hence Eq. (3.14) becomes the x component of the Galilean addition of velocities (1.7).

In order to get the inverse transformation of (3.14) we shall use an argument that is frequently useful to avoid unnecessary calculations. Because frames S and S' are on an equal footing—none of them is privileged at all—then the inverse transformation of (3.14) only differs from (3.14) in the change of V by $-V$. In fact, whereas S' moves relative to S with velocity $\mathbf{V} = V\,\hat{\mathbf{x}}$, S moves relative to S' with speed $\mathbf{V} = -V\,\hat{\mathbf{x}}$. Therefore the inverse transformation of (3.14) must be[2]

$$\gamma\,(u)\;u_x = \gamma\,(u')\;\gamma\,(V)\;(u'_x + V) \tag{3.15}$$

It is convenient to get the transformation of $\gamma(u)$, since this will allow us to achieve a direct relation between u'_x and u_x from Eqs. (3.14–3.15). Using Eq. (3.14) to substitute $\gamma(u')u'_x$ in (3.15), it results

$$\gamma\,(u)\;u_x = \gamma\,(u)\;\gamma\,(V\,)^2\;(u_x -\;V) + V\,\gamma\,(u')\;\gamma\,(V\,)$$

[2] Of course, the same argument applies to Galileo transformations, although their extreme simplicity makes the argument immaterial.

i.e.,

$$V\,\gamma(u')\,\gamma(V) = V\,\gamma(u)\,\gamma(V)^2 - \gamma(u)\,u_x\left(\gamma(V)^2 - 1\right)$$

By applying the identity

$$\gamma(V)^2 - 1 = \frac{V^2}{c^2}\,\gamma(V)^2 \tag{3.16}$$

it is obtained

$$\gamma(u') = \gamma(u)\,\gamma(V)\left[1 - \frac{u_x V}{c^2}\right] \tag{3.17}$$

whose inverse relation is

$$\gamma(u) = \gamma(u')\gamma(V)\left[1 + \frac{u'_x V}{c^2}\right] \tag{3.18}$$

Now we can rewrite Eqs. (3.14–3.15) by replacing Eqs. (3.17–3.18). Thus the transformation $u'_x \leftrightarrow u_x$ is obtained.

$$u'_x = \frac{u_x - V}{1 - \frac{u_x V}{c^2}} \qquad u_x = \frac{u'_x + V}{1 + \frac{u'_x V}{c^2}} \tag{3.19a–b}$$

Figure 3.5 shows the graph of the relativistic velocity transformation (3.19a). Notice that the inverse transformation (which can be regarded as the interchanging of axis) corresponds to replace V by $-V$.

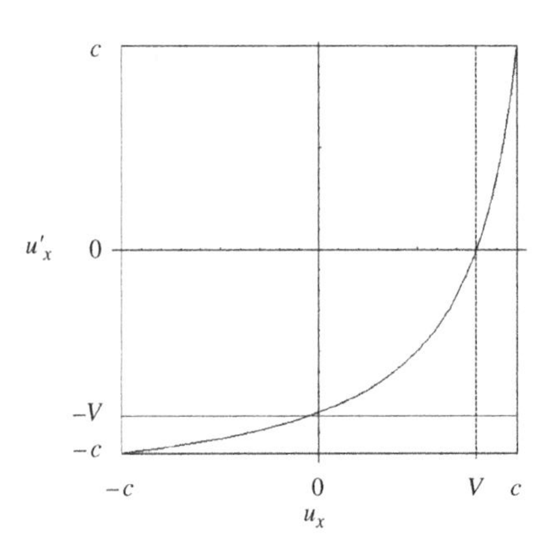

See § 4.6

Figure 3.5. Transformation of velocities (3.19), corresponding to $V = 0.8\,c$.

3.6. INTERPRETATION OF FIZEAU'S EXPERIMENT

Fizeau's 1851 experiment (§2.7) was considered until the beginning of the twentieth century as an experimental verification of the interaction between matter and ether. In particular, it confirmed with appropriate precision – improved by Michelson and Morley in 1886—the dragging coefficient proposed by Fresnel in 1818. In Fizeau's experiment it was believed that the water flow partially dragged the ether contained inside the water; the consequences of this dragging on the light velocity of propagation were detected by means of an interferometric device.

Since the theory of ether has been abandoned, it is natural to wonder how those results should be interpreted now. In Fizeau's experiment the light propagation in a medium (water) combines with the motion of the medium along a direction parallel to the direction of propagation. Let us then apply transformation (3.19b) to a light ray propagating inside a transparent substance of refractive index n. Light propagates with speed $u'_x = c/n$ relative to the transparent substance. According to Eq. (3.19b) the ray velocity in a frame S where the transparent substance moves with velocity V is

$$u_x = \frac{c}{n}\frac{1+\dfrac{nV}{c}}{1+\dfrac{V}{nc}} \qquad (3.20)$$

By approximating this result to the lowest order in V/c, it results

$$u_x \cong \frac{c}{n}\left(1+\frac{nV}{c}\right)\left(1-\frac{V}{nc}\right) \cong \frac{c}{n}\left(1+\frac{nV}{c}-\frac{V}{nc}\right) = \frac{c}{n}+\left(1-n^{-2}\right)V \qquad (3.21)$$

Fresnel's "dragging" is a relativistic composition of velocities

Let us compare (3.21) with (2.5b): both have the same aspect. However, u and V in Eq. (3.21) are the velocities of light and the transparent substance relative to an *arbitrary* frame S, respectively, whereas both velocities in Eq. (2.5b) are velocities relative to the universal ether. In Eq. (3.21) the speed of light is c/n in the frame fixed to the transparent substance[3] (the only frame with a natural privilege). Instead, in Eq. (2.5b) this is only true if the transparent substance is at rest relative to the universal ether. What Fizeau measured in 1851 was not the partial dragging of ether coming from the interaction between ether and the transparent substance. He measured a relativistic composition of velocities (it should be remarked that the first order result in Fizeau's experiment did not depend on "absolute" speeds, but it concerns with the speed of water *relative* to the laboratory, which is a well-defined velocity).

[3] In dispersive media the refractive index depends on the light frequency, so the value n to be used is the value corresponding to the frequency ν' measured in the system where the transparent substance is at rest.

3.7. TRANSVERSAL COMPONENTS OF THE VELOCITY

Those velocity components that are transversal to the relative movement between S and S' can be transformed by using the results of §3.4. The distance covered by the particle in the direction transverse to the relative motion $S' - S$ is equal in both frames:

$$u'_y \, \Delta t' = u_y \, \Delta t \qquad u'_z \, \Delta t' = u_z \, \Delta t \tag{3.22}$$

Equation (3.13) for the relation between $\Delta t'$ and Δt can also be applied here. In fact, Eq. (3.13) comes from Eqs. (3.12a,b) that express time dilatations for pairs of events happening along the trajectory of the particle. Factors γ depend only on the modulus of the relative speeds of the particle; it does not matter whether the speed directions coincide or not with a Cartesian axis. In other words, if Cartesian axes x, y, z were rotated in Figure 3.4, this action could not affect the relations between time intervals (3.12–3.13) even though the particle speed would acquire, in such case, non-vanishing components u_y and u_z. Then, replacing Eq. (3.13) in Eq. (3.22) one obtains:

$$\gamma(u') \, u'_y = \gamma(u) \, u_y \tag{3.23}$$

$$\gamma(u') \, u'_z = \gamma(u) \, u_z \tag{3.24}$$

If the speeds are much lower than c, then the factors γ approximate to 1 and Eqs. (3.23–3.24) become the transversal components of the Galilean addition of velocities (1.7). Using Eqs. (3.17–3.18):

$$u'_y = \frac{\sqrt{1 - \frac{V^2}{c^2}} \, u_y}{1 - \frac{u_x V}{c^2}} \qquad u_y = \frac{\sqrt{1 - \frac{V^2}{c^2}} \, u'_y}{1 + \frac{u'_x V}{c^2}} \tag{3.25a–b}$$

$$u'_z = \frac{\sqrt{1 - \frac{V^2}{c^2}} \, u_z}{1 - \frac{u_x V}{c^2}} \qquad u_z = \frac{\sqrt{1 - \frac{V^2}{c^2}} \, u'_z}{1 + \frac{u'_x V}{c^2}} \tag{3.26a–b}$$

3.8. THE NOTION OF SIMULTANEITY

So far we have examined several situations that involved pairs of events that allow a reference system where both events happen at the same position (proper frame). This means that the pair of events could be connected by a particle (traveling with velocity smaller than c). This will not always be the case. Let us consider two events happening at different positions, and *simultaneous* in a given reference system. In this case, there does not exist a proper frame where the events occur at the same position, because such condition would demand a proper frame moving with infinite speed relative to the frame where the events are simultaneous.

In classical physics the simultaneity of a pair of events has absolute character (independent of the reference system). The absolute character of the temporal coincidence of two events is a consequence of the as well absolute classical concept of time. In Relativity, however, the existence of an invariant finite speed deprives the time of its absolute character, so forcing us to review the notion of simultaneity. Unless both events occur at the same place—the only case where their simultaneity could not be disputed among different reference systems—the relativistic notion of simultaneity has to be examined in light of the existence of an invariant finite speed.

Figure 3.6 shows a vehicle having a configuration of back and front doors that are driven by light pulses detected by photoelectric sensors. A light source is placed halfway between the sensors. A light pulse is emitted by the source

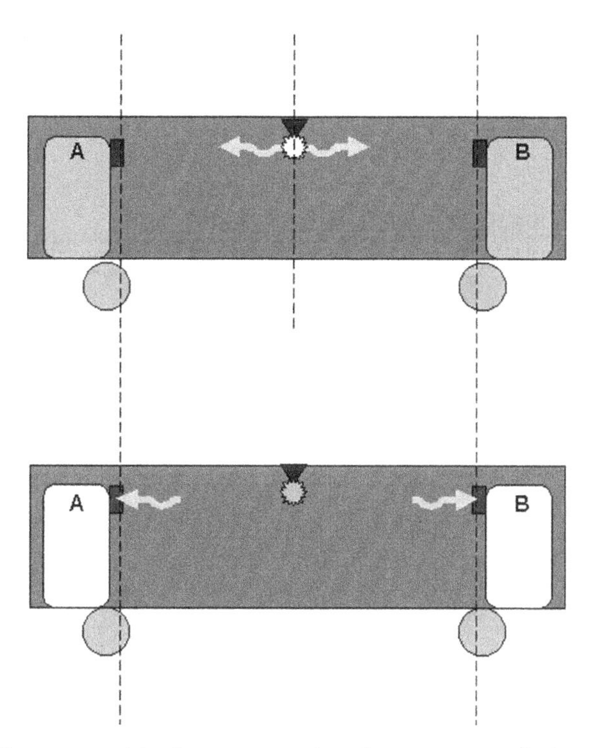

Figure 3.6. The openings of the doors are a pair of simultaneous events in the frame fixed to the vehicle, because the light covers with equal speed equal distances.

and covers, with equal speed in both directions, the equal distances between the source and the respective sensors. The opening of the doors is then a pair of simultaneous events in the vehicle frame.

Figure 3.7 shows the same situation as observed in a frame where the vehicle is moving. In this frame the light pulse also travels with the same speed c in both directions; however, whereas door A goes toward the pulse, door B moves away of the pulse. Therefore, in a reference system where the vehicle is moving, the back door opens before front door B: the events are not simultaneous.

simultaneity of events is a relative concept

It is immediately understood that the simultaneity of events is no longer an absolute concept (independent of the reference system), as a result of the

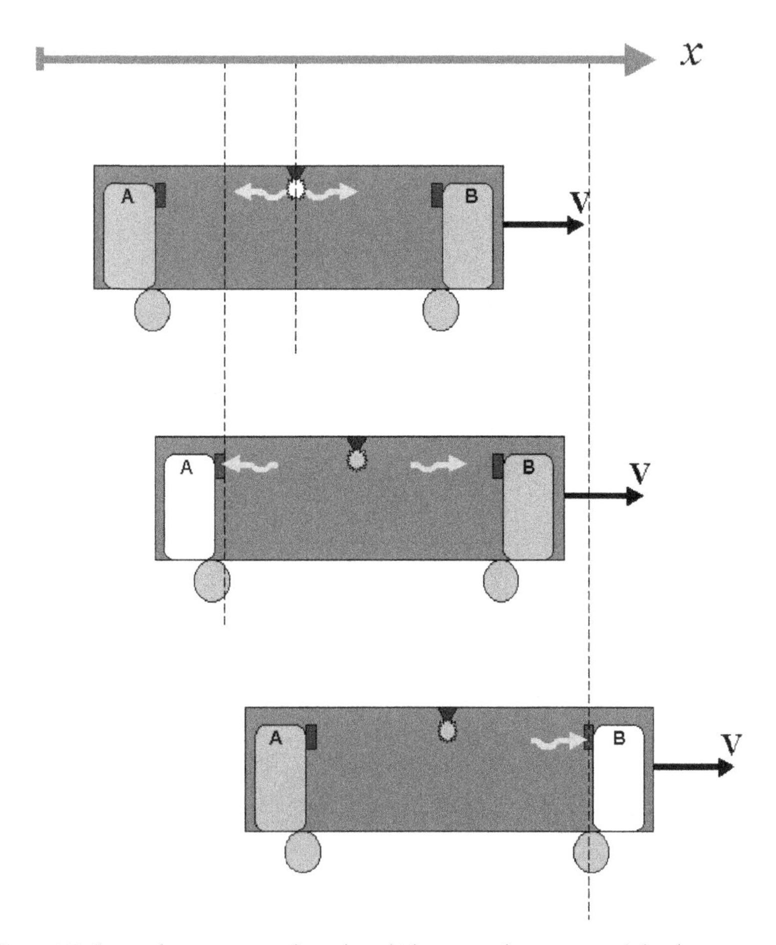

Figure 3.7. In a reference system where the vehicle moves, the openings of the doors are not simultaneous events because the light covers with equal speed different distances. The door going toward the light pulse opens first.

existence of an invariant finite speed. In fact, in Figure 3.7 the signal does not arrive simultaneously at each door: the doors have time to move from their positions equidistant from the source because the signal propagates with finite speed. If the signal propagated with infinite speed, the sensors of the doors would be reached by the signal "before" the vehicle (moving with finite speed V) could displace.

criterion of simultaneity

As a conclusion, we can establish a physical criterion of simultaneity of two events, which is based on the invariance of the speed of light. Two events are simultaneous in a reference system if they are coincident with the arrival of a light signal previously emitted from a position at equal distance from both events. In Figure 3.6, the light pulses coinciding with the events were emitted from a place equidistant from the positions where the events happen; the events are simultaneous in this reference system. In Figure 3.7, instead, the place where the signal is emitted is not equidistant of the positions where the events happen.

temporal coordinate of an event

The events that are simultaneous in a given reference system will be characterized by the same *temporal coordinate* in this frame. A way to measure the temporal coordinate of an event in a given reference system consists in distributing in the space a collection of clocks fixed to the reference system. The clocks are equally constituted: all of them change at the same rate; their locations in different positions do not modify this property because the space is homogeneous. The synchronization of the clocks (i.e., the procedure to manage that all clocks exhibit simultaneously the same reading) can be made using light signals. A pair of clocks A and B is synchronized if they are adjusted to a same reading (for instance, $t = 0$) at the moment when they receive a light pulse emitted from an equidistant position. After this procedure, the clocks A and B will remain synchronized, because the rates of the clocks are homogeneous. A third clock C can be synchronized with A (or B) in the same way. The procedure of interchanging light signals can be extended to synchronize an arbitrary number of clocks at relative rest. Once all clocks have been synchronized, the temporal coordinate of an event is the simultaneous reading of the clock located in the position where the event happens. Due to the relativity of the notion of simultaneity, this method of synchronization is only appropriate for the reference system where the clocks are fixed. Thus each reference system has its own way of assigning the temporal coordinate to the events.

synchronization of clocks

Any other way to use clocks to measure the temporal coordinate of an event is acceptable, whenever it is consistent with the criterion of simultaneity of events in the reference system (i.e., whenever the same temporal coordinate is assigned to simultaneous events in the reference system). For example, a unique clock can be used, provided that the time required by the information for traveling from the location of the event to the position of the clock is subtracted.[4]

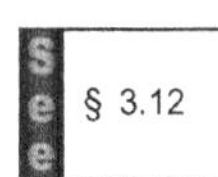
See § 3.12

[4] Usually distances are measured with rules and times with clocks; time and length units are considered independent. Nevertheless we could choose the value of the invariant c by definition, which would imply that time and length units would stop being independent (see Complement 1B). It is usual the non-dimensional choice $c = 1$, which means that the same unit is used for lengths and times.

3.9. EVENTS AND WORLD LINES

The position and time coordinates of an event in a given reference system S allow to represent the event as a point in a space-time diagram. Figure 3.8 shows the Cartesian coordinate x as the abscissa, and the temporal coordinate t times c as the ordinate (the Cartesian axes y and z were suppressed). Hence, both the abscissa and the ordinate have units of length.

The motion of a particle is a succession of events that draws a curve in a space-time diagram. This curve is called *world line*. If the motion is uniform (i.e., with a constant velocity $\mathbf{u}$), then the world line is a straight line. Figure 3.9 shows three world lines corresponding respectively to a particle at rest relative to S, a particle with uniform motion ($|\mathbf{u}| < c$), and a light ray. The three lines go through x_o at time $t = 0$. The particle at rest remains at position x_o for all time t; therefore its world line is a vertical straight line in the space-time diagram of the coordinates of the reference system S. Along a light ray, the elapsed times and the displacements fulfill $c\ \Delta t = \Delta x$; then the corresponding world line is a 45° straight line. Since the velocity of any particle is lower than c in any reference system, then the world lines of particles are curves whose slopes vary between the vertical (rest) case and the 45° straight line characterizing a light ray. In a space-time diagram, the tangent of the angle between the world line and the time axis is equal to $|\mathbf{u}|/c$.

The use of space-time diagrams is illustrated in Figure 3.10, where the problem studied in §3.2 is represented (see Figure 3.2). The movement of the bar with constant velocity V is shown in the space-time diagram by means of the world lines of the ends of the bar. These are two straight lines of equal slopes, because both ends move with the same velocity V. The round-trip of

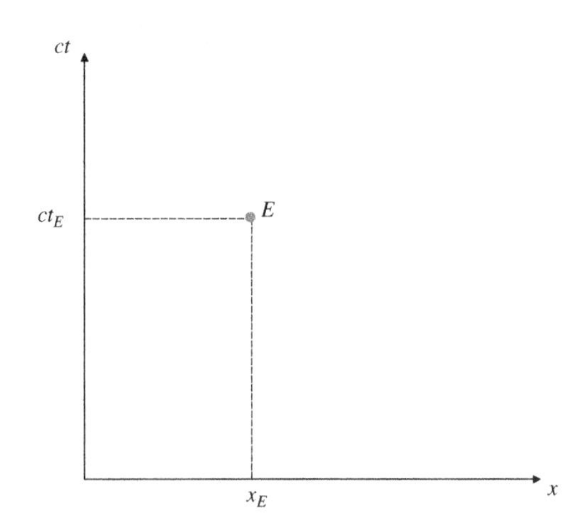

Figure 3.8. Representation in a space-time diagram of an event whose coordinates are (ct_E, x_E).

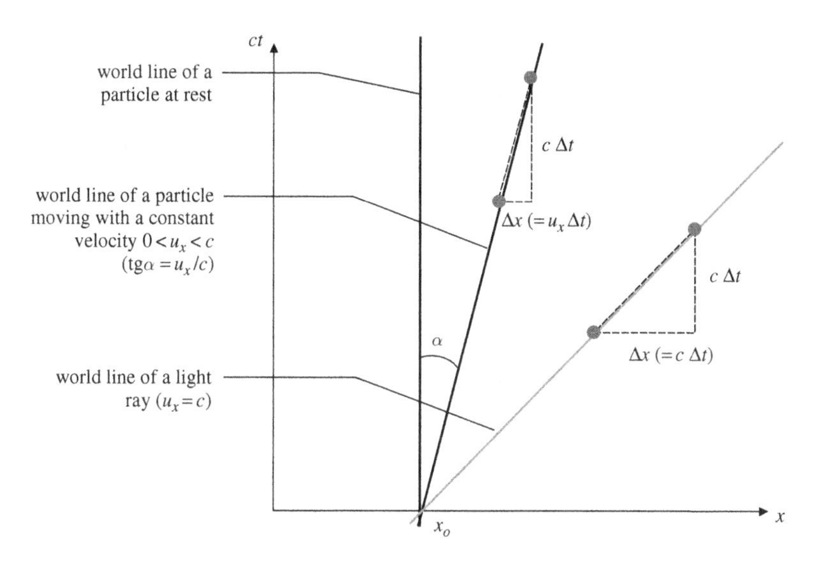

Figure 3.9. World lines of a particle at rest relative to S, a particle moving with a constant velocity, and a light ray.

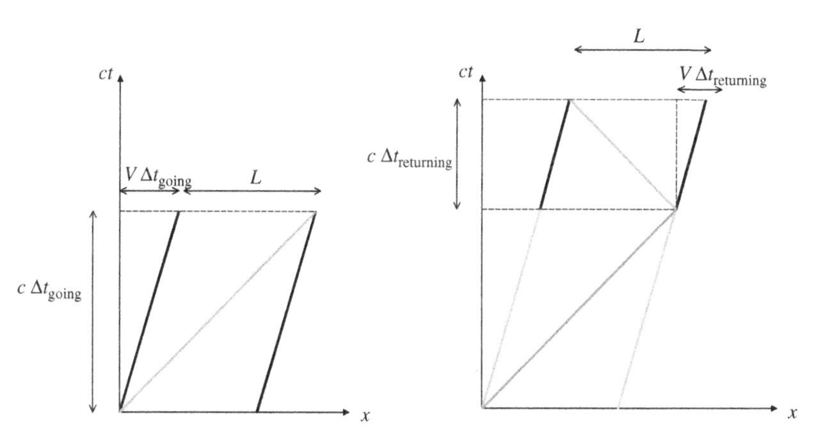

Figure 3.10. Outward (left) and return (right) journeys of a light ray between the extremes of a moving bar.

a light ray between the ends of the bar is represented by two 45° lines. These lines have opposite slopes because each one corresponds to the outgoing and returning rays, respectively. L in Eqs. (3.3) is the length of the bar in frame S; so L in the space-time diagram of Figure 3.10 is the distance between the world lines of the ends of the bar measured on a horizontal direction. In fact, each horizontal line in Figure 3.10 is made of a set of events that are

length and simultaneity

simultaneous in the reference system S associated with the coordinates of the diagram. Of course, the length L of the bar in a reference system S must be the distance between two simultaneous (in S) positions of the bar ends. Equations (3.3) appear in Figure 3.10 as the equality between the sides of 45° right triangles.

The relativity of the notion of simultaneity, which was analyzed in §3.8, is clearly displayed by using space-time diagrams. In Figure 3.6 let us call S' the reference system where the vehicle is at rest. The world lines of the source and the vehicle doors look as equidistant vertical straight lines when they are described with coordinates of frame S', because they are at rest relative to S'. The world lines of the light rays are, as always, 45° straight lines. In Figure 3.6 two rays travel in opposite directions, so they are represented with 45° world lines of opposite slopes. It then results in the space-time diagram shown in Figure 3.11, where the events 1 and 2 are the arrival of light pulses to the positions of the doors and the consequent simultaneous opening of the doors.

In §3.8 the same set of events was analyzed from another arbitrary reference system S, where the vehicle moved with velocity V (Figure 3.7). When coordinates of frame S are used, then the world lines of source and doors are equidistant straight lines of equal slope (equal velocity V). The world lines of light rays still are 45° straight lines, because of the invariance of the speed of light. Thus the space-time diagram shown in Figure 3.12 is obtained. The events 1 and 2 are not simultaneous in frame S, but the event 1 occurs first (opening of the back door).

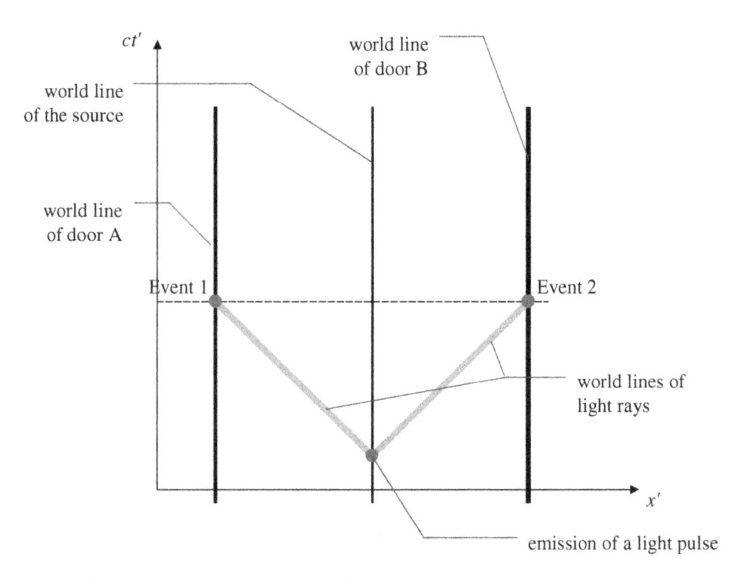

Figure 3.11. Space-time diagram for the problem analyzed in Figure 3.6.

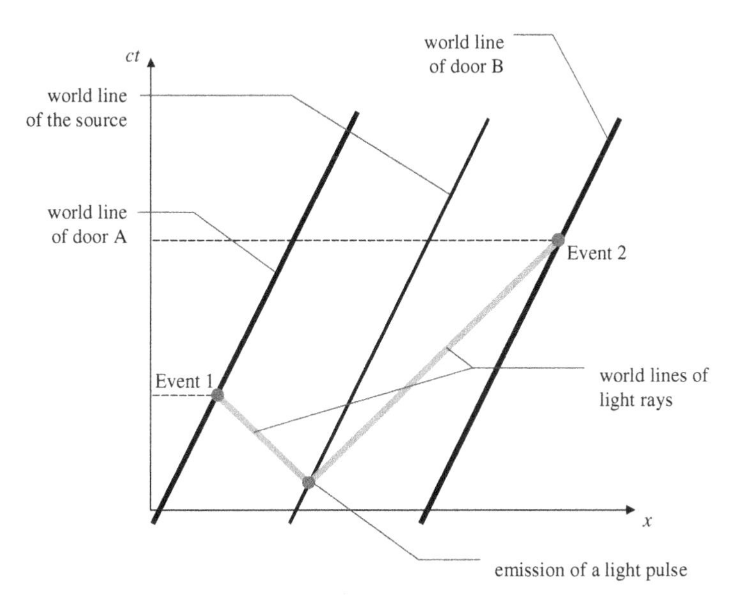

Figure 3.12. World lines and events taking part in Figure 3.11, as they look in coordinates of another reference system (space-time diagram for the problem analyzed in Figure 3.7).

3.10. COORDINATE LINES OF S' IN THE SPACE-TIME DIAGRAM OF S

Figure 3.13 shows *coordinate lines* belonging to two different reference systems S and S'. A coordinate line is a curve made up of points sharing a same value of all the coordinates except one of them. Along the vertical lines in Figure 3.13 the temporal coordinate varies while the Cartesian spatial coordinate remains

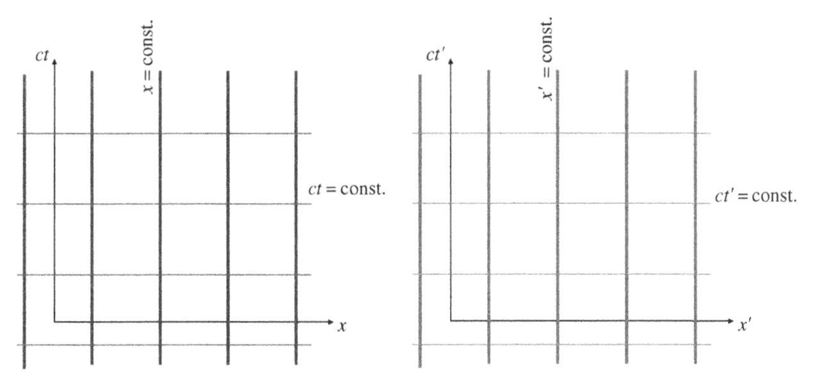

Figure 3.13. Coordinate lines of reference systems S and S' in their respective space-time diagrams.

unchanged (the rest of the Cartesian spatial coordinates are not represented but they should be considered likewise fixed). These vertical lines are world lines of particles at rest in each reference system. On the other hand, along the horizontal lines in Figure 3.13 one of the spatial coordinates varies while the rest of the coordinates remain fixed. The events lying on the horizontal lines are simultaneous events in the respective reference system.

We shall assume, as always, that the Cartesian axes x and x' have the direction of the relative movement between S and S'. We shall transfer the coordinate lines of one of the reference systems to the space-time diagram belonging to the other one. Let us begin with the vertical coordinate lines, which are, as already mentioned, world lines of particles at rest in the respective frame. This means that such lines look as world lines of moving particles in the space-time diagram belonging to another frame. If frame S' moves with respect to S toward increasing values of the x coordinate, then the lines $x' = constant$ look in the diagram of frame S as displayed in Figure 3.14a, while the lines $x = constant$ look in the diagram of frame S' as Figure 3.14b shows.

We shall now transfer the horizontal lines of Figure 3.13. These coordinate lines are made up of simultaneous events in the respective reference system, and they cannot be associated with movements of particles. Therefore we should imagine a physical procedure to build a set of events that results to be simultaneous in a given reference system. Then we shall see how such procedure is described in the space-time diagram of another reference system. Figure 3.15 shows a light ray traveling toward decreasing values of the coordinate x' of frame S'. When the ray goes through a set of equidistant positions $\ldots$, x'_4, x'_3, x'_2, x'_1, x'_0—marked by means of world lines of particles at rest relative to S'—light pulses are emitted toward the opposite direction. Afterwards the light pulses emitted at positions x'_n reach positions x'_{2n}, configuring a set of simultaneous events. The simultaneity of these events is guaranteed by the equidistance between consecutive positions x'_n , together with the independence of the speed

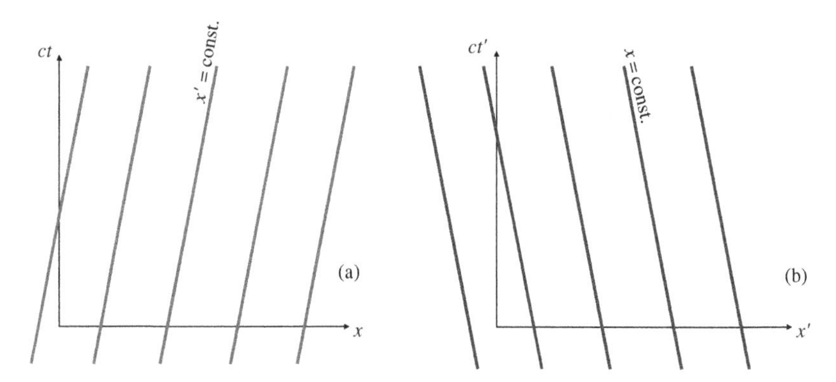

Figure 3.14. (a) Coordinate lines x'= const. represented in the space-time diagram of frame S. (b) Coordinate lines x = const. represented in the space-time diagram of frame S'.

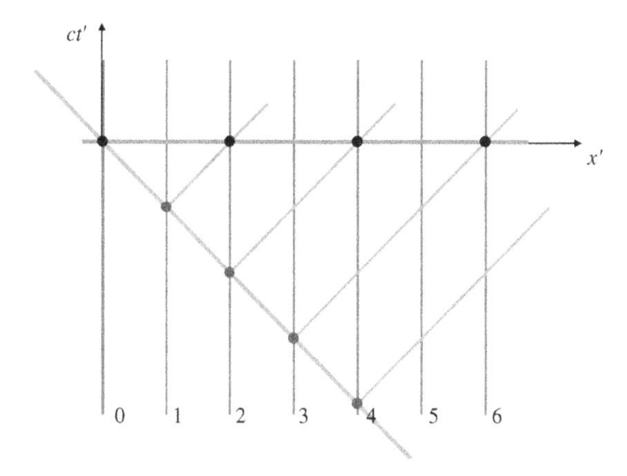

Figure 3.15. Building of a line of simultaneous events in a given frame S'.

of light with respect to the direction of propagation. In Figure 3.15 the coordinate $t' = 0$ was assigned to the instant where those events occur (choice of the origin of coordinate t'). Thus, the straight line joining the events is the x' axis of frame S'.

Figure 3.16 shows the procedure to draw the coordinate line $t' = 0$ in the space-time diagram of frame S. The freedom to choose the origin of coordinates of frame S was used to give coordinates $t = 0$, $x = 0$ in S to the event having coordinates $t' = 0$, $x' = 0$ in S'. Figure 3.16 shows that the bisectrix of the angle between the lines $t' =$ const. and $x' =$ const. is a light ray in frame S as well.

prescription for the coordinate origin

The same physical procedure involving light rays can be used to draw the lines $t =$ const. of frame S in the diagram of frame S'. In this case the lines of particles at rest in S will look as Figure 3.14b shows when they are transferred to the diagram of S'. Both results are displayed in Figure 3.17.

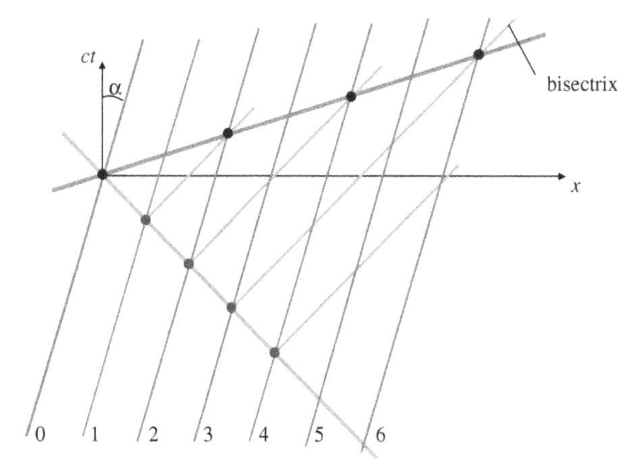

Figure 3.16. Simultaneous events in S', as they look in the space-time diagram of frame S. $\text{tg}\alpha = V/c$

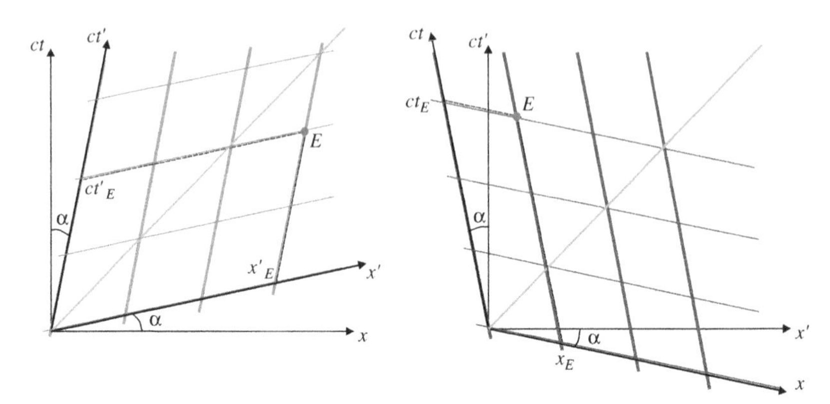

Figure 3.17. Aspect of coordinate lines of frame S' in the diagram of frame S (left), and coordinate lines of frame S in the diagram of frame S' (right). $\mathrm{tg}\alpha = V/c$

Figure 3.17 also shows the way of reading the coordinates of an event E belonging to the transferred reference system: of course, the coordinates of the event are those corresponding to the coordinate lines going through the event. However, in order that the coordinates assigned to an event in different reference systems can be compared on the same diagram, the axes of each frame must be properly calibrated since their scales will result to be different. In fact, Figure 3.18 shows a pair of events that happens at the same position ($x' = 0$) in

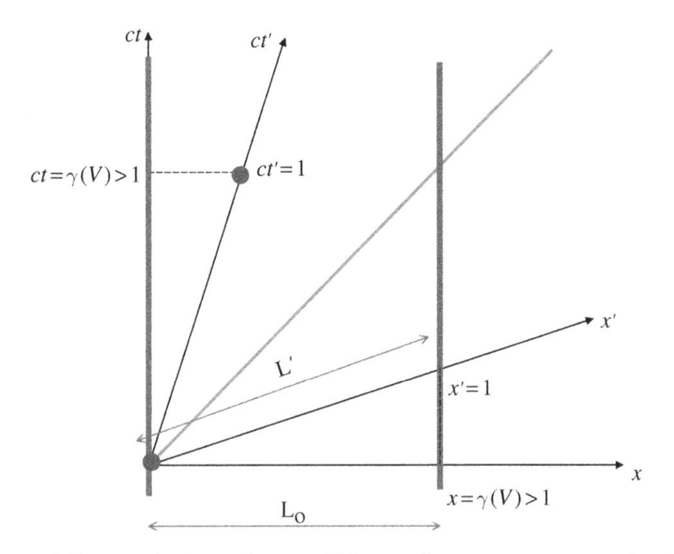

Figure 3.18. Calibration of scales on the axes of different reference systems represented in the same diagram.

frame S'. The elapsed time $\Delta t'$ is then a proper time, and Δt is larger than $\Delta t'$ according to (3.8). However, a crude look at the diagram of Figure 3.18 does not seem to reflect that relation, which is nothing but the evidence that we must not use equal scales on axes of different reference systems in the same diagram. These observations related with temporal axes are also valid for scales on axes x and x'. In Figure 3.18 the gray strokes are the world lines of the ends of a body at rest relative to frame S. According to (3.7), the body length (i.e., the distance between simultaneous positions of the ends) should be larger in S than in S'. Figure 3.18 clearly shows that the use of the same scale on axes x and x' would lead to an error. Therefore, the axes must be properly calibrated in order to express the relations (3.7–3.8). In Figure 3.18 an arbitrary unit was chosen on the axes of frame S'; then the scales of the axes of frame S were calibrated by using Eqs. (3.7–3.8).

axes belonging to different frames have different scales

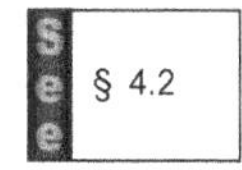

3.11. LORENTZ TRANSFORMATIONS

In §1.4 the transformation of the space-time coordinates of an event was built starting from the assumption of invariance of distances and times. As a consequence, Galileo transformations (1.4, 1.6) were obtained, together with the addition theorem of velocities (1.7) that derives from them. These transformations are incompatible with the postulate of invariance of the speed of light, so they have to be replaced by other transformations of coordinates of events that will be obtained from the subordination to this postulate. In §3.2 it was proved that the postulate of invariance of the speed of light leads to length contraction (3.7) and time dilatation (3.8). This behavior of distances and times is all we need to build the transformation of the coordinates of events—in the same way that distances and times invariance is sufficient to get Galileo transformations. In spite of this remark, in this section we shall build the transformation of coordinates as if we do not know yet the behavior of distances and times in Special Relativity. We shall review the arguments in §1.4 avoiding the prejudice about distances and times invariance; when necessary, we shall use the postulate of invariance of the speed of light.

Figure 1.5 and Eq. (1.2) will be again our starting point. As emphasized in §1.4, Eq. (1.2) is entirely correct—regardless of the behavior of distances and times—whenever all distances are measured in the same reference system, whether S or S':

$$d_{O'P} = d_{OP} - d_{OO'} \tag{3.27}$$

In frame S, the relations between distances involved in Eq. (3.27) and coordinates $(t\,,\,x)$ of the event considered are

$$d_{OO'}|_S = V\,t \tag{3.28}$$

(the origin of the temporal coordinate was chosen at the instant when O and O' are coincident) and

$$d_{OP}|_S = x \tag{3.29}$$

(by definition of the Cartesian coordinate x). On the other hand, the Cartesian coordinate x' is not equal to $d_{O'P}|_S$ but to $d_{O'P}|_{S'}$:

$$d_{O'P}|_{S'} = x' \tag{3.30}$$

The isotropy of space and the homogeneity of space and time guarantee that the relation between distances $d_{O'P}|_S$ and $d_{O'P}|_{S'}$ can only depend on the modulus $V = |\mathbf{V}|$ of the relative velocity between $S$and S'. In fact, the isotropy prevents a dependence on $\mathbf{V}$ direction, and the homogeneity prevents a dependency on the place and time of the event:

$$\frac{d_{O'P}|_{S'}}{d_{O'P}|_S} = \gamma(V) \tag{3.31}$$

where $\gamma(V)$ is an unknown function which will be determined applying the postulate of invariance of the speed of light.

By using Eq. (3.31), Eq. (3.27) is rewritten as

$$d_{O'P}|_{S'} \;=\; \gamma(V)\;(d_{OP}|_S - d_{OO'}|_S) \tag{3.32}$$

then replacing Eqs. (3.28–3.30):

$$x' = \gamma(V)\,(x - Vt) \tag{3.33}$$

The way leading from Eq. (3.27) to Eq. (3.33) can be analogously repeated to obtain the coordinate x as a function of coordinates (t', x'). Although this is an easy way, we shall substitute it by an argument that was already used in §3.5. We shall take into account that frames S and S' are on an equal footing. The only difference between them lies in the direction of the relative velocity: while S' moves with respect to S with velocity $\mathbf{V} = V\hat{\mathbf{x}}$, S moves with respect to S' with velocity $\mathbf{V} = -\,V\hat{\mathbf{x}}$. This difference does not affect the function γ, because γ depends just on the $\mathbf{V}$ modulus; it only affects the sign inside the parenthesis in Eq. (3.33). Therefore, the inverse transformation of (3.33) must be

$$x = \gamma(V)\,(x' + Vt') \tag{3.34}$$

The transformations (3.33–3.34) also contain the transformation of the temporal coordinate. In fact, the replacement of Eq. (3.33) in Eq. (3.34) leads to

$$x = \gamma\,(\gamma(x - V\,t) + V\,t') = \gamma^2 x - \gamma^2 V\,t + \gamma V\,t'$$

which can be solved for t' :

$$t' = \gamma t + \left(\gamma^{-1} - \gamma\right) V^{-1} x \tag{3.35}$$

The invariance of distances assumed in Galileo transformations corresponds to the adoption of $\gamma = 1$ in Eq. (3.31); in such a case Eqs. (3.33 and 3.35) become Galileo transformations (1.4, 1.6). Instead, function $\gamma(V)$ will be subordinated to the invariance of the speed of light. In order that this postulate can play its part, let us consider an event E belonging to the light ray illustrated in Figure 3.17 (the light ray going through the common origin of space-time coordinates of both reference systems, and propagating on the x axis direction). Any event E on this light ray has coordinates fulfilling $x = ct$ in frame S, and $x' = ct'$ in frame S', due to the invariance of the speed of light. Therefore, to subordinate the transformation of the coordinates of an event to the postulate of invariance of the speed of light, the transformation must be such that if $x = ct$ then it is $x' = ct'$ (and vice versa). We shall add this requirement in Eqs. (3.33–3.34) to obtain a result for function $\gamma(V)$. By replacing x and x' in (3.33–3.34):

$$ct' = \gamma(V)\,(ct - V\,t)$$

$$ct = \gamma(V)\,(ct' + V\,t')$$

Multiplying these two equations we get

$$c^2\,t' t = \gamma^2 (c - V) t (c + V) t' = \gamma^2 (c^2 - V^2)\, t\, t'$$

which leads to

$$\gamma(V) = \frac{1}{\sqrt{1 - \dfrac{V^2}{c^2}}} \tag{3.36}$$

in agreement with the results in §3.2. By replacing Eq. (3.36) in Eqs. (3.33, 3.35) we get Lorentz transformations of the coordinates t, x of any event:

Lorentz transformations

$$x' = \frac{x - V\,t}{\sqrt{1 - \dfrac{V^2}{c^2}}} \qquad x = \frac{x' + V\,t'}{\sqrt{1 - \dfrac{V^2}{c^2}}} \tag{3.37a–b}$$

$$t' = \frac{t - \dfrac{V}{c^2} x}{\sqrt{1 - \dfrac{V^2}{c^2}}} \qquad t = \frac{t' + \dfrac{V}{c^2} x'}{\sqrt{1 - \dfrac{V^2}{c^2}}} \tag{3.38a–b}$$

The coordinate lines of S' in Figure 3.17 result from fixing the values of x' and t' in Lorentz transformations (3.37a–3.38a). In particular, the x'axis ($t' = 0$) corresponds to $ct = Vc^{-1}\,x$, and the ct' axis ($x' = 0$) corresponds to $ct = V^{-1} c\,x$. Lorentz transformations go to Galileo ones in the limit $c \to \infty$.

Complement 3A: *Using Lorentz transformations*

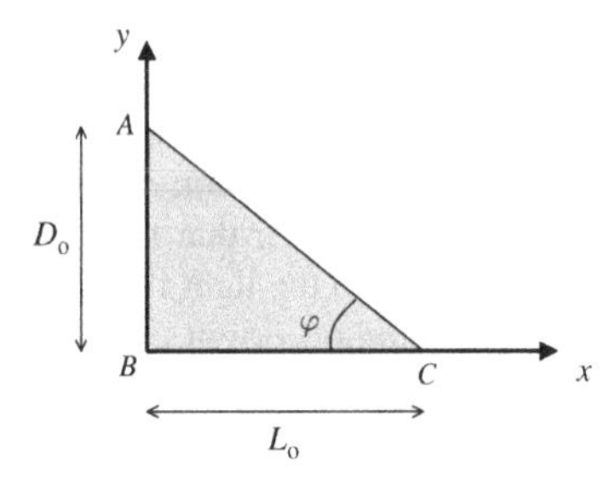

Problem 1. The triangle in the figure is at rest in a frame S. Due to the behavior of distances described in Eqs (3.7, 3.10), the dimensions and the shape of the triangle will be different in another frame S' moving with velocity $\mathbf{V} = V\,\hat{\mathbf{x}}$ relative to S. Use Lorentz transformations to know the shape of the triangle in frame S'.

Solution: In frame S the positions of vertices A, B, C do not vary in time. The world lines of the vertices are described by the equations:

$$x_A(t) = 0, \quad x_B(t) = 0, \quad x_C(t) = L_o$$
$$y_A(t) = D_o, \quad y_B(t) = 0, \quad y_C(t) = 0$$

If E_A , E_B , E_C are three events belonging to each one of these three world lines, then their coordinates in frame S' are, according to Eqs (3.37–3.39),

$$x'_A = -\gamma V\, t_A, \quad x'_B = -\gamma V\, t_B, \quad x'_C = \gamma(L_o - V t_C)$$
$$t'_A = \gamma\, t_A, \quad t'_B = \gamma\, t_B, \quad t'_C = \gamma(t_C - V\, c^{-2}\, L_o)$$
$$y'_A = D_o, \quad y'_B = 0, \quad y'_C = 0$$

Therefore the world lines of the vertices are described in frame S' by the equations:

$$x'_A(t') = -V\, t', \quad x'_B(t') = -V\, t', \quad x'_C(t') = \gamma\, L_o - V\, t' - \gamma V^2 c^{-2} L_o = \gamma^{-1} L_o - V\, t'$$
$$y'_A(t') = D_o, \quad y'_B(t') = 0, \quad y'_C(t') = 0$$

The dimensions and the shape of the triangle in frame S' result to be defined by the simultaneous positions of its three vertices. At each time t' the following relations are fulfilled:

$$L' = x'_C(t') - x'_B(t') = \gamma^{-1} L_o, \qquad \text{tg}\varphi' = \frac{D'}{L'} = \frac{D_o}{\gamma^{-1} L_o} = \gamma\ \text{tg}\varphi$$
$$D' = y'_A(t') - y'_B(t') = D_o$$

Problem 2. At time $t = 0$ two spacecrafts leave the Earth in opposite directions, with velocities V_1 and V_2. At a subsequent time t_o two light signals are sent from the Earth to both spacecrafts. Obtain the ratio between V_1 and V_2 in order that the reception of the signals can be simultaneous in the reference system fixed to the faster spacecraft.

Solution: Let us call events E_1 and E_2 the receptions of the light signals on spacecrafts 1 and 2 respectively. In the reference system fixed to the Earth, E_1 happens at a time t_1 such that:

$$c(t_1 - t_o) = V_1 t_1 \quad \Rightarrow \quad t_1 = \frac{c\, t_o}{c - V_1}$$

while the spatial coordinate of E_1 is the position $x_1 = V_1\, t_1$ of spacecraft 1 at that time. Thus:

$$E_1 \; coordinates: \; c\, t_1 = \frac{c^2 t_o}{c - V_1}, \qquad x_1 = \frac{c V_1 t_o}{c - V_1}$$

(we choose the x axis in such a way that $\mathbf{V}_1 = V_1\hat{\mathbf{x}}$ and $\mathbf{V}_2 = -V_2\hat{\mathbf{x}}$). Analogously:

$$E_2 \; coordinates: \; c\, t_2 = \frac{c^2\, t_o}{c - V_2}, \qquad x_2 = -\frac{c\, V_2\, t_o}{c - V_2}$$

Let spacecraft 1 be the faster one. The simultaneity of E_1 and E_2 in the frame S' fixed to spacecraft 1 implies

$$t'_1 = t'_2 \;\Rightarrow\; \gamma(V_1)(c\, t_1 - V_1 c^{-1} x_1) = \gamma(V_1)(c\, t_2 - V_1 c^{-1} x_2)$$

then

$$\frac{c^2 - V_1^2}{c - V_1} = \frac{c^2 + V_1 V_2}{c - V_2} \qquad \Rightarrow \qquad \frac{c}{V_2} - \frac{c}{V_1} = 2$$

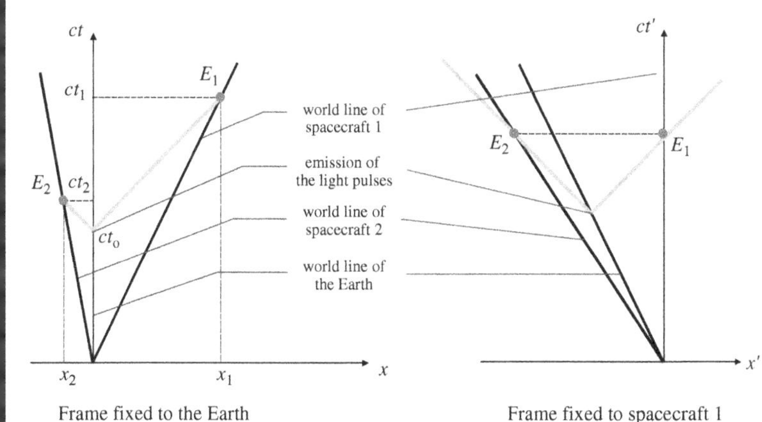

Frame fixed to the Earth

Frame fixed to spacecraft 1

In Euclidean geometry, the Cartesian coordinates y, z of a given event are the distances between the position where the event takes place and each one of two mutually perpendicular planes intersecting on the x axis. The y axis is contained in one of these planes—the "xy" plane—and the z axis is contained in the other—the "xz" plane (Figures 1.1 and 1.2). Since the orientations of y', z' axes were chosen like the ones of y, z axes, and, besides, x and x' axes are coincident, then plane xz coincides with plane $x'z'$, and plane xy coincides with plane $x'y'$. In this way, Cartesian coordinates y and y' measure the distance between a same point and a same plane in two different reference systems (idem for z and z'). We can repeat the argument used to write relation (3.31) and to state again that the relation between distances y and y' can only be a function of the modulus V of the relative velocity between S and S':

$$\frac{y'}{y} = \kappa(V)$$

i.e.,

$$y' = \kappa(V)y \qquad y = \kappa(V)^{-1}y'$$

The only admissible difference between direct and inverse transformations is the difference due to the change of direction of the relative velocity between both frames. Since function $\kappa(V)$ is not sensitive to the $\mathbf{V}$ direction, we conclude that $\kappa(V)^{-1} = \kappa(V)$; i.e., $\kappa(V) = 1$ (the other possibility, $\kappa(V) = -1$, corresponds to axes y and y' with opposite directions), in agreement with the result in §3.4 about the invariance of distances transversal to the direction of the relative movement. The same argument can be applied to the transformation of z coordinate. Then

$$\begin{aligned} y' &= y \\ z' &= z \end{aligned} \qquad (3.39)$$

It is used to call $\beta \equiv V/c$; then Lorentz transformations (3.37–3.38) have the form

See § 4.1 § 4.6

$$\begin{aligned} x' &= \gamma(x - \beta ct) \qquad & x &= \gamma(x' + \beta ct') \\ ct' &= \gamma(ct' - \beta x') \qquad & ct &= \gamma(ct' + \beta x') \end{aligned} \qquad (3.40a–b)$$

3.12. COMPARING CLOCKS IN DIFFERENT FRAMES

Time dilatation (3.8) can be immediately reobtained starting from Lorentz transformations.[5] In fact, if two events E_1 and E_2 happen at the same position in frame S'—i.e., $x'_1 = x'_2$ (proper frame) – then using Eq.(3.38b):

$$\Delta t = t_2 - t_1 = \gamma\left[(t'_2 - t'_1) + Vc^{-2}(x'_2 - x'_1)\right] = \gamma\Delta t'$$

[5] To obtain the length contraction by starting from Lorentz transformations see Complement 3A (Problem 1).

Since S' is the proper frame of the pair of events, then $\Delta t'$ is the proper time $\Delta\tau$ elapsed between the events. Therefore the result obtained corresponds to Eq. (3.8). Time dilatation is illustrated in Figure 3.19, which shows a pair of clocks synchronized in a frame S, and a third clock going past in front of the former ones. The encounters of the "moving" clock with each one of the clocks fixed in S are a pair of events which happen at the same position in the frame S' coming with the "moving" clock (proper frame). Therefore the time elapsed between events measured by the "moving" clock is the proper time between events, and has to be lower than the time measured in frame S.

The example in Figure 3.19 seems to display an inadmissible asymmetry between both reference systems, because time would pass more quickly in frame S than in frame S'. In order to analyze this point, it should be remarked that the chosen example itself is asymmetric. In fact, a unique S' clock has been compared with two S clocks. Therefore, a symmetric example is needed in order to give a satisfactory answer to the question. Then, let us compare two sets of synchronized clocks moving each one with respect to the other. Although each set is synchronized, this fact is only apparent in their respective proper frames, as a consequence of the relativity of the notion of simultaneity. The top of Figure 3.20 shows both sets of clocks at time $t = 0$ of frame S. Clock O occupies the origin of spatial coordinates of frame S. Since Lorentz transformations (3.40a–b) are written under the convention that the same event is the origin of space-time coordinates in both S and S', then clock O′—fixed to the origin of spatial coordinates of S'—goes through position O at time $t = 0$ when its hand reads zero. However, the rest of the clocks synchronized in S' do not read zero at time $t = 0$ in frame S. This circumstance can be understood with the help of Figure 3.21, where the world lines of clocks in S' and the coordinate line $t' = 0$

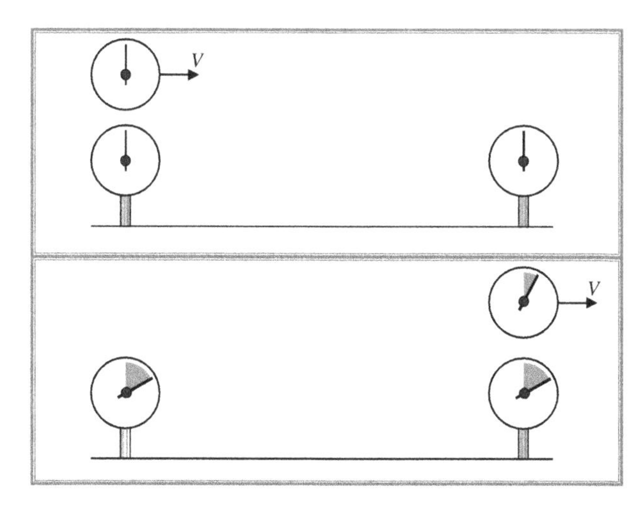

Figure 3.19. Clock moving with respect to two synchronized clocks.

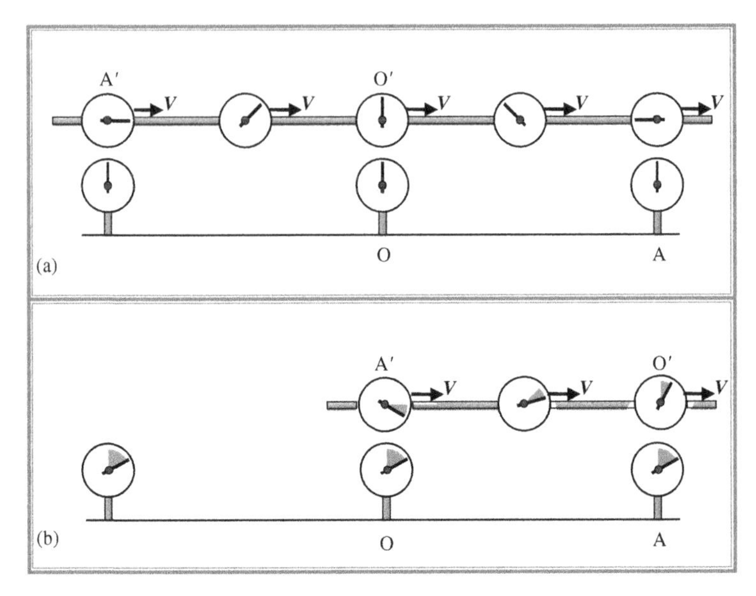

Figure 3.20. Two sets of synchronized clocks, as they are regarded in the proper frame belonging to one of them.

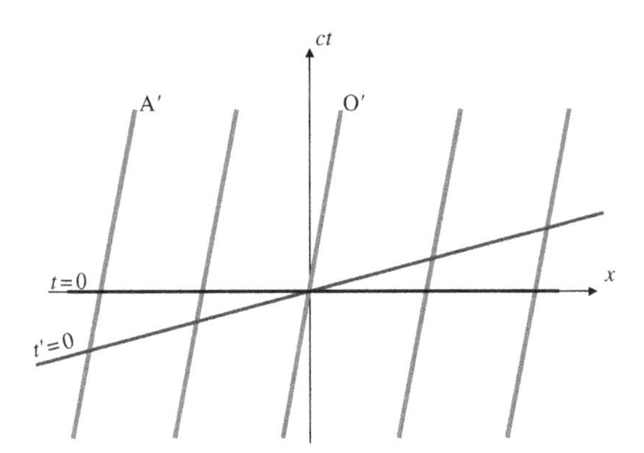

Figure 3.21. World lines of clocks of S' in the diagram of frame S .

are drawn in the space-time diagram of frame S. There it can be seen that clocks of S' placed on the left of O′ reach $t = 0$ after they have gone through $t' = 0$, while those on the right of O′ reach $t = 0$ before reaching $t' = 0$.

Although both sets of clocks are built in the same way—in particular, the proper distance between them is the same L_o in both cases—the distance between clocks of S' is contracted by the factor $\gamma(V)$ in frame S. Figure 3.20 is made with $\gamma(V) = 2$ (i.e., $\beta^2 = 3/4$); the value of L_o is such that L_o is traveled by clocks of S' in the time $\Delta t = L_o/V = 10\,\text{s}$, as can be regarded at the bottom of the figure. Let us use these values to compute the readings of clocks of S' at time $t = 0$ (Figure 3.20a). According to Eq. (3.38a) it is

$$t' = \gamma\left(t - \frac{V}{c^2}x\right)\Big|_{t=0} = -\gamma\frac{V}{c^2}x = -\gamma\beta^2\frac{x}{V} = -\frac{3}{2}\frac{x}{V}$$

Starting from A′, the x coordinates of clocks of S' at time $t = 0$ are $x = -L_o$, $-L_o/2$, 0, $L_o/2$, L_o, and the corresponding values of t' results to be 15 s, 7.5 s, 0 s, −7.5 s, −15 s, as displayed in Figure 3.20a. At $t = 10\,\text{s}$ (Figure 3.20b) the clocks of frame S' have displaced to positions $x = 0$, $L_o/2$, L_o, etc., and their readings result from

$$t' = \gamma\left(t - \frac{V}{c^2}x\right)\Big|_{t=10\,\text{s}} = \gamma 10\,\text{s} - \gamma\beta^2\frac{x}{V} = 20\,\text{s} - \frac{3}{2}\frac{x}{V}$$

i.e., $t' = 20\,\text{s}$, 12.5 s, 5 s, etc. Thus, after 10 s have passed in frame S, the clocks of S' have just advanced 5 s. In particular, clock O′ can be compared with the pair of synchronized clocks (O, A), as in Figure 3.19. Clocks O and O′ read zero at the time when they are in front; when O′ goes past in front of clock A (which is synchronized with O) the first one reads 5 s while A reads 10 s (dilatation factor $\gamma = 2$). Since both frames enter the examined problem in a completely symmetric way, and, on the other hand, there is no privilege at all of any reference system, then it should be equally verifiable that clocks of S also advanced a half of the time passed in S', when they are regarded from this last frame. To prove this statement, one clock of S, for instance O, should be compared with a pair of clocks synchronized S'. Again let us start at the time when O and O′ meet. According to Figure 3.20, when O goes past in front of clock A′ (synchronized with O′), O reads 10 s while A′ reads 20 s (the same dilatation factor $\gamma = 2$). Thus, the statement that the clocks of the other frame go more slowly is valid in both reference systems. The symmetry is complete because the dilatation factor is the same in both cases. This conclusion does not imply a logical contradiction only because the notion of simultaneity is not absolute. Each reference system can say that the clocks of the other frame go more slowly, basing such statement on its own notion of simultaneity.[6] Each reference system has synchronized its clocks by using its own notion of simultaneity, and the result is that the synchronization in S' is not acceptable in S (as evident in Figure 3.20); analogously, the synchronization in S is not acceptable in S'.

time dilatation does not imply privilege of any frame

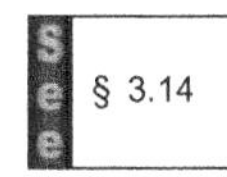
§ 3.14

[6] A similar discussion for length contraction can be found in §4.2.

3.13. VELOCITY AND ACCELERATION TRANSFORMATIONS

Let a particle be moving with velocity $\mathbf{u}(t) = \mathrm{d}\mathbf{r}/\mathrm{d}t$ relative to a frame S. According to Lorentz transformations (3.37–3.39), the coordinate differences along the world line of the particle in two distinct frames are related by (remember that V is constant):

$$\begin{aligned}
\mathrm{d}x' &= \gamma(V)(\mathrm{d}x - V\mathrm{d}t) = \gamma(V)(u_x - V)\mathrm{d}t \\
\mathrm{d}y' &= \mathrm{d}y \\
\mathrm{d}z' &= \mathrm{d}z \\
\mathrm{d}t' &= \gamma(V)(\mathrm{d}t - Vc^{-2}\mathrm{d}x) = \gamma(V)\left(1 - \frac{V\,u_x}{c^2}\right)\mathrm{d}t
\end{aligned} \tag{3.41}$$

The difference quotients of coordinates in S' lead to the components of velocity $\mathbf{u}'(t') = \mathrm{d}\mathbf{r}'/\mathrm{d}t'$ of the particle with respect to frame S'. In this way, the results (3.19, 3.25, 3.26) are recovered:

$$u'_x = \frac{\mathrm{d}x'}{\mathrm{d}t'} = \frac{u_x - V}{1 - \dfrac{Vu_x}{c^2}} \tag{3.42}$$

$$u'_y = \frac{\mathrm{d}y'}{\mathrm{d}t'} = \frac{\gamma(V)^{-1}u_y}{1 - \dfrac{V\,u_x}{c^2}} \qquad u'_z = \frac{\mathrm{d}z'}{\mathrm{d}t'} = \frac{\gamma(V)^{-1}u_z}{1 - \dfrac{V\,u_x}{c^2}} \tag{3.43}$$

The respective inverse transformations can be obtained from the replacement of V by $-V$, as it was explained in §3.5 and §3.7 (see Eqs. (3.19b) and (3.25b–3.26b)).

In the same way, we can obtain the transformations of acceleration $\boldsymbol{a}(t) = \mathrm{d}\mathbf{u}/\mathrm{d}t$. Differentiating Eqs. (3.42–3.43)

$$\mathrm{d}u'_x = \frac{\mathrm{d}u_x}{1 - \dfrac{V\,u_x}{c^2}} + \frac{u_x - V}{\left(1 - \dfrac{V\,u_x}{c^2}\right)^2}\frac{V}{c^2}\mathrm{d}u_x = \frac{\gamma(V)^{-2}}{\left(1 - \dfrac{V\,u_x}{c^2}\right)^2}\mathrm{d}u_x$$

$$\mathrm{d}u'_y = \frac{\gamma(V)^{-1}\,\mathrm{d}u_y}{1 - \dfrac{V\,u_x}{c^2}} + \frac{\gamma(V)^{-1}u_y}{\left(1 - \dfrac{V\,u_x}{c^2}\right)^2}\frac{V}{c^2}\mathrm{d}u_x$$

(together with an equation for $\mathrm{d}u'_z$ analogous to the last one) and taking the quotients with Eq. (3.41):

$$a'_x = \gamma(V)^{-3}\left(1 - \frac{V\,u_x}{c^2}\right)^{-3} a_x \tag{3.44}$$

$$a'_y = \gamma(V)^{-2}\left(1-\frac{V\,u_x}{c^2}\right)^{-2}\left(a_y+\frac{u_y\,V\,c^{-2}}{1-\dfrac{V\,u_x}{c^2}}a_x\right) \tag{3.45a}$$

the acceleration transformation

$$a'_z = \gamma(V)^{-2}\left(1-\frac{V\,u_x}{c^2}\right)^{-2}\left(a_z+\frac{u_z\,V\,c^{-2}}{1-\dfrac{V\,u_x}{c^2}}a_x\right) \tag{3.45b}$$

Far from being invariant, as under Galileo transformations, the acceleration changes under Lorentz transformations in a rather complicated way. The simplest case happens when the acceleration is transformed into a frame S_o whose velocity **V** is coincident with the particle velocity at the time when the transformation is performed. The equality $\mathbf{V}=\mathbf{u}$ forces to orient the axes $x-x'$ in the direction of **u**, which will be called the longitudinal direction. Then **u** in Eqs. (3.44–3.45) has only component x; besides it is $V=u_x$. Then acceleration $\boldsymbol{a}_o$ in frame S_o (*proper acceleration*) turns out to be

$$\boldsymbol{a}_{o_{\text{longitudinal}}} = \gamma(u)^3\,\boldsymbol{a}_{\text{longitudinal}} \tag{3.46}$$

proper acceleration

$$\boldsymbol{a}_{o_{\text{transversal}}} = \gamma(u)^2\boldsymbol{a}_{\text{transversal}} \tag{3.47}$$

In order to apply the result (3.46) we shall solve the unidimensional movement with constant proper acceleration. Equation (3.46) then reads

$$\text{constant} = \left(1-\frac{u^2}{c^2}\right)^{-3/2}\frac{\mathrm{d}u}{\mathrm{d}t}$$

Calling the constant a, and integrating the equation:

$$a(t-t_o) = \int_{u_o}^{u(t)}\left(1-\frac{u^2}{c^2}\right)^{-3/2}\mathrm{d}u = \left.\frac{u}{\sqrt{1-\dfrac{u^2}{c^2}}}\right|_{u_o}^{u(t)}$$

Choosing $t_o=0,\;\; u_o=u(t_o)=0$, and solving for u:

$$\frac{\mathrm{d}x}{\mathrm{d}t} = u = \frac{a\,t}{\sqrt{1+\dfrac{a^2t^2}{c^2}}} \tag{3.48}$$

Integrating once again:

$$x-x_o = \int_0^t \frac{a\,t}{\sqrt{1+\dfrac{a^2t^2}{c^2}}}\,\mathrm{d}t = \frac{c^2}{a}\sqrt{1+\frac{a^2t^2}{c^2}}-\frac{c^2}{a} \tag{3.49}$$

The world line corresponding to a motion with constant proper acceleration (3.49) is a hyperbola branch, as is clear when Eq. (3.49) is rewritten in the way

hyperbolic motion

$$\left(x - x_o + \frac{c^2}{a}\right)^2 - c^2 t^2 = \frac{c^4}{a^2} \tag{3.50}$$

The hyperbola branch associated with Eq. (3.49) is the right one if $a > 0$ (Figure 3.22), or the left one if $a < 0$.

The non-relativistic limit, $|u| << c$, corresponds to $|at| << c$ (see Eq. (3.48)). In this case the equation of motion (3.49) approximates to

$$x \cong x_o + \frac{1}{2} a t^2$$

while for $|at| >> c$ the velocity (3.48) goes to $\pm c$.[7]

In §4.2 it will be shown that any hyperbola that is asymptotic to light rays, as the one in Figure 3.22, keeps this aspect in all inertial reference systems.

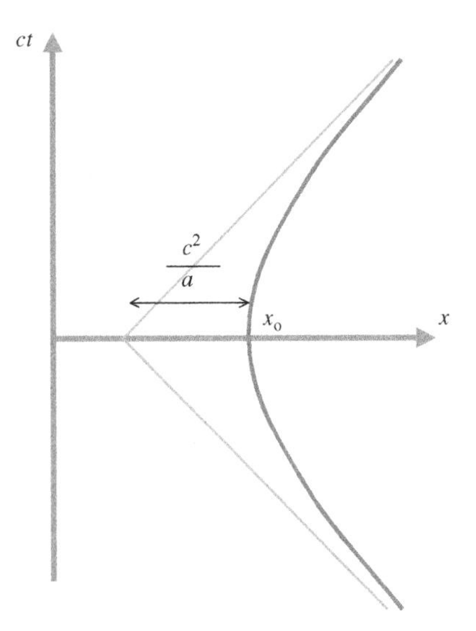

Figure 3.22. World line of a motion with constant proper acceleration.

[7] It must be emphasized that what is constant is not the acceleration in a given inertial frame (if this were the case, the particle would exceed the speed of light), but the value of the acceleration in the consecutive different inertial frames associated with the particle (proper frames) at each time t.

3.14. "PARADOXES": REMNANTS OF CLASSICAL THOUGHT

The learning of Special Relativity is full of puzzlements that come from an involuntary attachment to the classical notions of space and time. Most of them are connected with the relativity of the notion of simultaneity; others with the classical notion of rigid body, which is meaningless in Relativity. This section will show two typical examples of these false paradoxes. A paradox of different nature will be analyzed in §4.5.

In classical physics the difference between simultaneous position vectors of two points A and B results to be invariant under Galileo transformations: $\Delta\mathbf{r}' = \mathbf{r}'_{\mathrm{B}} - \mathbf{r}'_{\mathrm{A}} = (\mathbf{r}_{\mathrm{B}} - \mathbf{V}\,t) - (\mathbf{r}_{\mathrm{A}} - \mathbf{V}\,t) = \mathbf{r}_{\mathrm{B}} - \mathbf{r}_{\mathrm{A}} = \Delta\mathbf{r}$. This implies that the shape and orientation of an extensive body—which are determined by simultaneous positions of its points—are Galilean invariants. On the contrary, in Relativity the shape and orientation of an extensive body are modified under changes of reference system, as a consequence of length contraction and the relativity of the notion of simultaneity. In Complement 3A (Problem 1) it was explained how the body shape is modified when the body is regarded from a frame where the body moves, owing to the length contraction. In the "paradox" that we are going to examine, the crucial element will not be the change of shape but the change of orientation.

Figure 3.23 shows a ring moving along a direction perpendicular to its own plane, and a bar whose velocity is parallel to the plane of the ring. If the diameter of the ring is slightly larger than the length of the bar, then the bar can go through the ring provided that the initial conditions are suitable. The apparent paradoxical character of this situation appears when we consider it from the frame where the bar is at rest. In fact, in the proper frame of the bar the ring has an elliptical shape, due to the contraction of the dimension that is parallel to

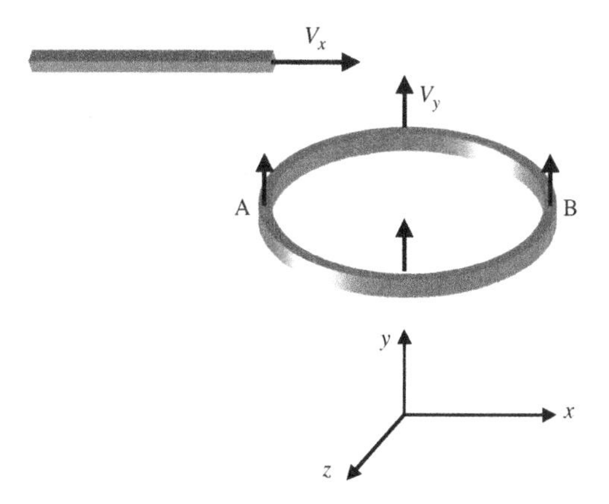

Figure 3.23. The ring–whose diameter slightly exceeds the length of the bar — can surround the bar. How is this fact understood in the bar proper frame, where the bar is longer and the ring is contracted?

the relative motion of the bar; besides, the length of the bar is maximum in this frame, because it is the proper length. Therefore it seems impossible that the bar can go through the ring in the bar frame. Of course, the issue about whether or not the bar goes through the ring must have a unique answer, independent of the reference system: it is an absolute fact.

The examined situation seems to be paradoxical because the argument forgot to take into account the change of orientation of the ring. In order to know the orientation of the ring in frame S' where the bar is at rest, we shall compare the simultaneous positions of points A and B on the ring. The world lines of A and B are described by the equations

$$x_A(t) = -R \qquad x_B(t) = R$$

$$y_A(t) = V_y\, t \qquad y_B(t) = V_y\, t$$

where R is the radius of the ring. In S' the coordinates of events happening at A and B are

$$x'_A = \gamma(V_x)\,(-R - V_x\, t_A) \qquad x'_B = \gamma(V_x)\,(R - V_x\, t_B)$$

$$y'_A = V_y\, t_A \qquad y'_A = V_y\, t_B$$

$$t'_A = \gamma(V_x)\,(t_A + \frac{V_x}{c^2} R) \qquad t'_B = \gamma(V_x)\,(t_B - \frac{V_x}{c^2} R)$$

A and B positions that are simultaneous in S' fulfill

$$t'_A = t'_B \qquad \Rightarrow \qquad t_B - t_A = \frac{2V_x R}{c^2}$$

Thus, at any time t' it is satisfied that

$$y'_B(t') - y'_A(t') = V_y\,(t_B - t_A) = \frac{2V_x V_y R}{c^2}$$

$$x'_B(t') - x'_A(t') = 2\gamma(V_x)\, R - \gamma(V_x)\; V_x\, \frac{2V_x R}{c^2} = 2\gamma(V_x)^{-1}\, R$$

the orientation of a body depends on the reference system

The orientation of the ring in S' differs from the one in S, since in S' it is $y'_B(t') \neq y'_A(t')$ (Figure 3.24). The angle between the plane of the ring and the x' axis is

$$\mathrm{tg}\varphi' = \frac{y'_B(t') - y'_A(t')}{x'_B(t') - x'_A(t')} = \gamma(V_x) \frac{V_x V_y}{c^2}$$

On the other hand, the velocity of the ring in $S'$is not perpendicular to the plane of the ring:

$$u'_x = -V_x, \qquad u'_y = \gamma(V_x)^{-1} V_y$$

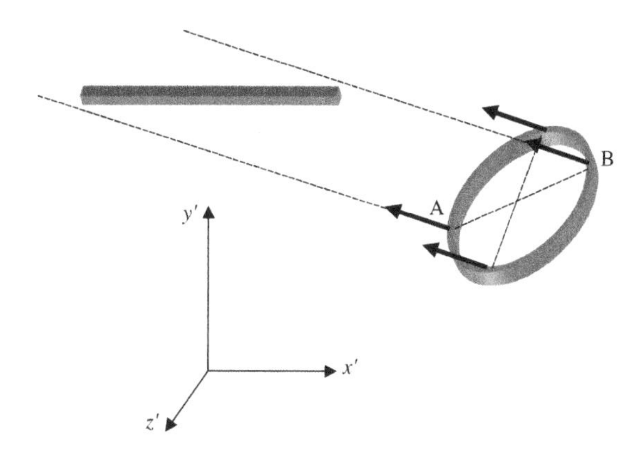

Figure 3.24. Situation described in Figure 3.23, as it occurs in frame S' where the bar is at rest.

As shown in Figure 3.24, the distinct orientation of the ring in each frame explains that it can be consistent to state that the bar goes through the ring in both frames.

The result of this problem can be enhanced by stating an important difference between Galileo and Lorentz transformations, which will be studied in more detail in §4.8. Let us imagine two different ways of transforming coordinates from the bar proper frame to the ring proper frame: (a) we can do it in two steps, first performing a transformation to the frame S in Figure 3.23 using velocity $V_x\hat{\mathbf{x}}$, and then applying a transformation with velocity $-V_y\hat{\mathbf{y}}$; (b) we can do it by means of only one transformation with velocity $-\mathbf{u}' = V_x\hat{\mathbf{x}} - \gamma(V_x)^{-1}V_y\hat{\mathbf{y}}$, opposite to the velocity of the ring in the bar frame (and equal to the relativistic composition of the velocities involved in the former case). Procedure (a) composes two Lorentz transformations; the ring is at rest in the arrival frame, and its orientation is shown in Figure 3.23. Instead, procedure (b) uses only one Lorentz transformation with a composed velocity; the ring is also at rest in the arrival frame, but its orientation is different from the one obtained in procedure (a) (this can be understood by contracting lengths). In §4.8 it will be shown that there is a spatial rotation connecting procedures (a) and (b), which is called Wigner rotation.

The next "paradox" is related to pairs of objects whose relative motion is such that, in a reference system, one of them is entirely enclosed by the other; this situation can seem inadmissible for other reference systems, where the dimensions of the objects are different. Concretely, Figure 3.25 shows a bar and a hole cylinder in relative motion. In the reference system of Figure 3.25 the cylinder is fixed and the bar moves. The dimensions of the bodies are such that the cylinder completely encloses the bar when this one passes through it. On the contrary, in the bar frame the cylinder is contracted and the bar length is maximum. Thus, in the bar frame the cylinder does not completely enclose the bar at any time (Figure 3.26). So far there is nothing paradoxical.

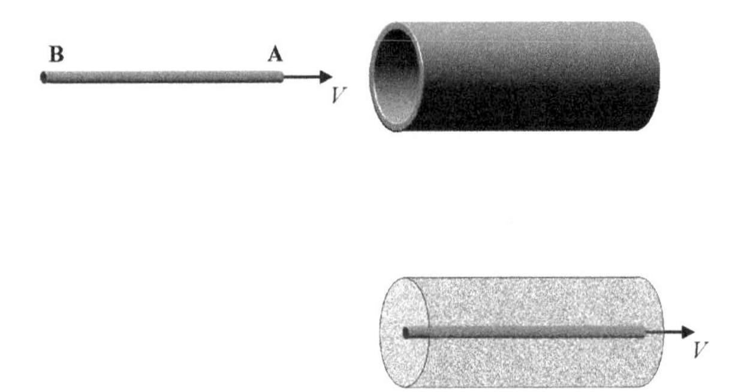

Figure 3.25. Relative motion between a bar and a cylinder, considered in the reference system fixed to the cylinder, where the dimensions of the bodies make possible that the bar is completely enclosed in the cylinder.

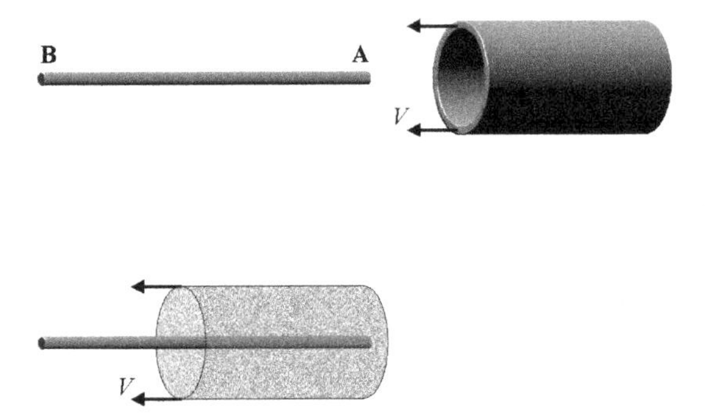

Figure 3.26. The same relative motion of Figure 3.25, as considered in the reference system where the bar is at rest.

Let us now suppose that the back of the cylinder is closed, so the bar cannot get out through the end of the cylinder. Figure 3.27 displays this situation: it shows a railway carriage going into a tunnel whose exit is closed by a buffer. In the frame fixed to the earth the tunnel and the carriage have the same length; this means that end B of the carriage arrives to the entrance of the tunnel at the time of the shock. Since the events are absolute, it has to be admitted that the arrival of end B of the carriage to the entrance of the tunnel also happens—although it is not simultaneous with the shock—in the carriage frame. But, how can this event be conceived in the carriage frame, where the tunnel is contracted and the carriage has maximum length (proper length)?

concept of rigidity is meaningless in Relativity

In this example, the perplexity appears as a consequence of considering the carriage as a rigid body. The notion of rigidity is meaningless in Relativity, because it implies that the information about the changes of the state of motion of a point belonging to an extensive body can be "instantaneously" communicated to the rest of the body. Thus, the notion of rigidity comes into conflict with the relativity of simultaneity and the impossibility of sending information at a

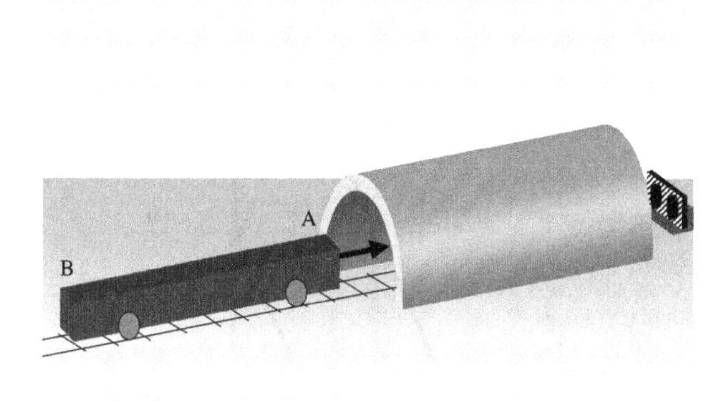

Figure 3.27. The carriage impacts on the buffer when its end B enters the tunnel. How can this be understood in the frame fixed to the carriage, where the tunnel is contracted and the carriage is longer?

speed higher than the speed of light. The notion of rigid body is only consistent within the context of classical physics, where the simultaneity is absolute, and the propagation of information at infinite velocity is compatible with its notions of space and time.

What does it happen, then, when the end A of the carriage impacts on the buffer? The information of the shock begins to propagate toward the rest of the carriage by means of elastic waves traveling with the speed of sound in the material the carriage is made of. Higher speeds of propagation can be obtained by increasing the "rigidity" of materials, but the speed of light will never be reached. In this way, the information about the shock will take a time until it arrives to the rest of the carriage; this time will always be larger than the time needed by a light ray for the same traveling. As long as the information about the shock does not reach a given point of the carriage, the point will continue its motion as if the shock had not happened. Therefore the end B of the carriage not only arrives to the entrance of the tunnel but it continues its motion and enters the tunnel compressing the length of the carriage. In the frame that follows the movement of the end B of the carriage (i.e., the frame that originally was the proper frame of all the points of the carriage) B enters the tunnel when the compression is in an advanced stage; at this time, the length of the carriage in this frame is no longer the original proper length. Figure 3.28 shows the world lines of the ends of the carriage in the coordinate system S fixed to the earth. In S, the end B of the carriage enters the tunnel at time $t = 0$, simultaneously with the shock. In the frame S' going with the end B, the entrance occurs at time $t' = 0$, after the beginning of compression. The light ray passing through the event of the shock was also drawn, in order to indicate the maximum speed

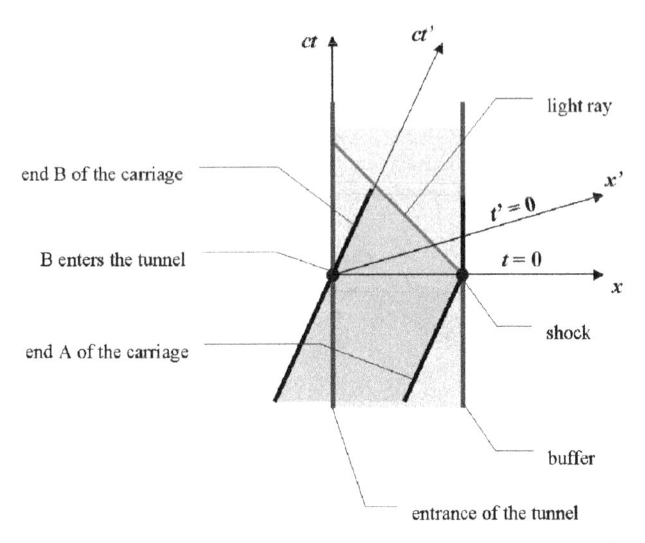

Figure 3.28. Space-time diagram for the movement of the railway carriage in the earth frame, and the light ray indicating the maximum speed for propagating the information about the shock.

for the transmission of the information to the rest of the carriage. The evolution of the points of the carriage after this information is received will depend on the elastic and dissipative properties of the material, and the boundary conditions imposed on the oscillations.

3.15. DOPPLER EFFECT

If observers move with respect to a source that emits signals with frequency ν_{src} then they will receive these signals with different frequencies ν_{obs}. This phenomenon is called *Doppler effect.* In this section, we shall first study the Doppler effect for mechanical waves in a material medium, in the classical context where the changes of reference systems are governed by Galileo transformations. The so-obtained result will be then amended to achieve the relativistic Doppler effect.

Let a mechanical wave be propagating in an isotropic and homogeneous medium. If the speed of propagation is c_s then the wave fronts emitted by a point-like source will be spheres of radii $c_s t$, where t is the time elapsed from the emission. In Figure 3.29 a source moves with velocity $\mathbf{u}_{\text{src}} (|\mathbf{u}_{\text{src}}| < c_s)$ relative to the medium; a pair of pulses emitted from the source is displayed as seen in the frame fixed to the medium. The 1st pulse was emitted from point A; after a time t, the pulse reaches an observer moving with velocity $\mathbf{u}_{\text{obs}}$ relative to the medium ($|\mathbf{u}_{\text{obs}}| < c_s$). In the meantime, the 2nd pulse was emitted at a time T_{src} after the emission of the first one (T_{src} is the *emission* period of the source), when the

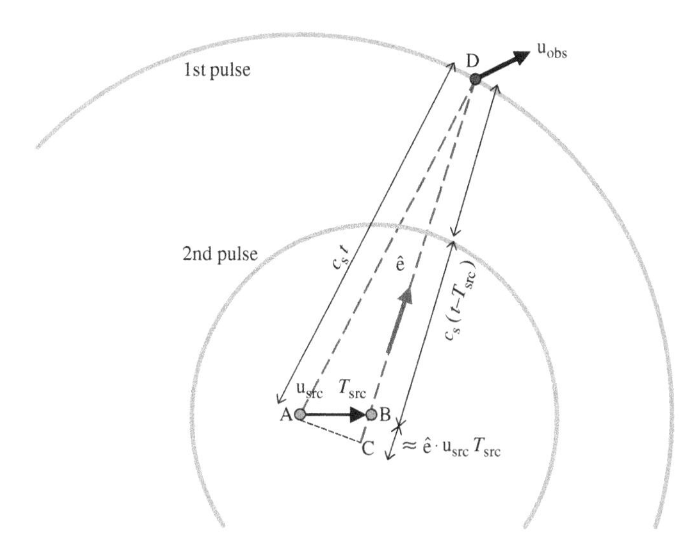

Figure 3.29. Propagation of a mechanical wave in the frame fixed to the medium.

source went through point B after having displaced the distance $\mathbf{u}_{src} T_{src}$. When the 1st pulse reaches the observer, the 2nd pulse has then developed the radius $c_s(t - T_{src})$. From the observer point of view, the period T_{obs} is the time elapsed between the *reception* of two consecutive pulses. Figure 3.29 shows that time T_{obs} is the time needed by the 2nd wave front to cover the sum of distance λ plus the displacement of the observer along the direction normal to the wave front during this time T_{obs}:

$$c_s T_{obs} = \lambda + \hat{\mathbf{e}} \cdot \mathbf{u}_{obs} T_{obs} \tag{3.51}$$

In this equation the curvature of the wave front has been considered negligible. The equation is exact for a plane wave; in another case the equation is a good approximation when the radius of the front is much larger than the displacement of the observer along the direction transversal to $\hat{\mathbf{e}}$ (it is sufficient to demand that $t >> T_{obs}$). On the other hand, Figure 3.29 shows an isosceles triangle ADC. If $\overline{AC}$ is small compared with the radius $c_s t$ (for it is enough that $t >> T_{src}$), then $\overline{BC}$ can be approximated by $\hat{\mathbf{e}} \cdot u_{src}\ T_{src}$; thus the equality of sides $\overline{AD} = \overline{CD}$ is expressed as

$$\hat{\mathbf{e}} \cdot \mathbf{u}_{src} T_{src} + c_s(t - T_{src}) + \lambda = c_s t$$

Solving for λ in both equations and equaling the results:

$$(c_s - \mathbf{u}_{obs} \cdot \hat{\mathbf{e}}) T_{obs} = (c_s - \mathbf{u}_{src} \cdot \hat{\mathbf{e}}) T_{src}$$

or, since the frequency ν is the inverse of period T:

classical Doppler effect

$$\frac{\nu_{\text{obs}}}{c_{\text{s}} - \mathbf{u}_{\text{obs}} \cdot \hat{\mathbf{e}}} = \frac{\nu_{\text{src}}}{c_{\text{s}} - \mathbf{u}_{\text{src}} \cdot \hat{\mathbf{e}}} \tag{3.52}$$

Equation (3.52) describes the classical Doppler effect. The frequency observed coincides with the source frequency only if $\mathbf{u}_{\text{obs}} \cdot \hat{\mathbf{e}} = \mathbf{u}_{\text{src}} \cdot \hat{\mathbf{e}}$, i.e., if the relative velocity $\mathbf{u}_{\text{obs}} - \mathbf{u}_{\text{src}}$ is perpendicular to the source–observer direction. Besides, Eq. (3.52) shows that the ratio between the source frequency and the one perceived by the observer separately depends on the source and observer velocities relative to the material medium. However, if both $\mathbf{u}_{\text{src}}$ and $\mathbf{u}_{\text{obs}}$ are much smaller than c_{s} then the classical Doppler effect only depends, at the lower order of approximation, on the relative velocity $\mathbf{u}_{\text{src}} - \mathbf{u}_{\text{obs}}$. In fact:

$$\begin{aligned} \nu_{\text{obs}} = \frac{1 - \dfrac{\mathbf{u}_{\text{obs}}}{c_{\text{s}}} \cdot \hat{\mathbf{e}}}{1 - \dfrac{\mathbf{u}_{\text{src}}}{c_{\text{s}}} \cdot \hat{\mathbf{e}}} \nu_{\text{src}} &\cong \left(1 - \frac{\mathbf{u}_{\text{obs}}}{c_{\text{s}}} \cdot \hat{\mathbf{e}}\right)\left(1 + \frac{\mathbf{u}_{\text{src}}}{c_{\text{s}}} \cdot \hat{\mathbf{e}}\right) \nu_{\text{src}} \\ &\cong \left(1 - \frac{\mathbf{u}_{\text{obs}} - \mathbf{u}_{\text{src}}}{c_{\text{s}}} \cdot \hat{\mathbf{e}}\right) \nu_{\text{src}} \end{aligned} \tag{3.53}$$

We wish to remark that all calculation was made using magnitudes measured in the frame fixed to the medium. Due to the classical invariance of times and distances, the times T_{src} and T_{obs} do not require a transformation to the source and observer proper frames to have the right of being called source period and observed period.

Let us now consider the Doppler effect for light signals propagating in vacuum, in a relativistic framework. In Relativity, light is not regarded as a mechanical phenomenon; the description of its propagation does not recognize privileged frame at all. Therefore the Doppler shift of light frequency cannot depend on two separated velocities but on a sole velocity: the velocity $\mathbf{V}$ of the observer relative to the source. On the other hand, light propagates with the same speed c along all directions whatever the reference system is. As a consequence, the light pulses emitted by a point-like source have spherical shape in all reference systems. Then it will be practical to work in the reference system fixed to the source. Figure 3.29 and previous calculation are still useful, but in the source frame it is $\mathbf{u}_{\text{src}} = 0$, $\mathbf{V} = \mathbf{u}_{\text{obs}} - \mathbf{u}_{\text{src}} = \mathbf{u}_{\text{obs}}$, and the light pulses are concentric spheres. In this frame Eq. (3.51) becomes

$$c\Delta t = \lambda + \mathbf{V} \cdot \hat{\mathbf{e}} \Delta t$$

where λ is the distance between consecutive pulses (λ is the wavelength) measured in the source frame: $\lambda = cT_{\text{src}} = \nu_{\text{src}}^{-1} c$, and Δt is the time elapsed between two consecutive pulse receptions at the observer positions. The former equation is exact if $\mathbf{V}$ is parallel to $\hat{\mathbf{e}}$; in other cases the transversal observer displacement should be much smaller than the radius of the spherical pulse (i.e., the wave front could be approximated by a plane). Solving for Δt:

$$\Delta t = \frac{1}{1 - \dfrac{\mathbf{V}}{c} \cdot \hat{\mathbf{e}}} \nu_{\mathrm{src}}^{-1}$$

In the observer proper frame the pair of events involved in Δt— the two consecutive pulse receptions by the observer – happen at the same position. Therefore, the time elapsed in the observer proper frame – which is what the observer calls period – is a proper time: $\nu_{\mathrm{obs}}^{-1} = T_{\mathrm{obs}} = (1 - V^2/c^2)^{1/2} \Delta t$. Then

relativistic Doppler effect for light propagating in vacuum

$$\nu_{\mathrm{obs}} = \frac{1 - \dfrac{\mathbf{V}}{c} \cdot \hat{\mathbf{e}}}{\sqrt{1 - \dfrac{V^2}{c^2}}} \nu_{\mathrm{src}} \qquad (3.54)$$

If $V << c$, Eq. (3.54) approximates to Eq. (3.53). While in the classical Doppler effect there is not frequency shift if $\mathbf{V} = \mathbf{u}_{\mathrm{obs}} - \mathbf{u}_{\mathrm{src}}$ is perpendicular to $\hat{\mathbf{e}}$ (see Eq. (3.52)), in Relativity there is frequency shift even if the relative velocity $\mathbf{V}$ is perpendicular to $\hat{\mathbf{e}}$, as it results from the presence of the factor $\gamma(V)$ coming from time dilatation. This *transversal* Doppler effect —$\nu_{\mathrm{obs}} = \gamma(V)\ \nu_{\mathrm{src}}$—has been experimentally verified, which must be taken as a direct proof of time dilatation.[8]

As an application of transformation (3.54), let us consider the case where the direction of propagation is parallel to $\mathbf{V}$. Then $\mathbf{V} \cdot \hat{\mathbf{e}} = \pm V$ for equal and opposite directions. Replacing in Eq. (3.54) it results

$$\frac{\nu_{\mathrm{obs}}}{\nu_{\mathrm{src}}} = \sqrt{\frac{1 - \dfrac{V}{c}}{1 + \dfrac{V}{c}}} \text{ if } \mathbf{V} \cdot \hat{\mathbf{e}} = V, \qquad \frac{\nu_{\mathrm{obs}}}{\nu_{\mathrm{src}}} = \sqrt{\frac{1 + \dfrac{V}{c}}{1 - \dfrac{V}{c}}} \text{ if } \mathbf{V} \cdot \hat{\mathbf{e}} = -V \qquad (3.55\text{a–b})$$

Therefore if the observer moves away from the source, then $\nu_{\mathrm{obs}} < \nu_{\mathrm{src}}$ and it is said that light is "redshifted" (because the red is the lowest frequency of the visible spectrum). If the observer moves closer to the source, then $\nu_{\mathrm{obs}} > \nu_{\mathrm{src}}$ and it is said that light is "blueshifted."

The space-time diagram in Figure 3.30 explains the result (3.55a). A set of light pulses propagating along the $\hat{\mathbf{x}}$ direction ($\mathbf{V} \cdot \hat{\mathbf{e}} = V$) is shown; the period is T or T' depending on the measurement is performed in S or S'. The text of Figure 3.30 describes the kinematical study leading to Eq. (3.55).

[8] The first experimental verification of the presence of factor $\gamma(V)$ in the law of Doppler effect was performed in 1938 by H.E. Ives and G.R. Stilwell.

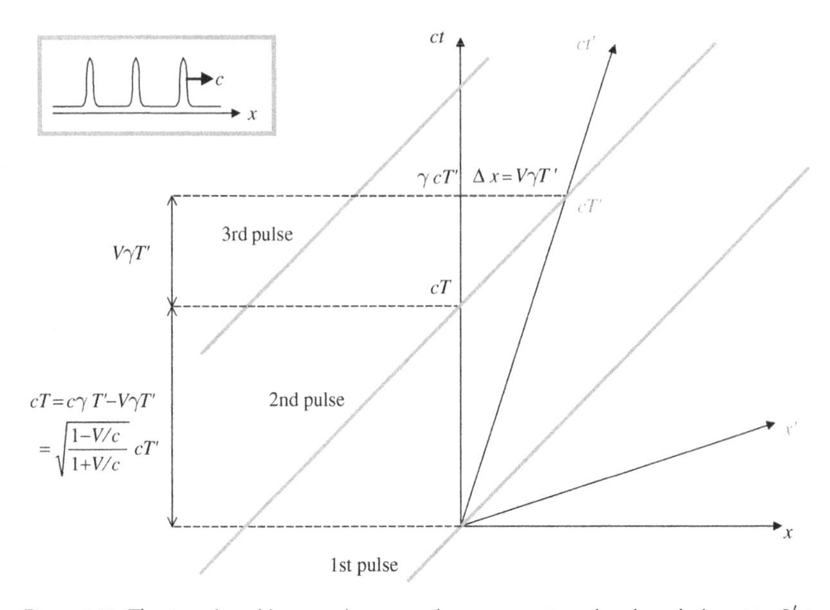

Figure 3.30. The time elapsed between the passes of two consecutive pulses through the origin O' is T' in S' (by definition of period) and $\Delta t = \gamma T'$ in S (because of time dilatation). During this time, the distance $d_{OO'}$ measured in S increases $\Delta x = V\Delta t = V\gamma T'$. The period T is calculated by subtracting the time used by the light to cover the distance $V\gamma T'$ from $\gamma T'$.

3.16. TRANSFORMATION OF LIGHT RAYS

A periodic plane light-wave propagating in vacuum is characterized by its frequency ν and its direction of propagation $\hat{\mathbf{n}}$ (perpendicular to the plane wave fronts). Whenever a wave is plane in a frame S, it will also be plane in another frame S' uniformly translating with respect to S, because the coordinate transformations are linear. However, its frequency and direction of propagation will be different in each reference system.

The frequency transformation has been already studied in the previous section, where we emphasized that the reasoning leading to Eq. (3.54) is exact if the wave is plane. Substituting $\hat{\mathbf{e}}$ with $\hat{\mathbf{n}}$ in (3.54), it results

frequency transformation for light in vacuum

$$\nu' = \frac{1 - \frac{\mathbf{V}}{c} \cdot \hat{\mathbf{n}}}{\sqrt{1 - \frac{V^2}{c^2}}} \nu \qquad \nu = \frac{1 + \frac{\mathbf{V}}{c} \cdot \hat{\mathbf{n}}'}{\sqrt{1 - \frac{V^2}{c^2}}} \nu' \tag{3.56a–b}$$

The inverse transformation (3.56b) was obtained by maintaining the structure of (3.56a) and replacing $\mathbf{V}$ by $-\mathbf{V}$, since neither S nor S' are privileged in any way.

Let us consider the transformation of the direction of propagation (aberration of light). It was already mentioned that the ray velocity transforms as a particle velocity. In vacuum the ray direction coincides with the direction of propagation in any reference system, so the ray velocity is $\mathbf{u} = c\hat{\mathbf{n}}$. If, as usual, the x axis is defined by $\mathbf{V}$ direction, and y axis is chosen in such a way that the ray propagates in the x–y plane, then $u_x = c\cos\theta$, $u_y = c\sin\theta$, $u_z = 0$, where θ is the angle from $\mathbf{V}$ to $\mathbf{u}$ taken in a counterclockwise direction. By applying the relativistic transformation of velocities (3.42–3.43) it is obtained:

$$\cos\theta' = \frac{\cos\theta - \dfrac{V}{c}}{1 - \dfrac{V}{c}\cos\theta} \qquad \sin\theta' = \frac{\sqrt{1 - \dfrac{V^2}{c^2}}\,\sin\theta}{1 - \dfrac{V}{c}\,\cos\theta} \tag{3.57a–b}$$

aberration of light in vacuum

or

$$\mathrm{tg}\theta' = \frac{\sqrt{1 - \dfrac{V^2}{c^2}}\,\sin\theta}{\cos\theta - \dfrac{V}{c}} \tag{3.57c}$$

Equation (3.57c) can be compared with the classical result (2.22). Nevertheless it should be remarked that θ in Eq. (2.22) measures the ray direction in the ether frame, while θ in Eqs. (3.57) measures the ray direction in an arbitrary frame S. In Relativity, the change of the ray direction implies the change of orientation of the wave fronts, since the ray in vacuum must be perpendicular to the wave front in any reference system. On the contrary, in the ether theory the ray was perpendicular to the wave front only in the ether frame, and the aberration of light did not imply change of orientation of the wave fronts (see Figure 2.9).

It is enlightening to show that the change of the light direction of propagation in vacuum can also be obtained without invoking the identity with the ray direction or the transformation of particle velocities. In fact, we can start from Eqs. (3.56), where only the undulatory features of light played a role. Multiplying Eqs. (3.56a) and (3.56b) the frequencies cancel out, and a relation between the directions of propagation emerges:

$$1 - \frac{V^2}{c^2} = \left(1 - \frac{\mathbf{V}}{c}\cdot\hat{\mathbf{n}}\right)\left(1 + \frac{\mathbf{V}}{c}\cdot\hat{\mathbf{n}}'\right)$$

Replacing $\mathbf{V}\cdot\hat{\mathbf{n}} = V\cos\theta$ and $\mathbf{V}\cdot\hat{\mathbf{n}}' = V\cos\theta'$, it yields

$$-\frac{V}{c} = \cos\theta' - \cos\theta - \frac{V}{c}\cos\theta'\cos\theta$$

Solving for $\cos\theta'$, one obtains the result of Eq. (3.57a).

3.17. TRANSFORMATION OF A PLANE WAVE

Instead of considering a light wave propagating in vacuum, we can try with a periodic plane wave of different nature. Then, for a complete description, the phase velocity w must be added to the frequency ν and the direction of propagation $\hat{\mathbf{n}}$ (perpendicular to the plane wave fronts). Besides, we have to distinguish the angle θ between the ray and the x axis from the angle ϑ between the direction of propagation and x axis.

Reviewing the way of obtaining Eqs. (3.56a–b) in §3.15, we notice that they will remain valid for any periodic plane wave if we replace c by phase velocity w:

frequency transformation for a plane wave

$$\nu' = \frac{1 - \dfrac{\mathbf{V}}{w} \cdot \hat{\mathbf{n}}}{\sqrt{1 - \dfrac{V^2}{c^2}}}\, \nu \qquad \nu = \frac{1 + \dfrac{\mathbf{V}}{w'} \cdot \hat{\mathbf{n}}'}{\sqrt{1 - \dfrac{V^2}{c^2}}}\, \nu' \tag{3.58a–b}$$

(of course, the coefficient c in factor $\gamma(V)$, which comes from the time dilatation, should not be replaced).

In order to obtain the phase velocity transformation, let us begin by studying the case where both w and w' are smaller than c. In this case there will exist a frame $\bar{S}$ moving with velocities $w\, \hat{\mathbf{n}}$ and $w'\, \hat{\mathbf{n}}'$ relative to S and S' respectively, where the wave fronts are at rest. In this frame the wavelength $\bar{\lambda}$ is a proper length, so it is related with λ and λ' through the equations

$$\lambda' = \sqrt{1 - \frac{w'^2}{c^2}}\,\bar{\lambda}, \qquad \lambda = \sqrt{1 - \frac{w^2}{c^2}}\,\bar{\lambda} \qquad \Rightarrow \qquad \gamma(w')\lambda' = \gamma(w)\lambda$$

Since $\lambda = \nu^{-1} w$, it results

$$\gamma(w') w' \nu'^{-1} = \gamma(w) w \nu^{-1} \tag{3.59}$$

By multiplying Eqs. (3.58a) and (3.59) we obtain the relativistic transformation of phase velocity:

$$\gamma(w') w' = \gamma(V) \gamma(w)\, (w - V \cos \vartheta) \tag{3.60a}$$

the inverse transformation being

$$\gamma(w) w = \gamma(V) \gamma(w')\, (w' + V \cos \vartheta') \tag{3.60b}$$

We can solve for w' in Eq. (3.60a) by inverting and squaring:

transformation of phase velocity

$$\frac{c^4}{w'^2} = c^2 + \frac{(c^2 - V^2)(c^2 - w^2)}{(w - V \cos \vartheta)^2} \tag{3.61}$$

We wish to remark that both w and w' in Eq. (3.61) are either smaller than c or larger than c. Actually, transformation (3.61) is also valid when w and w'

are larger than c,[9] since Eq. (3.59) is meaningful even in this case. In fact, if the phase velocity is larger than the speed of light, then there will exist a frame $\underline{S}$, moving with velocities $c^2/w\,\hat{\mathbf{n}}$ and $c^2/w'\,\hat{\mathbf{n}}'$ relative to S and S' respectively, where the period $\underline{T}(=\underline{\nu}^{-1})$ is maximum. As shown in Figure 3.31, the relations between periods T, T', and $\underline{T}$ resemble those of wavelengths λ, λ', and $\bar{\lambda}$; then it is obtained:

$$\gamma(c^2/w')\nu'^{-1} = \gamma(c^2/w)\nu^{-1}$$

which is only a different way of writing Eq. (3.59).

The transformation of the direction of propagation can be obtained from the multiplication of (3.58a) and (3.58b), or (3.60a) and (3.60b) as well. The result is

$$-V = \frac{c^2}{w'}\cos\vartheta' - \frac{c^2}{w}\cos\vartheta - \frac{c^2\,V}{w\,w'}\cos\vartheta'\cos\vartheta$$

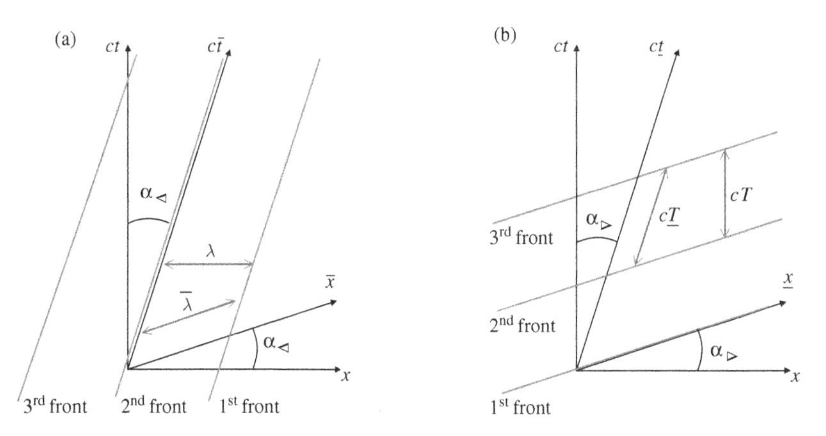

Figure 3.31. (a) A plane wave propagates in frame S with phase velocity $w < c$. In frame $\bar{S}$ (tg $\alpha_\triangleleft = w/c$), the wave fronts are at rest, and the wavelength is maximum because it is a proper length. (b) When phase velocity is $w > c$, there exists a frame $\underline{S}$ such that tg $\alpha_\triangleright = c/w$ (i.e., $\underline{S}$ moves with velocity c^2/w relative to S), where the period is maximum. In fact, the relation $\underline{T}\,/\,T$ has the same nature as $\bar{\lambda}/\lambda$, since diagrams (a) and (b) only differ by the exchange of spatial and temporal coordinates.

[9] Phase velocities bigger than c happen when light propagates in a bounded region (*wave guides*) or in *dispersive media*. In these cases phase velocity $w = \lambda\nu$ varies with frequency. As a consequence, different monochromatic components of a light pulse have different phase velocities, what produces the deformation of the pulse during the propagation. This characteristic makes difficult the study of the energy velocity carried by the pulse (see Brillouin, 1960). However, if the pulse is composed by components of similar wavelengths $\lambda \sim \lambda_o$, and the medium does not exhibit "anomalous dispersion" for these wavelengths, then the energy velocity carried by the pulse is well characterized by the *group velocity* $u \equiv d\nu/d\lambda^{-1}(\lambda_o)$, which turns out to be smaller than c.

i.e.,

$$\frac{c^2}{w'}\cos\vartheta' = \frac{\dfrac{c^2}{w}\cos\vartheta - V}{1 - \dfrac{V}{w}\cos\vartheta} \tag{3.62}$$

Squaring Eq. (3.62) and subtracting it from Eq. (3.61), it results

$$\frac{c^2}{w'}\sin\vartheta' = \frac{\sqrt{1 - \dfrac{V^2}{c^2}}\dfrac{c^2}{w}\sin\vartheta}{1 - \dfrac{V}{w}\cos\vartheta} \tag{3.63}$$

Finally, the quotient of Eqs. (3.63) and (3.62) leads to the transformation of the direction of propagation:

transformation of the direction of propogation

$$\mathrm{tg}\vartheta' = \frac{\sqrt{1 - \dfrac{V^2}{c^2}}\sin\vartheta}{\cos\vartheta - \dfrac{wV}{c^2}} \tag{3.64}$$

In the Galilean limit ($c \to \infty$) the direction of propagation does not change, and the phase velocity transforms as $w' = w - V\cos\vartheta$.

On the other hand, the ray velocity transforms as a particle velocity. Therefore, writing $u'^2 = u'^2_x + u'^2_y$ and using $u_x = u\cos\theta,\ u_y = u\sin\theta$ in Eqs. (3.42–3.43), it results that the modulus of ray velocity transforms according to

transformation of ray velocity

$$u'^2 = c^2 + \frac{(c^2 - V^2)\left(c^2 - \dfrac{c^4}{u^2}\right)}{\left(\dfrac{c^2}{u} - V\cos\theta\right)^2} \tag{3.65}$$

while the change of the ray direction yields

transformation of ray direction

$$\mathrm{tg}\theta' = \frac{u'_y}{u'_x} = \frac{\sqrt{1 - \dfrac{V^2}{c^2}}\sin\theta}{\cos\theta - \dfrac{V}{u}} \tag{3.66}$$

Equations (3.61–3.66) exhibit the following remarkable property: if phase velocity w and ray velocity u fulfill the relation $u = c^2/w$, then transformations (3.61) and (3.65) will keep this relation in any reference system, and transformations (3.64) and (3.66) will say that the direction of propagation transforms, in such case, as the ray direction. If a pulse is built in such a way that *group velocity* $\mathrm{d}\nu/\mathrm{d}\lambda^{-1}\lambda_o)$ properly describes the energy propagation (see Note 9), then the ray velocity will be identified with group velocity, and the relation $u = c^2/w$

a special dispersion relation

will emerge when the dispersion relation is $\nu\ (\lambda) = (c^2\lambda^{-2} + constant)^{1/2}$.[10] In particular, if the *constant* is zero it results the dispersion relation for light propagating in vacuum, where the condition $u = c^2/w$ is trivially satisfied because both u and w are equal to c. The examined property is the consequence of the existence of an invariant finite velocity; therefore, it lacks the Galilean analogue (in particular the direction of propagation is a Galilean invariant, but the ray direction changes under Galileo transformations).

3.18. PROPAGATION OF LIGHT IN MATERIAL MEDIA

Transformations (3.61, 3.64–3.66) can be applied to light propagation inside an isotropic transparent substance of refractive index n. Phase velocity is $w_o = c/n$ in the frame S_o where the substance is at rest; this value coincides with the ray velocity in S_o if the substance is non-dispersive. On the other hand, the ray direction is equal to the direction of propagation in S_o, therefore it is not necessary to distinguish them. In a reference system S where the substance moves with velocity $\mathbf{V} = V\hat{\mathbf{x}}$, phase velocity and the direction of propagation are given by the inverse transformations of (3.61) and (3.64):

$$\frac{c^4}{w^2} = c^2 + \frac{(c^2 - V^2)(n^2 - 1)}{\left(1 + \frac{nV}{c}\cos\theta_o\right)^2} \tag{3.67}$$

$$\mathrm{tg}\vartheta = \frac{\sqrt{1 - \frac{V^2}{c^2}}\sin\theta_o}{\cos\theta_o + \frac{V}{nc}} \tag{3.68}$$

while the ray velocity and direction in S turn out to be

$$u^2 = c^2 - \frac{(c^2 - V^2)(1 - n^{-2})}{\left(1 + \frac{V}{nc}\cos\theta_o\right)^2} \tag{3.69}$$

$$\mathrm{tg}\theta = \frac{\sqrt{1 - \frac{V^2}{c^2}}\sin\theta_o}{\cos\theta_o + \frac{nV}{c}} \tag{3.70}$$

---

[10] This dispersion relation happens in Quantum Mechanics for the relativistic particle as a consequence of the identification $E \to h\nu$, $p \to h\lambda^{-1}$ proposed by L. de Broglie, where h is Planck constant (see Eq. (6.25)). We remark that a dispersion relation such that $u = c^2/w$ implies that phase velocity is infinite in the frame moving with group velocity $\mathbf{u}$.

The frequency transformation results from (3.58b):

$$\nu = \frac{1 + \frac{n\,V}{c}\cos\theta_o}{\sqrt{1 - \frac{V^2}{c^2}}}\,\nu_o \qquad (3.71)$$

Velocity V in Eqs. (3.67–3.71) is not a relative velocity between two arbitrary reference systems, but the velocity of the transparent substance relative to frame S; θ_o and ν_o do not indicate the ray direction and the frequency in an arbitrary frame but in the substance frame S_o. By no means this implies the abandonment of the spirit of Relativity, because the presence of a material medium naturally privileges the substance frame from the rest of the frames. What matters is velocity V in (3.67–3.71) is not an absolute property of frame S_o but its velocity relative to an *arbitrary* frame S.

Snell's law in Relativity

In order to apply these transformations to ray refraction, it should be noticed that Snell's law (2.1) in electromagnetism comes from applying Maxwell's laws for continuous media in the frame where the transparent substance is at rest (*whichever* this frame is). Therefore the use of Eq. (2.1) is only valid in the substance proper frame S_o.[11] Instead, in classical physics Snell's law was applicable only if the substance was at "absolute" rest, since the wave equation was valid only in the ether frame. As emphasized in Chapter 2, the results of the experiments by Arago and Airy are nothing but the confirmation that Snell's law is verified in the transparent substance proper frame. Nevertheless, the classical notions of space and time forced to interpret these experimental results by resorting to a complicated combination of refraction and partial dragging of ether.

[11] In the same way, reflection law is valid only in the frame S_o fixed to the reflective surface.

4

Geometric Structure of Space-Time

4.1. INTERVAL

Galileo transformations are the consequence of assuming the invariance of distance (1.1) – which is built with the squared differences of Cartesian coordinates of a pair of points in the space. Also Lorentz transformations have an associated invariant whose structure resembles a distance but includes the temporal coordinates of the involved pair of events. In fact, let there be two events E_1 and E_2 having respectively coordinates (t_1, x_1, y_1, z_1) and (t_2, x_2, y_2, z_2) in the frame S. Let us call

$$\Delta t = t_2 - t_1, \quad \Delta x = x_2 - x_1, \quad \Delta y = y_2 - y_1, \quad \Delta z = z_2 - z_1 \tag{4.1}$$

and study the behavior of the combination $c^2 \Delta t^2 - \Delta x^2$ under transformations (3.40). Since Lorentz transformations are linear, coordinate differences (4.1) transform like the respective coordinates. Thus, using (3.40b),

$$\begin{aligned} c^2 \Delta t^2 - \Delta x^2 &= \gamma^2 (c \Delta t' + \beta \Delta x')^2 - \gamma^2 (\Delta x' + \beta c \Delta t')^2 \\ &= \gamma^2 (1 - \beta^2)(c^2 \Delta t'^2 - \Delta x'^2) = c^2 \Delta t'^2 - \Delta x'^2 \end{aligned}$$

Therefore, the value of the combination of coordinates $c^2 \Delta t^2 - \Delta x^2$ is not altered by transformations (3.40): it is invariant. Taking into account transformations (3.39), we can include coordinates y, z in the invariant. Let us define the *interval* Δs of the pair of events E_1 and E_2 as

$$\Delta s^2 \equiv c^2 \Delta t^2 - \Delta x^2 - \Delta y^2 - \Delta z^2 \tag{4.2}$$

interval

Since the combination $|\Delta \mathbf{r}|^2 = \Delta x^2 + \Delta y^2 + \Delta z^2$ is invariant under spatial rotations, then the interval (4.2) is invariant both under Lorentz transformations (3.39–40) and spatial rotations.

While in classical physics distance $|\Delta \mathbf{r}|$ and the lapse of time Δt were regarded as absolute relations happening separately in space and time, in Relativity the sole absolute relation between two events happens in *space-time*. Although the interval is not positive definite, it is a sort of "distance" in space-time, so conferring it a geometrical structure. Because the form of the interval differs from the Euclidean distance (1.1) by the unequal signature of its terms, the space-time geometry is said to be *pseudo-Euclidean*. Hermann Minkowski (1864–1909) is the author of this unified vision of space-time, where space and time are nothing but mere shadows – according to their own words – of a unique four-dimensional geometrical structure. Minkowski translated relativistic physics into a "four-vectorial" language (see Chapter 7), which is the natural language for the pseudo-Euclidean space-time geometrical structure (4.2). Minkowski taught the importance of associating physical concepts with magnitudes having geometrical meaning (independent of the coordinate system), like the invariant interval.

See Chap. 7

4.2. CALIBRATION HYPERBOLA

As remarked in §3.10, whenever two reference systems are represented in a same space-time diagram the scales of their axes are different and require calibration. Figure 3.18 showed a calibration procedure; in this section we are going to see another procedure. Let us consider all events E having $\Delta s^2 = -1$ (in the chosen unit of length) with respect to the space-time coordinate origin. Such events have then coordinates satisfying the equation

$$c^2\, t^2 - x^2 - y^2 - z^2 = -1 \tag{4.3}$$

which is the equation for a one-sheeted hyperboloid in space-time. In Figure 4.1 (left) the hyperboloid is displayed without the dimension associated to z coordinate. Events having $t = 0$ make up a circle of radius 1 in the x–y plane (a sphere of radius 1 in space). In particular the hyperboloid intersects the spatial Cartesian axes at the point corresponding to the unit of measure.

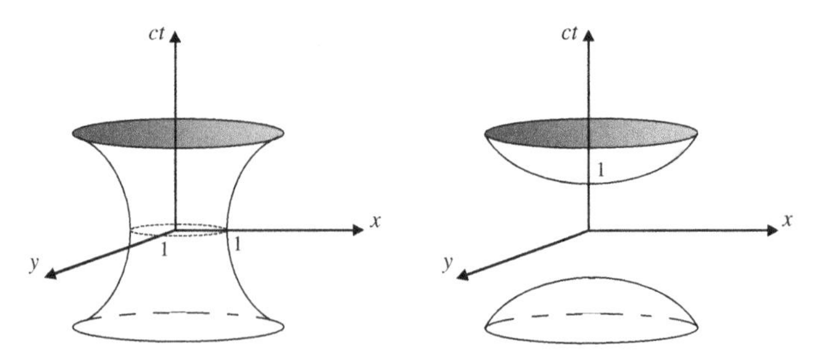

Figure 4.1. One- and two-sheeted hyperboloids defined by the set of events having respectively intervals $\Delta s^2 = -1$ and $\Delta s^2 = 1$ with the event at the coordinate origin. They intersect Cartesian axes at the unit of measure.

In the same way, we can consider those events E having interval $\Delta s^2 = 1$ with respect to the coordinate origin. These events are defined by the equation

$$c^2 t^2 - x^2 - y^2 - z^2 = 1 \tag{4.4}$$

which describes the two-sheeted hyperboloid represented in Figure 4.1 (right). The vertex of each sheet has coordinates $x = y = z = 0$, and so it is $ct = \pm 1$. Thus the vertex indicates the unit of measure on the temporal axis.

On the other hand, the invariance of the interval implies that the events belonging to each one of these hyperboloids have coordinates in S' satisfying the equation

$$c^2 t'^2 - x'^2 - y'^2 - z'^2 = -1$$

or

$$c^2 t'^2 - x'^2 - y'^2 - z'^2 = 1$$

(notice that the event at the space-time coordinate origin has coordinates (0, 0, 0, 0) in any reference system). This means that the same hyperboloids calibrate the unit of measure of the coordinate axes of any other reference system.

Figure 4.2 displays in a same space-time diagram the axes of two different reference systems, together with the *calibration hyperbolas* (the projections of the hyperboloids on *ct*–*x* plane). The units of measure have been marked on each axis at the points where the hyperbolas intersect the axes.

In §3.11 it was explained that there is not contradiction at all if each one of two different frames states that the clocks belonging to the other one go slower: time dilatation is verified in both frames because each one uses a different notion of simultaneity to synchronize its clocks. The relativity of the notion of

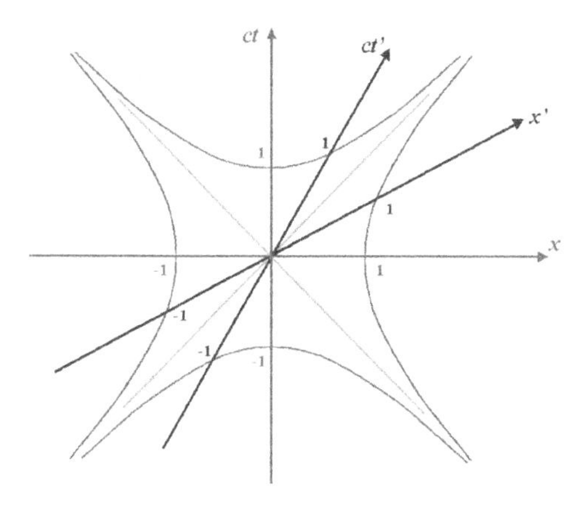

Figure 4.2. Calibration of coordinate axes by means of hyperbolas.

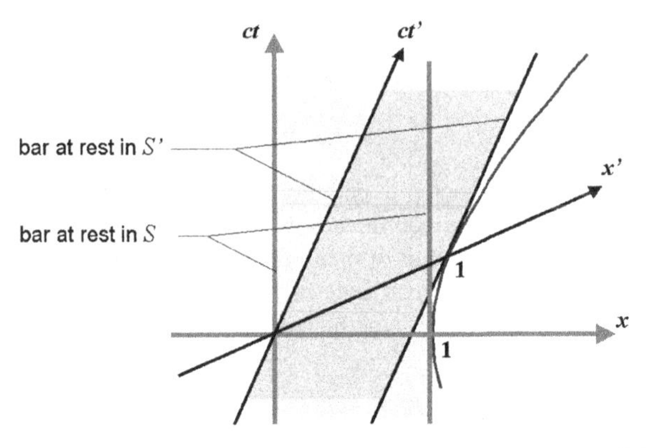

Figure 4.3. World lines of the ends of two bars of proper length equal to the unit, and their respective proper frames. The length of a bar results from the determination of simultaneous positions of its ends. The notion of simultaneity is different in each frame. This explains why there is not contradiction if each frame states that the bodies at rest in the other one are contracted.

length contraction does not imply privilege of a frame

simultaneity is also the reason why there is not contradiction if each frame states that the bodies at rest in the other frame are contracted. Figure 4.3 shows the world lines of the ends of two equal bars: one of them is at rest in frame S, while the other is at rest in frame S'. Proper lengths of both bars are equal to the unit, as certified by the calibration hyperbola. In fact, each bar is placed between the coordinate origin and the unit of measure of the respective spatial coordinate axis. In order to determine the bar length in an arbitrary reference system, we have to establish the *simultaneous* positions of its ends. Thus the determination of the length in S of the bar at rest in S' implies the choice of a pair of events, belonging to the world lines of each end, having the same t coordinate; for instance, we can use the events placed on the x axis (i.e., $t = 0$). Figure 4.3 shows that the world lines of the ends of the bar fixed to S' cross the x axis belonging to S at $x = 0$ and $x < 1$ (contraction). In the same way, the world lines of the ends of the bar fixed to S cross the x' axis belonging to S' (i.e., $t' = 0$) at $x' = 0$ and $x' < 1$. The contraction is verified in both frames.

4.3. LIGHT CONE

The invariance of the interval allows an absolute (independent of the reference system) classification of pairs of events. A pair of events is said to be

timelike separated	if $\Delta s^2 > 0$
spacelike separated	if $\Delta s^2 < 0$
lightlike separated	if $\Delta s^2 = 0$

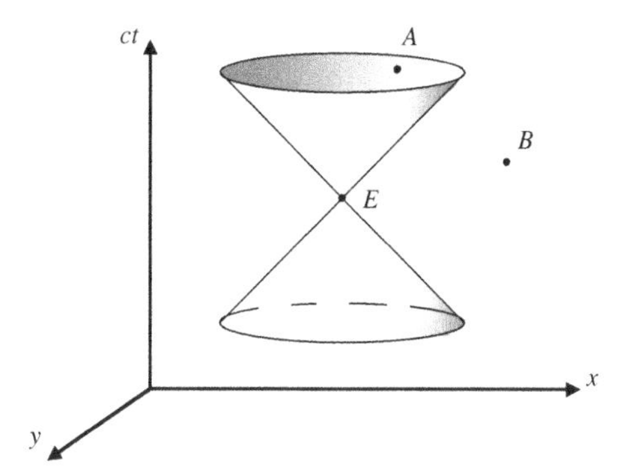

Figure 4.4. Light cone of event E. The events belonging to the interior of the cone, as the event A, are "timelike" separated from E. The events being outside the cone, as the event B, are "spacelike" separated from E.

If a pair of events are timelike separated then the term $c^2\Delta t^2$ in (4.2) dominates on the spatial terms forming $|\Delta\mathbf{r}|^2$. Instead, if they are spacelike separated it happens the contrary, what explains the reason for each type of separation denomination.

Given an arbitrary event E, the set of events being lightlike separated from E are those whose coordinates fulfill the equation

$$c^2(t-t_E)^2-(x-x_E)^2-(y-y_E)^2-(z-z_E)^2=0 \tag{4.5}$$

light cone

which is the equation of a cone with vertex at E. Any event on this conical surface satisfies that $|\mathbf{r}-\mathbf{r}_E|^2=c^2(t-t_E)^2$, which is the equation of a light ray going through E.Therefore the cone is generated by the world lines of the light rays going through the event E. Due to the invariance of the interval, this conclusion is also true in other reference systems uniformly moving with respect to S. Figure 4.4 displays the *light cone* of event E.

4.4. TIMELIKE-SEPARATED EVENTS

Figure 4.4 shows that the events belonging to the interior of the light cone of E are timelike separated from E, because $|\mathbf{r}-\mathbf{r}_E|^2<c^2(t-t_E)^2$. If a pair of events E_1 and E_2 are timelike separated, then they can be joined by the world line of a particle traveling with constant velocity $\mathbf{u}=(\mathbf{r}_2-\mathbf{r}_1)/|t_2-t_1|$, smaller than the speed of light. Thus, there exists a frame where both events happen at the same place (the frame fixed to this particle). In Figure 4.5, S' is the proper frame of a timelike separated pair of events, as it looks in the space-time diagram of an arbitrary reference system S (the axes x–x' were chosen along the direction defined by the locations where the events happen). Since in frame S' it is $\mathbf{r}'_2=\mathbf{r}'_1$, then the interval between events E_1 and E_2 takes the form

$$\Delta s^2=c^2\,\Delta t'^{\,2}$$

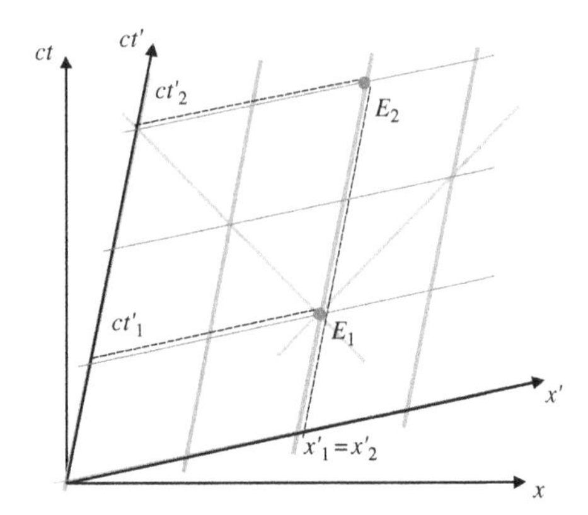

Figure 4.5. Building of the proper frame S' for a timelike separated pair of events E_1 and E_2.

On the other hand $\Delta t'$ is, in this case, the proper time $\Delta\tau$ elapsed between the events (because both events occur at the same place in reference system S'). Therefore, the interval for timelike separated events is the measure of the proper time between them:

timelike separation: relation between proper time and interval

$$\Delta\tau = \frac{\Delta s}{c} \tag{4.6}$$

We are going to apply Eq. (4.6) to a pair of events belonging to the world line of a particle. Since no particle can reach the speed of light, then a world line of particle is always interior to the light cone of each one of its events (Figure 4.6). This kind of world lines is said to be *timelike*. In fact, any pair of events belonging to a world line of particle is timelike separated. Let us choose two events such that their coordinates slightly differ: the coordinates of E_1 and E_2 in a reference system S are (t, x, y, z), and $(t+dt, x+dx, y+dy, z+dz)$ respectively. So the interval is infinitesimal and has the value

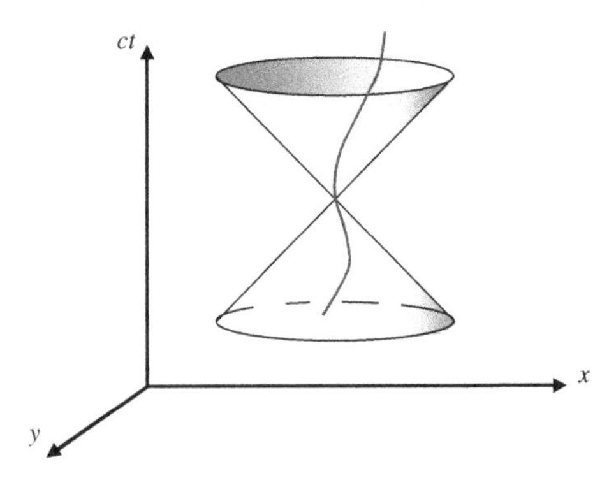

Figure 4.6. The world line of a particle is contained in the light cone of each one of its events.

$$ds^2 = c^2 dt^2 - |d\mathbf{r}|^2$$

$d\mathbf{r}$ in the former equation is the particle displacement during the time dt. Then $|d\mathbf{r}| = u(t)\,dt$, and the proper time between events turns out to be

$$d\tau = \frac{ds}{c} = \frac{1}{c}\sqrt{c^2dt^2 - u(t)^2dt^2} = \sqrt{1 - \frac{u(t)^2}{c^2}}\,dt = \gamma\,(u(t))^{-1}\,dt \qquad (4.7)$$

$d\tau$ is the time measured in the frame where both events happen at the same position, i.e., the frame moving relative to S with the same velocity $\mathbf{u}(t)$ of the particle at the considered time t (compare with (3.12a)). Then, it is the time measured by a clock going with the particle. Equation (4.7) can be integrated to get the time read by a clock fixed to the particle (proper time along the world line of a particle)

proper time along the world line of a particle

$$\Delta\tau = \int dt\sqrt{1 - \frac{u(t)^2}{c^2}} \qquad (4.8)$$

4.5. TWIN PARADOX

In the previous section, "ds" does not denote an "exact" differential form, but it is just a symbol to indicate an infinitesimal magnitude. In fact, the integral of ds (Eq. (4.8)) does not result to be the difference of the values taken by a function "s" at the ends of the integration path. Instead, we are going to see that the integral of ds depends on the world line along which the integral is made. Figure 4.7 shows two different world lines joining two timelike-separated events E_i and E_f: one of them is a straight line and the other is an arbitrary world line of particle. The space-time diagram corresponds to the proper frame of the events, because both events have equal spatial coordinates. A given dt defines on each line a pair of events infinitesimally separated. The interval between them is $ds_1 = c\ dt$ on the straight line, while on an arbitrary line it is $ds_2 = (c^2dt^2 - {u_2}^2\,dt^2)^{1/2} \leq ds_1$. Therefore, the integral of the interval along each line will be maximum on the straight line. The integral of the interval divided by c is the proper time along the world line (Eq. (4.8)). On the straight line, this integral coincides with the proper time between the events at the ends. By comparing the intervals on each line, it results that the time read by a clock going with a particle that travels between two given events is maximum if the world line is straight. Similarly to the fact that the length for joining two points depends on the path traveled (being minimum, in an Euclidean geometry, on the straight path), in Special Relativity the time read by a *unique* clock joining two events depends on the path followed by the clock, being maximum if the world line is straight (i.e., if the motion is rectilinear and uniform). Although the diagram in Figure 4.7 corresponds to the proper frame of the events at the ends

maximization of proper time along a world line joining a given pair of events

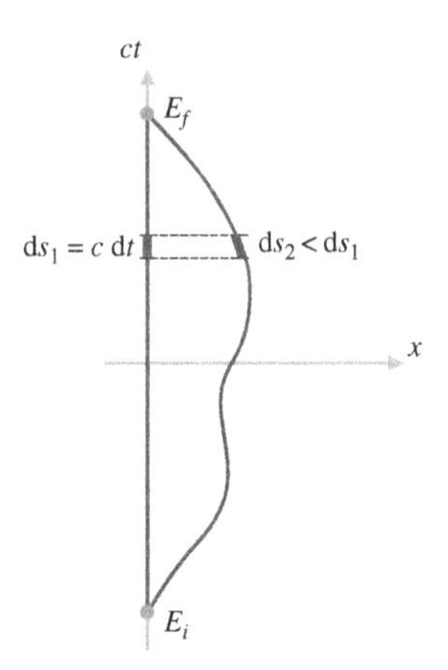

Figure 4.7. Comparison between intervals, for a given d*t*, along two different world lines joining a same pair of events E_i and E_f.

of the world line, the result would be the same in other reference systems moving uniformly with respect to the former one, because the interval is invariant under Lorentz transformations.

The question examined in the previous paragraph is usually introduced as "the twin paradox". One of the twins is astronaut and undertakes a space journey. At the departure the twins check their clocks. When the astronaut comes back the twins compare their clocks again and realize that the astronaut's clock went slower than that of his brother on Earth. Not only the clocks exhibit the discrepancy, but the astronaut came back younger than his brother. The reason for describing this result as "paradoxical" lies in the fact that, from the astronaut's point of view, he would have the right to consider that the traveler is not him but his brother; if so, his conclusion would be that his brother should be younger when they meet again. Thus we arrive to two contradictory conclusions, depending on whether the problem is analyzed from the Earth or the astronaut frames.

The nature of the question differs from that examined in §3.12, where the rate of two synchronized clocks was compared with a unique clock moving with respect to the others. In this case, the same time dilatation factor $\gamma(V)$ was obtained either if the pair of synchronized clocks belonged to one or the other of the two frames involved. This result does not imply contradiction at all because each reference system synchronizes its clocks according to its own notion of simultaneity. But now we are comparing the rate of only two clocks in relative motion. Therefore the synchronization of clocks in different frames is not implicated, and we cannot admit an ambivalent result. So, it should be elucidated which is the clock that goes slower, and why the clocks do not experiment a symmetrical behavior in spite of the apparent symmetry of the problem. Does one of the clocks have some privilege?

As already shown in this Chapter, in Special Relativity space-time has a pseudo-Euclidean geometrical structure, which was inherited from the hypothesis that space is Euclidean and time is homogeneous (§1.2, 3.1), together with the postulate of invariance of the speed of light. This geometric framework confers a privilege to the straight world line (i.e., to the rectilinear uniform movement). In the same way that Euclidean geometry states that the distance is minimum along a straight line, the pseudo-Euclidean geometry of Special Relativity states that the time elapsed between two events, as read by a unique clock, is maximum along a straight world line. So, there is no symmetry in the behavior of the astronaut and his brother clocks because their relative movement must be non-uniform in order that the clocks can go away and meet again: if one of the clocks has a rectilinear uniform movement (straight world line), *necessarily the other will not have it*. Like in classical physics, in Special Relativity there are also privileged world lines: the straight timelike world lines corresponding to inertial trajectories of particles. The privilege of these trajectories is conferred by an underlying geometrical structure existing independently of the physical phenomena, and playing the same role that Euclidean absolute space and absolute time play in classical physics. The privilege of inertial world lines on the others explains the asymmetry between the rates of the clocks carried by the astronaut and his brother.

privileged world lines

See Chap. 8

Figure 4.8 displays a space-time diagram for three different world lines joining the same initial and final events: (1) the inertial line, (2) a polygonal line corresponding to a going travel with velocity v and a return with opposite velocity, and (3) the hyperbolic motion described in Eq. (3.48). The proper time along each world line is computed by using the formula (4.8):

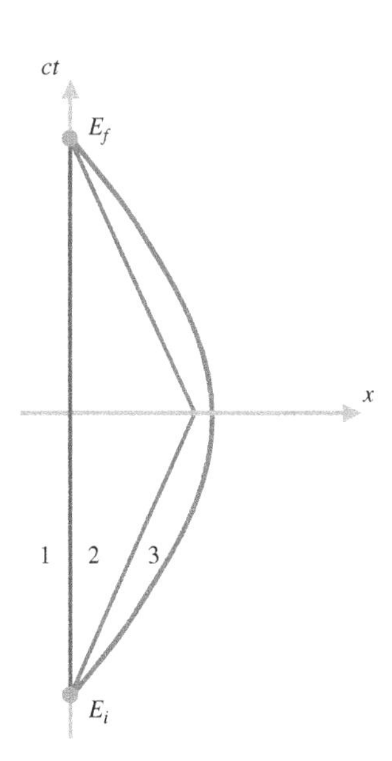

Figure 4.8. Three different movements joining the same pair of events: (1) inertial motion, (2) going and return with constant velocity, and (3) hyperbolic motion.

$$\Delta\tau_1 = \int_{t_i}^{t_f} dt\sqrt{1-\frac{u_1(t)^2}{c^2}} = \int_{t_i}^{t_f} dt = t_f - t_i = \Delta t$$

$$\Delta\tau_2 = \int_{t_i}^{t_f} dt\sqrt{1-\frac{u_2(t)^2}{c^2}} = \int_{-\Delta t/2}^{0} dt\sqrt{1-\frac{v^2}{c^2}} + \int_{0}^{\Delta t/2} dt\sqrt{1-\frac{v^2}{c^2}} = \sqrt{1-\frac{v^2}{c^2}}\Delta t$$

$$\Delta\tau_3 = \int_{t_i}^{t_f} dt\sqrt{1-\frac{u_3(t)^2}{c^2}} = \int_{-\Delta t/2}^{\Delta t/2} \frac{dt}{\sqrt{1+\frac{a^2t^2}{c^2}}} = \frac{c}{a}\,\mathrm{arcsinh}\,\frac{at}{c}\Big|_{-\frac{\Delta t}{2}}^{\frac{\Delta t}{2}} = \frac{2c}{a}\mathrm{arcsinh}\frac{a\Delta t}{2c}$$

The result $\Delta\tau_2 = \gamma(v)^{-1}\Delta t$ is the well-known delay of a clock moving with respect to clocks synchronized in the Earth frame (Eq. (3.8) and Figure 3.19). However, would it not be licit to think that the astronaut traveling along world line 2 would also observe an equal delay for the terrestrial clock, since this is in motion relative to the clocks synchronized in the astronaut frame? The answer is "yes and no." Motion 2 decomposes in two stages associated to two *different* inertial proper frames S' and S''. The answer is "yes" during each one of the inertial stages. However, when the astronaut goes through the return event E_c (see Figure 4.9) his velocity relative to Earth changes from v to $-v$, and his proper frame changes from S' to S''. This change causes a discontinuity of his notion of simultaneity. Figure 4.9 includes the coordinate lines marked with t'_c and t''_c which correspond to events that are simultaneous with E_c in frames S' and S'' respectively. It can be seen that the change of the astronaut's notion of simultaneity skips a finite part of the evolution of the terrestrial clock (the part covered between the lines t'_c and t''_c). The delay of the terrestrial clock during the inertial stages is compensated and exceeded by the sudden advance registered when the astronaut passes the event E_c where his acceleration is infinite. Then the answer is "no" whenever the complete motion is considered, and the result $\Delta\tau_2 = \gamma(v)^{-1}\Delta t$ is understandable even from the astronaut's point of view. The

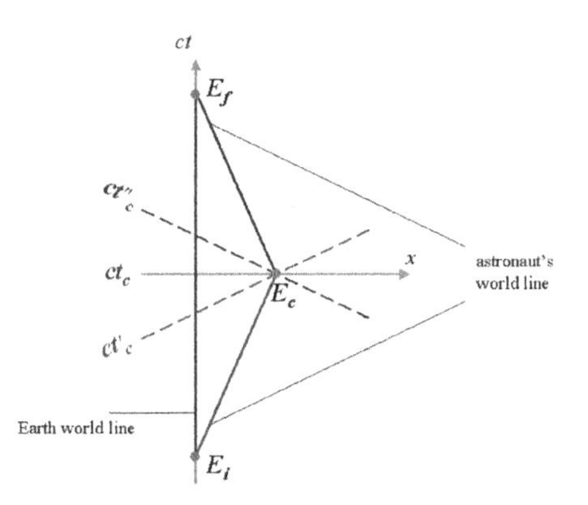

Figure 4.9. Discontinuity of the astronaut's notion of simultaneity.

skip of the astronaut's notion of simultaneity does not mean that a part of the Earth history is not accessible for him; in Figure 4.9 we can verify that the history passed between the departure (event E_i) and the arrival (event E_f) can be communicated to the astronaut in a continuous way by means of light signals, for instance. On the other hand, the discontinuity of the notion of simultaneity can be eliminated by smoothing the movement, as it happens in motion 3, hence obtaining a gradual change of the notion of simultaneity.[1]

4.6. SPACELIKE-SEPARATED EVENTS

In Figure 4.4 those events that are external to the light cone of E are spacelike separated from E, because $|\mathbf{r}-\mathbf{r}_E|^2 > c^2\ (t-t_E)^2$. If two events are spacelike separated, there always exists a frame where both events are simultaneous. This fact can be understood by means of the diagram in Figure 4.10, where the x–x' axes have been chosen in the direction defined by the places where the events happen. Since in the frame S' it is $t'_2 = t'_1$, then the interval between both events is

$$\Delta s^2 = -|\Delta \mathbf{r}'|^2$$

Therefore, the interval between spacelike-separated events is the measure of the spatial distance between them, in the frame where the events are simultaneous (*proper distance*).

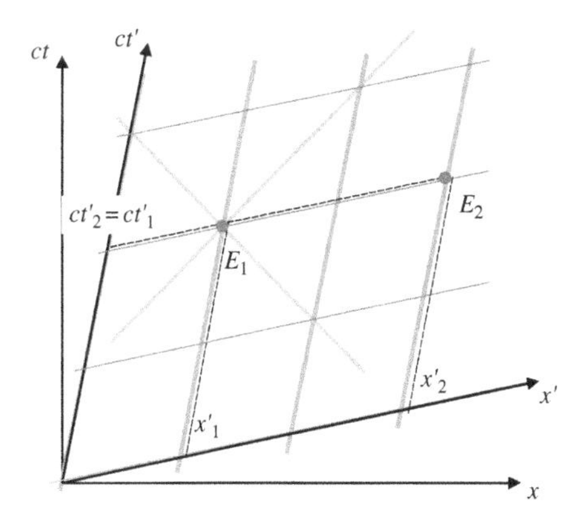

Figure 4.10. Construction of frame S' where two spacelike-separated events are simultaneous.

[1] Time differences between clocks that meet again after undergoing different accelerations have been measured by comparing four atomic clocks flying around the planet with reference clocks on earth (Hafele and Keating, 1972). The observed differences of -59 ± 10 (eastward flights) and 273 ± 7 (westward flights) nanoseconds include the effect of gravity that will be studied in Chapter 8.

A distinctive feature of a pair of spacelike-separated events is that its temporal order depends on the reference system. In the example of Figure 3.6, the opening of the doors was a pair of simultaneous events in the frame fixed to the vehicle. However, in a frame where the vehicle moved to the right, the opening of the left door would happen before the opening of the other door (Figure 3.7). Analogously, the right door would open first in a frame where the vehicle moved to the left. Figure 4.11 displays these alternatives in a space-time diagram: E_1 and E_2 are two spacelike-separated events, and S is the frame where they are simultaneous. For simplicity E_1 was chosen as the space-time coordinate origin; in particular, the temporal coordinate of E_1 is zero in all reference systems under consideration. In frame S'' it is $t''_2 > 0$, and so E_2 happens after E_1. On the contrary, in frame S' it is $t'_2 < 0$, and E_2 occurs before E_1.

causal structure of space-time

Contrary to what could be suspected, the alteration of the temporal order is not in conflict with the notion of causality. In fact, one could wonder what could happen if an event E_2 is caused by the event E_1; in such a case the inversion of the temporal order would lead to a paradoxical situation where the effect happens before the cause. However, in order that two events can be causally related the communication between them has to be possible, either through material bodies or through light signals. But this communication cannot be performed to a speed higher than the speed of light. The communication is not possible between spacelike-separated events, so such pair of events cannot establish causal relation at all. The only events that can be causally related with E_1 are those events being either on the light cone of E_1 (null separation; i.e., communication by means of light rays) or inside the light cone (timelike separation; i.e., communication by means of particles). If a pair of events can be causally related, then Lorentz transformations will not invert its temporal order.

absolute past and future

In Figure 4.4 all events belonging to the upper cone and its interior happen after E in any reference system; then they make up the *absolute future* of E. In the same way, the events belonging to the lower cone and its interior make up the *absolute past* of event E, because they happen before E in any reference

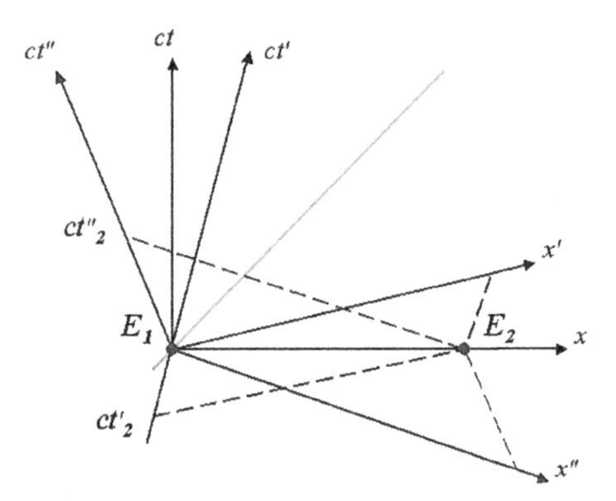

Figure 4.11. The temporal order of two spacelike-separated events depends on the reference system. While E_1 and E_2 are simultaneous in S, E_2 happens before E_1 in S', but after E_1 in S''.

system. The invariance of the interval – i.e., the invariance of the structure of space-time light cones – together with the concept that the speed of light cannot be surpassed guarantee the preservation of the causal structure of space-time.

4.7. VELOCITY PARAMETER: RAPIDITY

Galileo transformations commute. This means that the result of successively applying two Galileo transformations is independent of which transformation applies first. In fact, the application of two Galileo transformations (1.5), G_1 and G_2 with velocities $\mathbf{V}_1$ and $\mathbf{V}_2$ (which not necessarily share the direction), is equivalent to perform a unique Galileo transformation with a velocity that turns out to be the Galilean composition of the velocities associated to G_1 and G_2: $\mathbf{V} = \mathbf{V}_1 + \mathbf{V}_2$,

$$\begin{array}{ccccc} \mathbf{r} & \xrightarrow{G_1} & \mathbf{r}' = \mathbf{r} - \mathbf{V}_1 t & \xrightarrow{G_2} & \mathbf{r}'' = \mathbf{r}' - \mathbf{V}_2 t = \mathbf{r} - (\mathbf{V}_1 + \mathbf{V}_2)\, t \\ \mathbf{r} & \xrightarrow{G_2} & \mathbf{r}' = \mathbf{r} - \mathbf{V}_2 t & \xrightarrow{G_1} & \mathbf{r}'' = \mathbf{r}' - \mathbf{V}_1 t = \mathbf{r} - (\mathbf{V}_2 + \mathbf{V}_1)\, t \end{array} \tag{4.9}$$

Galilean composition of velocities (1.7), where the roles played by the composed velocities are interchangeable, exhibits then the mark of the commutativity of Galileo transformations. On the contrary, in relativistic composition of velocities, each velocity plays a different role, as evident in (3.43) because of the presence of factor $\gamma(V)^{-1}$ associated to time dilatation. This means that Lorentz transformations in different directions do not commute.

On the other hand, it is quite plain that Galileo transformations do not mix up coordinates, but each Cartesian coordinate of S' is only a function of the corresponding Cartesian coordinate of S (time is just a parameter because it does not transform). Instead Lorentz transformation (3.40) – which is called a *boost* – combines the temporal coordinate with a spatial coordinate, so acting in a space-time plane. In this way the action of a boost evokes spatial rotations, which also acts on a plane (the plane that is orthogonal to the rotation axis). Spatial rotations do not commute either (Figure 4.12). However, both for boosts and spatial rotations, there is only a case where the result of applying successive transformations is independent of their order: when all act on the same plane. This means that boosts velocities have the same direction or that spatial rotations share the rotation axis.

If two rotations of angles θ_1 and θ_2 successively act on a same plane (i.e., they rotate around the same axis), the result is equal to a sole rotation of parameter $\theta = \theta_1 + \theta_2$. Like it happens with the Galilean composition of velocities, the interchangeable roles of θ_1 and θ_2 in the equivalent rotation parameter reflects the commutativity of rotations around the same axis. Relativistic composition of velocities of equal direction (3.19b) also displays both velocities with interchangeable roles, as a consequence of the commutativity of boosts of equal direction. But, different from the other examined transformations, relativistic

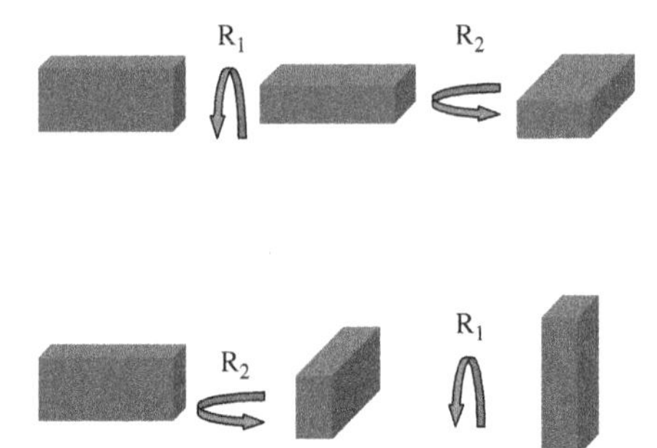

Figure 4.12. Rotations acting on different planes do not commute. The result of successively applying R_1 and R_2 depends on the order of application.

composition of velocities of equal direction does not lead to the addition of velocities. This only means that the velocity is not the more natural parameter characterizing a Lorentz boost. The natural *velocity parameter* Θ_V can be easily found by taking into account the similarity existing between the composition of velocities (3.19b) and the expression for the hyperbolic tangent of the sum of two arguments:

$$\mathrm{th}\,(\Theta_1 + \Theta_2) = \frac{\mathrm{th}\,\Theta_1 + \mathrm{th}\,\Theta_2}{1 + \mathrm{th}\,\Theta_1\,\mathrm{th}\,\Theta_2}$$

Then it becomes clear that the natural velocity parameter is Θ_V such that

velocity parameter

$$\mathrm{th}\,\Theta_V = \frac{V}{c} = \beta_V \tag{4.10}$$

In fact, using Θ_{u_x}, $\Theta_{u'_x}$, and Θ_V, composition (3.19b) can be written as

$$\mathrm{th}\,\Theta_{u_x} = \frac{u_x}{c} = \frac{\dfrac{u'_x}{c} + \dfrac{V}{c}}{1 + \dfrac{u'_x V}{c^2}} = \frac{\mathrm{th}\,\Theta_{u'_x} + \mathrm{th}\,\Theta_V}{1 + \mathrm{th}\,\Theta_{u'_x}\,\mathrm{th}\,\Theta_V} = \mathrm{th}\,(\Theta_{u'_x} + \Theta_V)$$

Therefore, the velocity parameter Θ is additive under composition:

$$\Theta_{u_x} = \Theta_{u'_x} + \Theta_V \tag{4.11}$$

Lorentz transformations (3.40) take an elegant form when it is written in terms of the velocity parameter (4.10). Noticing that

$$\gamma(V) = \frac{1}{\sqrt{1 - \beta_V^2}} = \frac{1}{\sqrt{1 - th^2\Theta_V}} = \cosh\,\Theta_V \tag{4.12}$$

$$\gamma(V)\,\beta_V = \frac{\beta_V}{\sqrt{1 - \beta_V^2}} = \frac{\mathrm{th}\,\Theta_V}{\sqrt{1 - \mathrm{th}^2\Theta_V}} = \sinh\,\Theta_V \tag{4.13}$$

then Lorentz transformations have the form[2]

$$\begin{aligned} ct' &= ct \cosh \Theta_V - x \sinh \Theta_V \\ x' &= -ct \sinh \Theta_V + x \cosh \Theta_V \end{aligned} \tag{4.14}$$

On the other hand, the elapsed time $c\,\mathrm{d}t$ and the displacement $|\mathrm{d}\mathbf{r}| = u\,\mathrm{d}t$ corresponding to an interval $\mathrm{d}s$ along the world line of a particle (see Eq. (4.7)), can be written, using (4.12–13), as a sort of "projections" of $\mathrm{d}s$ in terms of the particle *rapidity* Θ_u:

$$c\,\mathrm{d}t = \mathrm{d}s \cosh \Theta_u \qquad u\,\mathrm{d}t = \mathrm{d}s \sinh \Theta_u \tag{4.15}$$

4.8. WIGNER ROTATION

In the previous section we remarked some analogies and differences among Galileo transformations, spatial rotations, and Lorentz boosts, which were related to the commutativity of the compositions of transformations. Another important property of composition is the one defining the *group* structure of the set of transformations. While the successive application (or composition) of two Galileo transformations *is* a Galileo transformation (see Eq. (4.9), and the successive application of two rotations *is* a rotation, instead the successive application of two Lorentz boosts is only equal to a Lorentz boost if the boosts are performed along the same direction. This means that, different from the other examined transformations, Lorentz boosts themselves do not constitute a group.[3]

The Lorentz boost (3.39–3.40) is a linear transformation that can be summarized in a unique matrix equation:

$$\begin{pmatrix} ct' \\ x' \\ y' \\ z' \end{pmatrix} = \begin{pmatrix} \gamma & -\gamma\beta & 0 & 0 \\ -\gamma\beta & \gamma & 0 & 0 \\ 0 & 0 & 1 & 0 \\ 0 & 0 & 0 & 1 \end{pmatrix} \begin{pmatrix} ct \\ x \\ y \\ z \end{pmatrix} \tag{4.16}$$

matrix form of a boost

Equation (4.16) shows that the matrix corresponding to the boost, which will be called $B_{(x)}$, is symmetric and unimodular (its determinant is equal to 1). Of course, $B_{(x)}$ is not the most general boost, but it is just a boost along the x

---

[2] If regarded as an "active" transformation, a rotation "moves" points on spheres, because the distance of the point to the origin – the sphere radius – is left invariant. Analogously, a Lorentz boost "moves" points on hyperboloids like the ones of Figure 4.1, because the interval is left invariant. This feature is highlighted by using *Rindler coordinates* (ξ, Θ) which resemble polar coordinates: $ct = \pm\xi \cosh \Theta$, $x = \pm\xi \sinh \Theta$ if $\pm ct > |x|$, or $ct = \pm\xi \sinh \Theta$, $x = \pm\xi \cosh \Theta$ if $|ct| < \pm x$. Under Lorentz boosts, coordinate ξ (i.e., $|ct^2 - x^2|^{1/2}$) does not change, while Θ changes to $\Theta' = \Theta - \Theta_V$.

[3] A set of transformations is a group G if it has an associative operation – composition or multiplication – that is *closed* (the composition of two elements of G belongs to G), has unit element, and such that each transformation of G has an inverse belonging to G.

direction (with velocity $\boldsymbol{\beta} = \beta\hat{\mathbf{x}}$). However, a boost along an arbitrary direction can always be obtained by applying a spatial rotation to matrix $B_{(x)}$: $B = R\, B_{(x)} R^{-1}$. Matrix B is still unimodular (because det $R = 1$) and symmetric. The symmetry of the arbitrary boost B—which will play an important role in this section— comes from the symmetry of $B_{(x)}$, together with the property that rotations are orthogonal matrices (the inverse R^{-1} is equal to the transpose R^T). In fact, $B^T = (R^{-1})^T B^T_{(x)}\, R^T = R\, B_{(x)} R^{-1} = B$.

Now we shall show that the result of composing two boosts whose velocities $\boldsymbol{\beta}_1$ and $\boldsymbol{\beta}_2$ are not parallel is not a boost. In spite of this, the composition will result to be equivalent to a boost whose velocity is the relativistic composition of velocities $\boldsymbol{\beta}_1$ and $\boldsymbol{\beta}_2$ (as one could suspect), followed by a spatial rotation. This spatial rotation is called *Wigner rotation*, and its existence implies that Lorentz boosts along arbitrary directions – each one of them being a transformation that does not rotate spatial Cartesian axes – must join to spatial rotations for reaching the structure of a group (which is called *Lorentz group*).

For simplicity we shall consider the composition of two boosts $B_{(x)}$ and $B_{(y)}$ along mutually perpendicular directions $\boldsymbol{\beta}_1 = \beta_1\hat{\mathbf{x}}$, $\boldsymbol{\beta}_2 = \beta_2\hat{\mathbf{y}}$. The first one is the transformation (4.16); the second one results from (4.16) by exchanging the roles played by x and y. The successive application of two Lorentz transformations can be summarized in a unique matrix equation where the matrix performing the coordinate transformation turns out to be the product of the matrices performing each individual transformation:

$$B_{(y)}(\beta_2)\, B_{(x)}(\beta_1) = \begin{pmatrix} \gamma_2 & 0 & -\gamma_2\beta_2 & 0 \\ 0 & 1 & 0 & 0 \\ -\gamma_2\beta_2 & 0 & \gamma_2 & 0 \\ 0 & 0 & 0 & 1 \end{pmatrix} \begin{pmatrix} \gamma_1 & -\gamma_1\beta_1 & 0 & 0 \\ -\gamma_1\beta_1 & \gamma_1 & 0 & 0 \\ 0 & 0 & 1 & 0 \\ 0 & 0 & 0 & 1 \end{pmatrix}$$
$$= \begin{pmatrix} \gamma_2\gamma_1 & -\gamma_2\gamma_1\beta_1 & -\gamma_2\beta_2 & 0 \\ -\gamma_1\beta_1 & \gamma_1 & 0 & 0 \\ -\gamma_2\gamma_1\beta_2 & \gamma_2\gamma_1\beta_2\beta_1 & \gamma_2 & 0 \\ 0 & 0 & 0 & 1 \end{pmatrix} \tag{4.17}$$

As is evident in Eq. (4.17), the matrix resulting from the composition of two boosts is not symmetric; therefore it cannot be a boost. However it is easy to prove that result (4.17) can be written as the successive application of a unique boost B_c and a spatial rotation R:

$$B_{(y)}\, B_{(x)} = R\, B_c \tag{4.18}$$

Rotation R necessarily acts on the x–y plane, because coordinate z did not play any role. Then we shall write R, without lack of generality, as

$$R(\varpi) = \begin{pmatrix} 1 & 0 & 0 & 0 \\ 0 & \cos\varpi & \sin\varpi & 0 \\ 0 & -\sin\varpi & \cos\varpi & 0 \\ 0 & 0 & 0 & 1 \end{pmatrix} \tag{4.19}$$

where ϖ is an unknown angle that we are going to solve (note in Eq. (4.16) that the rows and columns acting on the x–y plane are the second and third ones).

With the aim of finding ϖ, the unimodular matrix $B_c = R(\varpi)^{-1} B_{(y)}\, B_{(x)}$ will be required to be symmetric, since this is a necessary condition in order that B_c be a boost.

$$B_c = R(\varpi)^{-1}\, B_{(y)}\, B_{(x)} = \begin{pmatrix} 1 & 0 & 0 & 0 \\ 0 & \cos\varpi & -\sin\varpi & 0 \\ 0 & \sin\varpi & \cos\varpi & 0 \\ 0 & 0 & 0 & 1 \end{pmatrix} \begin{pmatrix} \gamma_2\gamma_1 & -\gamma_2\gamma_1\beta_1 & -\gamma_2\beta_2 & 0 \\ -\gamma_1\beta_1 & \gamma_1 & 0 & 0 \\ -\gamma_2\gamma_1\beta_2 & \gamma_2\gamma_1\beta_2\beta_1 & \gamma_2 & 0 \\ 0 & 0 & 0 & 1 \end{pmatrix}$$
$$= \begin{pmatrix} \gamma_2\gamma_1 & -\gamma_2\gamma_1\beta_1 & -\gamma_2\beta_2 & 0 \\ \cdots & \cdots & -\gamma_2 sin\varpi & 0 \\ \cdots & \gamma_1 sin\varpi + \gamma_2\gamma_1\beta_2\beta_1 \cos\varpi & \cdots & 0 \\ 0 & 0 & 0 & 1 \end{pmatrix} \tag{4.20}$$

B_c is symmetric whenever the angle ϖ satisfies

$$\gamma_1 \sin\varpi + \gamma_2\gamma_1\beta_2\beta_1 \cos\varpi = -\gamma_2 \sin\varpi$$

(it is not difficult to verify that the non-written components in the former matrix product are also symmetric if ϖ fulfills this equation). Therefore the angle ϖ of Wigner rotation associated to the composition of boosts along mutually perpendicular directions is given by

$$\mathrm{tg}\varpi = -\frac{\gamma_2\gamma_1\beta_2\beta_1}{\gamma_2 + \gamma_1} \tag{4.21}$$

Wigner rotation

(if the velocities are much smaller than c, then ϖ approximates to $-\beta_2\beta_1/2$).

Boost B_c can be computed by replacing the Wigner angle (4.21) in Eq. (4.20). Actually, B_c is completely characterized by its velocity $\boldsymbol{\beta}_c$. The velocity $\boldsymbol{\beta}_c$ is already visible in Eq. (4.20), because the boost velocity appears in the first row of its matrix. In fact, the first row of a boost $B(\boldsymbol{\beta})$ has always the form $(\gamma, -\gamma\beta_x, -\gamma\beta_y, -\gamma\beta_z)$, since in that way Eq. (4.16) will state $ct' = \gamma(ct - \boldsymbol{\beta}.\mathbf{r})$, which generalizes the transformation of the temporal coordinate in Eq. (3.40a) (the transformation of the temporal coordinate must be insensitive to spatial rotations; this requirement is guaranteed by the scalar product of vectors because it is invariant under spatial rotations). Then Eq. (4.20) gives

reading β in the boost matrix

$$\gamma_c = \gamma_2\gamma_1, \qquad \boldsymbol{\beta}_c = \beta_1\,\hat{\mathbf{x}} + \gamma_1^{-1}\beta_2\,\hat{\mathbf{y}} \tag{4.22}$$

Complement 4A: *Thomas precession and the atomic spectra*
In relativistic Quantum Mechanics, the energy of an electron moving in a central electrostatic potential $\phi(r)$—for instance, the one of atomic nucleus—contains, at order u^2/c^2, a *spin-orbit coupling* term:

$$U_{LS} = \frac{e}{2\,m_e^2\,c^2}\,\mathbf{L}\cdot\mathbf{S}\,\frac{1}{r}\,\frac{\mathrm{d}\phi}{\mathrm{d}r} \tag{4A.1}$$

where $\mathbf{L} = m_e\, r\,\hat{\mathbf{r}} \times \mathbf{u}$ is the orbital angular momentum and $\mathbf{S}$ is the spin. The electric field is $\mathbf{E} = -\mathrm{d}\phi/\mathrm{d}r\,\hat{\mathbf{r}}$, so U_{LS} is equal to

$$U_{LS} = -\frac{e}{m_e}\mathbf{S}\cdot\left(\mathbf{E}\times\frac{\mathbf{u}}{2c^2}\right) \tag{4A.2}$$

The spin is a dynamical variable with characteristics of intrinsic angular momentum; this property, together with the electric charge, gives rise to a magnetic momentum, which is equal to $\mathbf{m} = e/m_e\mathbf{S}$ $(e < 0)$ for the electron. Then U_{LS} resembles the energy $U_B = -\mathbf{m}\cdot\mathbf{B}$ of a magnetic momentum in the presence of a magnetic field $\mathbf{B}$. The energy U_B describes the dynamics produced by the torque exerted by the magnetic field $\mathbf{B}$ on the momentum $\mathbf{m}$. If $\mathbf{m}$ is associated with an angular momentum—as it is the case for the electron– the dynamics will be like that of a top turning in a gravitational field: the torque produced by the top weight with respect to the leaning point does not make the top fall but produces a swinging (precession) around the direction of the gravitational field. Analogously, the electron spin precesses in a magnetic field; and it does it with angular velocity $\boldsymbol{\omega}_P = -e/m_e\mathbf{B}$.

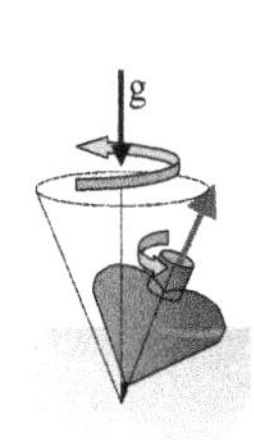

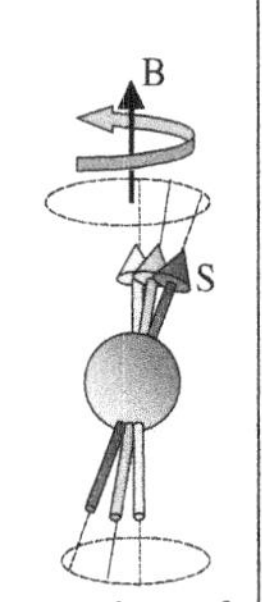

The analogy with U_B shows that energy U_{LS} involves the precession of spin $\mathbf{S}$ around $\mathbf{L}$ direction with a velocity $\boldsymbol{\omega} = -e/m_e\,\mathbf{E}\times\mathbf{u}/(2c^2)$. However, since the electron is just in an electric field, where does the torque come from? It should be remarked that the expression (4A.1) is valid in the laboratory frame S, where the field is purely electrostatic. Instead, in the electron proper frame S_o there is also a magnetic field owing to the fact that the source of the electric field (the atomic nucleus charge) is moving. This magnetic field is $\mathbf{B} \cong -\mathbf{u}\times\,\mathbf{E}/c^2$ (see §5.2) and makes the spin to precess at a rate $\boldsymbol{\omega}_o = -e/m_e\,\mathbf{E}\times\mathbf{u}/c^2$. (In §5.7 we shall show that this torque is regarded in S as the action of $\mathbf{E}$ on the dipolar electric momentum associated with $\mathbf{m}$).

As it can be seen, $\boldsymbol{\omega}_o = 2\boldsymbol{\omega}$. The difference between ω_o and ω reveals that the electron proper frame precesses with respect to the laboratory frame at a rate $\boldsymbol{\omega} - \boldsymbol{\omega}_o = e/m_e\,\mathbf{E}\times\mathbf{u}/(2c^2) = (\mathbf{F}/m_e)\times\mathbf{u}/(2c^2) = \mathbf{a}\times\mathbf{u}/(2c^2)$, which, according to Eqs. (4.23), is Thomas precession.

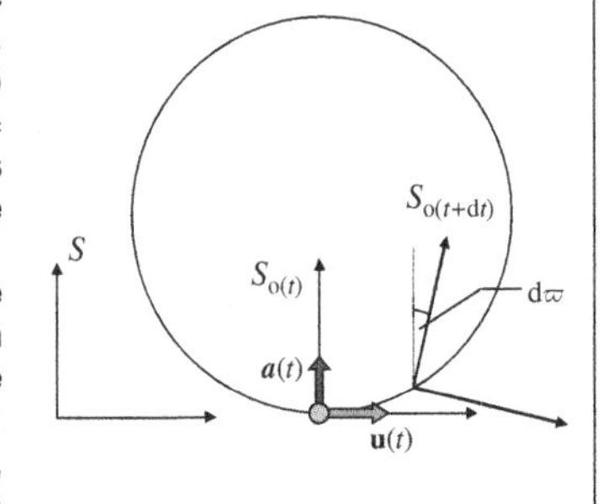

The experimental verification of (4A.1), by means of the analysis of the fine structure of atomic spectra, can be considered as a proof of Thomas precession.

Therefore, velocity $\boldsymbol{\beta}_c$ of boost B_c is the relativistic composition of velocities $\boldsymbol{\beta}_1$ and $\boldsymbol{\beta}_2$.[4] In summary, the composition of two mutually perpendicular Lorentz boosts is equal to a unique boost whose velocity is the relativistic composition (4.22), followed by a spatial rotation in the plane defined by the velocities to be composed, whose rotation angle ϖ is given by Eq. (4.21). [5]

Wigner rotation is a relativistic effect which has not a Galilean analog. One of the consequences of Wigner rotation is the precession of the series of proper frames of an accelerated particle. The inertial frame that coincides with the proper frame of an accelerated particle – the one where the particle is at rest – only keeps that quality for an instant. If $S_{o(t)}$ is the particle proper frame at time t, then at time $t+\mathrm{d}t$ the particle is moving with velocity $\boldsymbol{a}_{o(t)}\mathrm{d}\tau$ relative to $S_{o(t)}$ ($\boldsymbol{a}_{o(t)}$ is the particle proper acceleration at time t, and $\mathrm{d}\tau$ is the proper time along the world line of the particle). The proper frame at time $t+\mathrm{d}t$ is then defined as the one that results from applying a Lorentz boost with velocity $\boldsymbol{a}_{o(t)}\mathrm{d}\tau$ to the coordinates of frame $S_{o(t)}$. The axes of the so-obtained frame $S_{o(t+\mathrm{d}t)}$ will have the orientation of those of $S_{o(t)}$, but they will be rotated with respect to the axes of a "laboratory" frame S chosen to have the same orientation than $S_{o(t)}$. In fact $S_{o(t+\mathrm{d}t)}$ is reached from S through a composition of boosts: the first boost transforms the S coordinates into the $S_{o(t)}$ ones, while the second boost transforms the $S_{o(t)}$ coordinates into the $S_{o(t+\mathrm{d}t)}$ ones. The first boost has the velocity $\mathbf{u}(t)$ of the particle in the laboratory S; the second boost has a velocity $\boldsymbol{a}_{o(t)}\mathrm{d}\tau$. Unless both velocities are parallel (i.e., except for the case where proper acceleration is parallel to the velocity), the transformation directly leading from S coordinates to $S_{o(t+\mathrm{d}t)}$ coordinates involves a Wigner rotation. In this case, the Wigner angle is infinitesimal. Equation (4.21) can be used to compute it when the velocities to be composed are perpendicular, i.e., when $\boldsymbol{a}_{o(t)}$ is perpendicular to $\mathbf{u}(t)$:

$$\mathrm{d}\varpi = -\frac{\gamma(u)\, a_o \mathrm{d}\tau\, u}{c^2\,(1+\gamma(u))}$$

Using Eqs. (3.47) and (4.7) to transform the acceleration and elapsed time to the values measured in the laboratory frame S, it results

$$\mathrm{d}\varpi = -\frac{\gamma(u)^2\, a\, \mathrm{d}t\, u}{c^2\,(1+\gamma(u))} \tag{4.23}$$

---

[4] It is easy to verify that $\gamma_c = \gamma_2\gamma_1$ is really equal to $(1-\beta_c{}^2)^{1/2}$. In fact $1-\gamma_c{}^{-2} = 1-\gamma_2{}^{-2}\gamma_1{}^{-2} = 1-(1-\beta_2{}^2)(1-\beta_1{}^2) = \beta_1{}^2+\gamma_1{}^{-2}\,\beta_2{}^2 = \beta_c{}^2$. The proof that B_c is a boost is completed by direct inspection of the resulting matrix, or checking that B_c does not change the spatial lengths that are perpendicular to β_c.

[5] The composition of two boosts along arbitrary directions can be analyzed by decomposing one boost in two mutually perpendicular directions – inverting the process described in this section – chosen in such a way that one of them is coincident with the direction of the other boost. Thus, the parallel boosts can be composed first and the result is then composed with the remaining perpendicular boost. More details about this derivation of Wigner rotation and the result for the composition of boosts along arbitrary directions can be consulted in *R. F. and M. Thibeault, European Journal of Physics* **20** (1999), 143–151.

If $u << c$, then the precession angular velocity of the accelerated particle proper frame is

$$\frac{\mathrm{d}\varpi}{\mathrm{d}t} = -\frac{1}{2}\frac{a\,u}{c^2}$$

This result can be expressed in a vector way, which will be valid even if $\boldsymbol{a}$ were not perpendicular to $\mathbf{u}$ (but under the approximation $u << c$):

$$\boldsymbol{\omega}_{\mathrm{T}} = \frac{1}{2}\frac{\boldsymbol{a} \times \mathbf{u}}{c^2} \tag{4.24}$$

The precession $\boldsymbol{\omega}_{\mathrm{T}}$ of the proper frame of an accelerated particle that moves with respect to the laboratory with velocity $\mathbf{u}$ and acceleration $\boldsymbol{a}$ is called *Thomas precession* (see also Complement 4A).[6]

[6] The spatial rotation R is not the rotation underwent by a direction fixed to $S_{o(t)}$ when it is regarded from the laboratory frame S, because also boost B_c distorts the spatial directions. In the case examined in Complement 4A, the contribution from B_c to the change of the spin orientation is averaged to zero during each orbit. This issue is tackled in §7.8.

5

Transformation of the Electromagnetic Field

.1. THE ELECTROMAGNETIC PLANE WAVE

\s was remarked in Complement 1B, Maxwell's laws (1B.1–2) entail the wave quations (1B.5) for the electric and magnetic fields in the absence of charges nd currents. The velocity of propagation for waves described by equations 1B.5) is the constant $c = (\mu_0 \varepsilon_0)^{-1/2}$—the speed of light—which has been raised ɔ the rank of invariant by Special Relativity postulates, as a kinematic condition eeded for preserving the form of Maxwell's laws in all inertial reference ystems (Principle of relativity for electromagnetic phenomena). This *covariance* f Maxwell's laws requires not only the substitution of Galileo transformations vith Lorentz transformations; but fields must properly transform in order that the ɔrm of Maxwell's laws be really preserved under reference system changes. We hall begin the analysis of this issue by studying the properties of the simplest olution of a wave equation: *the plane wave*.

Any function having the form $\psi(\mathbf{r}, t) = \psi(\hat{\mathbf{n}} \cdot \mathbf{r} - ct)$—where $\hat{\mathbf{n}}$ is an rbitrary direction—solves a wave equation whose velocity of propagation is constant c. To prove it, we shall choose—without lack of generality—the 'artesian x axis to be coincident with the $\hat{\mathbf{n}}$ direction, so the argument in ψ is educed to $x - ct$. Then we can easily verify that the wave equation is identically ılfilled no matter what the function ψ is:

$$\left(c^2 \frac{\partial^2}{\partial x^2} - \frac{\partial^2}{\partial t^2} \right) \psi(x - ct) \equiv 0 \tag{5.1}$$

Figure 5.1 shows the behavior of a function $\psi(x - ct)$: after time Δt has lapsed, the only change underwent by the perturbation ψ is the displacement $x = c\,\Delta t$. In fact, $\psi(x + \Delta x, t + \Delta t)|_{\Delta x = c\Delta t} = \psi(x - ct)$; so the graphic of function at two different times repeats the same form although displaced in $c\,\Delta t$. 'herefore, the wave propagates along the direction $\hat{\mathbf{n}}$ (the x axis in the present ase) with speed c.

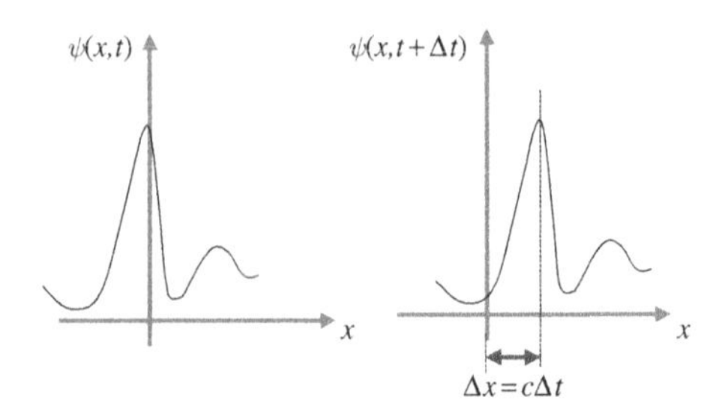

Figure 5.1. Temporal evolution of a function $\psi(x,t)=\psi(x\text{-}ct)$.

A solution of the wave equation having the form $\psi(\hat{\mathbf{n}}\cdot\mathbf{r}-ct)$ is called plane wave because, at each time t, the perturbation ψ has the same value at all the points belonging to a plane perpendicular to $\hat{\mathbf{n}}$ (as projection $\hat{\mathbf{n}}\cdot\mathbf{r}$ is equal for all of them). These planes are the wave fronts.

transversal character of electromagnetic waves

If a magnetic field has the form $\mathbf{B}=\mathbf{B}(x-ct)$, then the zero value of its divergence (see Eq. (1B.1a)) implies that the component B_x can only be a uniform "background": $0=\nabla\cdot\mathbf{B}(x-ct)=\partial B_x/\partial x$. After eliminating this background by means of a boundary condition, it results that the perturbation $\mathbf{B}$ is orthogonal to the propagation direction $\hat{\mathbf{n}}=\hat{\mathbf{x}}$. The same conclusion is valid for an electric field having the form $\mathbf{E}=\mathbf{E}(x-ct)$ in the absence of sources, because it is divergenceless as well (see Eq.(1B.2a)). This means that electromagnetic waves are *transversal*.

On the other hand, the rotor of the examined fields is also perpendicular to the propagation direction of each field. Since both fields mix in the sector of Maxwell's laws involving rotors, then $\mathbf{E}$ and $\mathbf{B}$ are compelled to share the propagation direction. The functional form of fields in a plane wave allows the substitution of $\partial/\partial t$ for $-c\,\partial/\partial x$. In that way equations (1B.1b) and (1B.2b) become

$$\frac{\partial}{\partial x}\left(E_z+cB_y\right)=0 \qquad \frac{\partial}{\partial x}\left(E_y-cB_z\right)=0$$

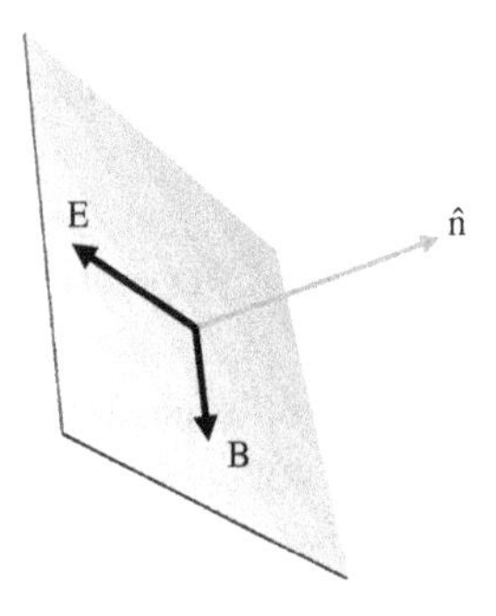

Figure 5.2. Relation among the propagation direction $\hat{\mathbf{n}}$ and fields E , B in an electromagnetic plane wave.

After eliminating uniform fields by means of boundary conditions, it is concluded that perturbations **E** and **B**, besides being each of them perpendicular to the propagation direction $\hat{\mathbf{n}}$, are mutually orthogonal (Figure 5.2) and their modulus are proportional. The properties obtained are summarized in the expression

$$\mathbf{B} = c^{-1}\hat{\mathbf{n}} \times \mathbf{E} \tag{5.2}$$

5.2. TRANSFORMATION OF E AND B

A wave that is plane in a reference system is also plane in any other reference systems: since Lorentz transformations are linear, then the plane wave fronts remain plane under a Lorentz transformation, although with a different propagation direction. An electromagnetic plane wave fulfills Eq. (5.2) as a consequence of Maxwell's laws. According to the principles of Special Relativity, Maxwell's laws have to be valid in any inertial reference system. It is concluded that whenever relation (5.2) is accomplished in some reference system (at an event), then the relation remains valid in any inertial reference system:[1]

$$\mathbf{B}(\mathbf{r}, t) = c^{-1}\hat{\mathbf{n}} \times \mathbf{E}(\mathbf{r}, t) \qquad \Leftrightarrow \qquad \mathbf{B}'(\mathbf{r}', t') = c^{-1}\hat{\mathbf{n}}' \times \mathbf{E}'(\mathbf{r}', t') \tag{5.3}$$

Conclusion (5.3) will guide us through the searching of field transformations; its observance is equivalent to guarantee that Maxwell's laws be satisfied in all inertial reference systems. Since the linearity of relation (5.2) has to be preserved, we shall propose a linear transformation for **E** and **B**; besides, the transformation must be homogeneous in order that the vanishing of both fields in a frame implies the vanishing of fields in any other reference system. The only ingredient that can enter the transformation, besides the fields, is the velocity **V** of frame S' relative to the frame S. The transformation cannot depend on the event $(\mathbf{r}, t)$ where it is evaluated, because space-time is homogeneous.

Therefore, fields $\mathbf{E}'$ and $\mathbf{B}'$ have to be written as combinations of vectors built with **E**, **B**, and **V** that must be linear in the fields. In order to combine such vectors, we have to take into account their behavior under spatial reflections. As is well known, while **E** and **V** are genuine vectors (or *polar* vectors) because they do not change under spatial reflections, **B** is a pseudovector (or *axial* vector) because its orientation depends on the orientation of the triad of Cartesian versors.[2] $\mathbf{E}$, $\mathbf{V} \times \mathbf{B}$, $\mathbf{V}(\mathbf{V} \cdot \mathbf{E})$, $\mathbf{V} \times (\mathbf{V} \times \mathbf{E})$ are examples of polar vectors.

[1] The speed of propagation c enters Eq. (5.2) as a universal constant contained in the universal Maxwell's laws. Gaussian System substitutes **B** for $\mathbf{B}/c$, so relation (5.2) is even simpler: $\mathbf{B} = \hat{\mathbf{n}} \times \mathbf{E}$.

[2] In a spatial reflection ($x \to -x$, $\hat{\mathbf{x}} \to -\hat{\mathbf{x}}$) a right triad changes to a left triad. A polar vector **P** does not change ($\mathbf{P} \to \mathbf{P}$, $P_x \to -P_x$, $P_y \to P_y$, $P_z \to P_z$), but an axial vector **M** inverts its orientation ($\mathbf{M} \to -\mathbf{M}$, $M_x \to M_x$, $M_y \to -M_y$, $M_z \to -M_z$). The vector product of two vectors of equal type results to be an axial vector, while the vector product of two vectors of opposite type results to be polar. The rotor of an axial vector is a polar vector, and vice versa. It is then clear in (1B.1b) that **E** and **B** have opposite behavior under reflections.

$\mathbf{B}$, $\mathbf{V}\times\mathbf{E}$, $\mathbf{V}\,(\mathbf{V}\cdot\mathbf{B})$, $\mathbf{V}\times(\mathbf{V}\times\mathbf{B})$ are examples of axial vectors. These examples of polar and axial vectors linear in the fields will be the building blocks of field transformations. It is not necessary to consider other possibilities (for instance, vector $\mathbf{V}\times(\mathbf{V}\times(\mathbf{V}\times\mathbf{E}))$ is proportional to $\mathbf{V}\times\mathbf{E}$). It will be useful to decompose fields in their projections parallel and perpendicular to $\mathbf{V}$; thus, we can write $\mathbf{V}\times(\mathbf{V}\times\mathbf{E}) = \mathbf{V}\times(\mathbf{V}\times\mathbf{E}_\perp) = -V^2\mathbf{E}_\perp$, and $\mathbf{V}(\mathbf{V}\cdot\mathbf{E}) = V^2\mathbf{E}_{||}$. This shows that each projection of the fields can enter the transformation with different coefficients. In sum, polar vector $\mathbf{E}'$ is equal to some linear combination of $\mathbf{E}_{||}$, $\mathbf{E}_\perp$ and $\mathbf{V}\times\mathbf{B}_\perp$ with coefficients that could depend on V:

$$\mathbf{E}' = e_1(V)\,\mathbf{E}_{||} + e_2(V)\,\mathbf{E}_\perp + e_3(V)\,\mathbf{V}\times\mathbf{B}_\perp \tag{5.4}$$

An analogous combination will be valid for axial vector $\mathbf{B}'$:

$$\mathbf{B}' = b_1(V)\,\mathbf{B}_{||} + b_2(V)\,\mathbf{B}_\perp + b_3(V)\,\mathbf{V}\times\mathbf{E}_\perp \tag{5.5}$$

The last two terms at the right sides of these equations are perpendicular to $\mathbf{V}$. Therefore the parallel projections of Eqs. (5.4-5) say that

$$\mathbf{E}'_{||} = e_1\;\mathbf{E}_{||} \qquad \mathbf{B}'_{||} = b_1\;\mathbf{B}_{||}$$

In order that the inverse transformations have the same form than the direct ones, the coefficients e_1 and b_1 should be equal to 1; this value guarantees the absence of privilege among different inertial frames:

$$\mathbf{E}'_{||} = \mathbf{E}_{||} \qquad \mathbf{B}'_{||} = \mathbf{B}_{||} \tag{5.6}$$

On the other hand, after a suitable redefinition of coefficients e_2 and e_3, the projection of transformation (5.4) on the direction perpendicular to $\mathbf{V}$ becomes

$$\mathbf{E}'_\perp = \Gamma\,(\mathbf{E}_\perp + \kappa\,\mathbf{V}\times\mathbf{B}_\perp) \tag{5.7}$$

where Γ and κ are non-dimensional and depend at least on V/c. The inverse transformation should have the same form, apart from the change of $\mathbf{V}$ for $-\mathbf{V}$:

$$\mathbf{E}_\perp = \Gamma\,(\mathbf{E}'_\perp - \kappa\,\mathbf{V}\times\mathbf{B}'_\perp) \tag{5.8}$$

In order to obtain the coefficients Γ and κ we shall make use of requirement (5.3). First, let us consider a plane electromagnetic wave propagating along the $\mathbf{V}$ direction; in this case it is $\hat{\mathbf{n}} = \mathbf{V}/V = \hat{\mathbf{n}}'$ since Eq. (3.57) states that no aberration exists for this propagation direction. Replacing Eq. (5.3) in (5.7) and (5.8) it results

$$\mathbf{E}'_\perp = \Gamma\left(\mathbf{E}_\perp + \kappa\,\mathbf{V}\times\frac{(\mathbf{V}\times\mathbf{E}_\perp)}{V\,c}\right) = \Gamma\left(1-\kappa\frac{V}{c}\right)\mathbf{E}_\perp$$

$$\mathbf{E}_\perp = \Gamma\left(\mathbf{E}'_\perp - \kappa\,\mathbf{V}\times\frac{(\mathbf{V}\times\mathbf{E}'_\perp)}{V\,c}\right) = \Gamma\left(1+\kappa\frac{V}{c}\right)\mathbf{E}'_\perp$$

Fields can be eliminated in these two equations to get

$$1 = \Gamma^2 \left(1 - \kappa^2 \frac{V^2}{c^2}\right) \tag{5.9}$$

It is easy to establish that κ is equal to 1. If $\mathbf{E}' = 0$ then Lorentz force (1B.4) on a charge at rest in S' will vanish. The vanishing of the force in frame S' implies the vanishing of the force in any inertial frame S, because if the charge is at rest in S' it will move with constant velocity $\mathbf{V}$ with respect to S. In frame S, Lorentz force on the charge is proportional to $\mathbf{E} + \mathbf{V} \times \mathbf{B}$; thus it is concluded that $\mathbf{E} + \mathbf{V} \times \mathbf{B} = 0 \Leftrightarrow \mathbf{E}' = 0$. Therefore, it is $\kappa = 1$ in (5.7), and $\Gamma = \gamma(V)$ in (5.9). In spite of this simple demonstration, it is not our aim to obtain the transformations of the electromagnetic field by resorting to charges or forces. We pretend to use only Maxwell's laws properties in the absence of sources. In harmony with Special Relativity postulates, we want to reach the transformations starting from the properties of light propagation in vacuum. So, in order to obtain a second equation that, together with (5.9), will allow to solve κ and Γ let us also consider a wave propagating along a direction perpendicular to $\mathbf{V}$ in frame S ($\hat{\mathbf{n}} \perp \mathbf{V}$; see Figure 5.3). Replacing $\theta = \pi/2$ in the Eq. (3.57c) for aberration of light, it results

$$\operatorname{tg} \theta' = -\frac{c}{\gamma\, V}$$

The wave chosen has $\mathbf{B}_{||} = 0$, $\mathbf{E}_{\perp} = 0$ ($\mathbf{B}' = \Gamma\, \mathbf{B}$). As Figure 5.3 shows, tg θ' can be obtained from the components of the electric field in frame S':

$$\operatorname{tg} \theta' = -\frac{|\mathbf{E}'_{||}|}{|\mathbf{E}'_{\perp}|} = -\frac{|\mathbf{E}_{||}|}{|\Gamma\, \kappa\, \mathbf{V} \times \mathbf{B}_{\perp}|} = -\frac{E}{\Gamma\, \kappa\, V\, B} = -\frac{c}{\Gamma\, \kappa\, V}$$

This result and the former one coincides only if $\kappa = 1$ and, then, $\Gamma = \gamma(V)$.

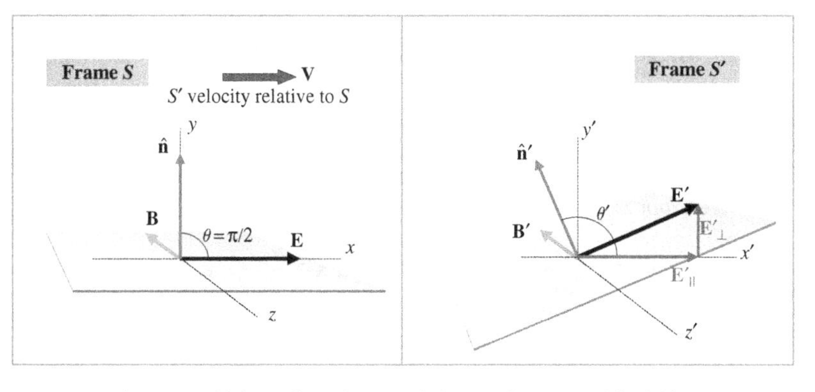

Figure 5.3. Aberration of light, and its relation with the transformation of the field components.

To complete the field transformations, we shall now consider the transformation of $\mathbf{B}_\perp$. With this aim, we shall vectorially multiply Eqs. (5.7) and (5.8) by $\mathbf{V}$. Observing that $\mathbf{V} \times (\mathbf{V} \times \mathbf{B}_\perp) = -V^2 \mathbf{B}_\perp$ we obtain the equations

$$\mathbf{V} \times \mathbf{E}'_\perp = \gamma(V)(\mathbf{V} \times \mathbf{E}_\perp - V^2 \mathbf{B}_\perp) \qquad \mathbf{V} \times \mathbf{E}_\perp = \gamma(V)(\mathbf{V} \times \mathbf{E}'_\perp + V^2 \mathbf{B}'_\perp)$$

where $\mathbf{B}'_\perp$ can be solved as a function of $\mathbf{E}_\perp$ and $\mathbf{B}_\perp$ by replacing the first one in the second one.

In sum, the field projections that are perpendicular to the relative velocity $\mathbf{V}$ between frames S and S' transform in the following way:

$$\mathbf{E}'_\perp = \frac{\mathbf{E}_\perp + \mathbf{V} \times \mathbf{B}_\perp}{\sqrt{1 - \frac{V^2}{c^2}}} \qquad \mathbf{E}_\perp = \frac{\mathbf{E}'_\perp - \mathbf{V} \times \mathbf{B}'_\perp}{\sqrt{1 - \frac{V^2}{c^2}}}$$

$$\mathbf{B}'_\perp = \frac{\mathbf{B}_\perp - c^{-2} \mathbf{V} \times \mathbf{E}_\perp}{\sqrt{1 - \frac{V^2}{c^2}}} \qquad \mathbf{B}_\perp = \frac{\mathbf{B}'_\perp + c^{-2} \mathbf{V} \times \mathbf{E}'_\perp}{\sqrt{1 - \frac{V^2}{c^2}}} \qquad (5.10\text{a–b})$$

It is not difficult to verify that field transformations (5.6) and (5.10), together with the coordinate transformations (3.39–3.40), do not modify the form of Maxwell's laws in the absence of sources.

5.3. CHARGE AND CURRENT TRANSFORMATIONS

Maxwell's laws (1B. 2a–b) tell us that charge density $\rho(\mathbf{r}, t)$ and current density $\mathbf{j}(\mathbf{r}, t)$ must transform as $\nabla \cdot \mathbf{E}$ and $\nabla \times \mathbf{B}$ respectively.[3] However, it is more practical to find out the transformations of charge and current densities by examining some simple configurations of sources and fields. With this aim, let us consider the stationary configuration proposed in Figure 5.4: in a frame S, an infinite plane slab of thickness d is uniformly charged with density ρ, and carries a uniform current density $\mathbf{j}$. The fields due to this source distribution at places outside the slab result to be uniform. According to Maxwell's laws, their values in the upper semi-space are

$$\mathbf{E} = \frac{\rho\, d}{2\,\varepsilon_0} \hat{\mathbf{y}} \qquad \mathbf{B} = -\frac{\mu_0}{2} j_z\, d\, \hat{\mathbf{x}} + \frac{\mu_0}{2} j_x\, d\, \hat{\mathbf{z}},$$

[3] $\rho(\mathbf{r}, t)$ is the electric charge per unit of volume. $\mathbf{j}(\mathbf{r}, t) \cdot \delta\mathbf{S}$ is the current, i.e., the electric charge going through the surface $\delta\mathbf{S}$ per unit of time.

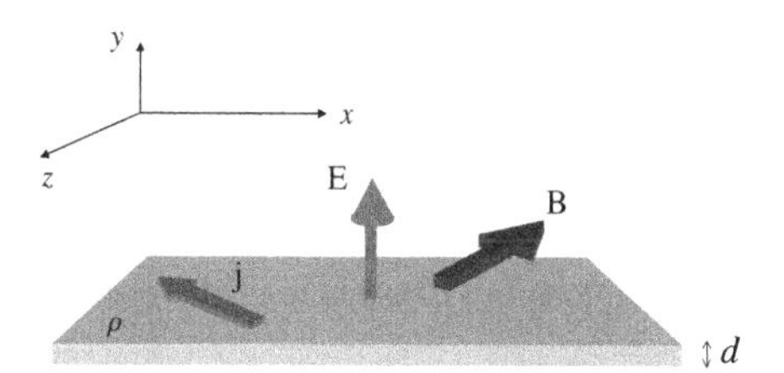

Figure 5.4. A plane slab with charge and current uniformly distributed creates a field **E** orthogonal to the slab, and a field **B** parallel to the slab and perpendicular to the current.

while in the lower semi-space the **E** and **B** components have opposite sign. In another frame S' moving with respect to frame S along a direction parallel to the slab, the charge and current configuration will be qualitatively similar, but with different values ρ' and $\mathbf{j}'$. According to the postulates of Special Relativity, Maxwell's laws have the same form in all frames. Therefore, in frame S' the fields keep the same relation with their sources:

$$\mathbf{E}' = -\frac{\rho' d'}{2\,\varepsilon_o}\hat{\mathbf{y}} \qquad \mathbf{B}' = -\frac{\mu_o}{2} j'_z\, d'\, \hat{\mathbf{x}} + \frac{\mu_o}{2} j'_x\, d'\, \hat{\mathbf{z}}$$

The charge transformation will result from replacing these fields values in transformations (5.6) and (5.10) (choosing as always $\mathbf{V} = V\hat{\mathbf{x}}$, the symbol $_{||}$ will allude to component x). Taking into account that $d = d'$, because it is a length perpendicular to the direction of the relative movement, and being $\mu_o \varepsilon_o = c^{-2}$ (see Eq. (1B.5)), it is obtained

$$\rho' = \frac{\rho - \frac{V}{c^2} j_x}{\sqrt{1 - \frac{V^2}{c^2}}} \qquad j'_x = \frac{j_x - V\rho}{\sqrt{1 - \frac{V^2}{c^2}}} \qquad \mathbf{j}'_\perp = \mathbf{j}_\perp \tag{5.11}$$

It is easy to verify that these charge and current transformations, together with Lorentz transformations for coordinates, do not change the form of the continuity equation (1B.3) which expresses the local charge *conservation*.

Transformations (5.11) can be used to prove the *invariance of charge*. Let us consider a charge and current distribution allowing a frame S_o where the current density vanishes. In such a case, the charge and current densities in another frame S, moving with velocity $\mathbf{V} = \mathbf{u}$ relative to S_o, become (using the inverse transformations of Eq. (5.11))

charge invariance

$$\rho = \frac{\rho_o}{\sqrt{1 - \frac{u^2}{c^2}}} = \gamma(u)\,\rho_o \qquad \mathbf{j} = \frac{\rho_o\,\mathbf{u}}{\sqrt{1 - \frac{u^2}{c^2}}} = \rho\,\mathbf{u} \tag{5.12}$$

which is the form of a purely convective current. The charge contained in a given volume is equal to the charge density multiplied by the volume (assuming, for simplicity, a uniform density). If v_o is a volume at rest in frame S_o, then in the frame S it is $v = \gamma(u)^{-1} v_o$ as a consequence of length contraction along the direction of the relative movement.[4] In the arbitrary frame S the charge contained in the volume v is $Q = \rho v = \gamma(u) \rho_o \gamma(u)^{-1} v_o = \rho_o v_o = Q_o$. This means that the charge contained in the volume has the same value in any reference system. In particular, this result guarantees the invariance of the charge of a particle (S_o is the particle proper frame).

The results in this section provide the relativistic explanation for the interaction between the line charge distributions in Figure 1.6, when observed from the proper frame of positive charges. Each individual electric charge keeps its value under reference system changes; but the distances among them are not invariant under Lorentz transformations. The modification of the distances among charges causes an alteration of densities when the frame is changed, in such a way that the upper line charge is not neutral in the frame going with the positive charges. In this frame the positive charges are at rest, and distances among them are proper lengths; i.e., in this second frame the distance between positive charges is maximum. On the contrary, the distance between negative charges, which are at rest in the first frame, are contracted in the second frame. The result is that the upper line charge is negatively charged in the second frame (see Figure 5.5), so it undergoes an electrostatic attraction with the lower line charge. Thus, a magnetostatic interaction in the first frame is regarded as an electrostatic interaction in the second frame. As already seen, fields **E** and **B** intermingle under reference system changes; so the electric or magnetic character of an interaction is a relative concept. The sole absolute thing is the existence of the interaction.

In the former argument we have used the invariance of charge to state that each individual charge has the same value in any frame. However, Figure 5.5

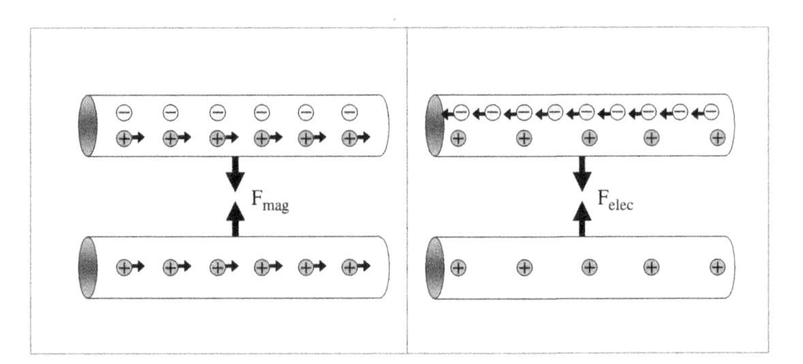

Figure 5.5. A same electromagnetic interaction between parallel line charges, as it is regarded from the negative charges proper frame (left) and the positive charges proper frame (right). The interaction is magnetostatic in the first frame and electrostatic in the second one.

[4] Thus the volume associated with particles moving with velocity u fulfills $\gamma(u')\, v' = \gamma(u)\, v$.

suggests that the total positive charge is lower in the second frame, and the total negative charge seems to be higher in this frame; this situation would contradict the charge invariance. It is clear that, due to the ideal nature of the configuration in Figure 5.5—the line charges are infinite—the total positive and negative charges are *equally* infinite in both frames. For this reason, the configuration of infinite parallel line charges is not suitable for analyzing this situation; we should rather study a finite distribution. Figure 5.6 shows a current and charge configuration distributed in a finite volume with a circuit shape. In frame S_o where the circuit is fixed (left) the charge density is null everywhere, and current is stationary (i.e. current i_o is the same along the circuit: $\nabla_o \cdot \mathbf{j}_o = 0$). Current i_o could come from a movement of positive charges along the $\mathbf{j}_o$ direction or a movement of negative charges along the opposite direction, or a combination of both. In a frame S where the circuit moves with velocity $\mathbf{u} = u\,\hat{\mathbf{x}}$, the values of charge and current densities will be (using the inverse transformations of (5.11))

$$\rho = \gamma(u)\frac{u}{c^2}\, j_{ox} \qquad j_x = \gamma(u)\, j_{ox} \qquad j_y = j_{oy}$$

Thus, in frame S there will be a positive charge density in the circuit upper branch (where j_{ox} is positive), and a negative charge density in the lower branch. In spite of this, the total charge remains null.

Current i in frame S is the flux of density $\mathbf{j}$. Care must be taken in calculating this flux, because the circuit section is contracted for a factor $\gamma(u)^{-1}$ in those branches that are transversal to the relative motion. Figure 5.6 displays the current in each branch: it is evident that the current is not stationary. This behavior is explained with the help of Figure 5.7, where a volume surrounding one of the corners of the circuit has been drawn. The difference between incoming and outgoing currents is $\gamma\, i_o - \gamma^{-1}\, i_o = \gamma(1-\gamma^{-2})\, i_o = \gamma(u)\, i_o\, u^2/c^2$. This means that a net charge $\Delta Q = \gamma(u)\, i_o\, u^2\, \Delta t/c^2$ enters the volume in time Δt. The result, obtained from a balance of incoming and outgoing charge flux,

meaning of the continuity equation

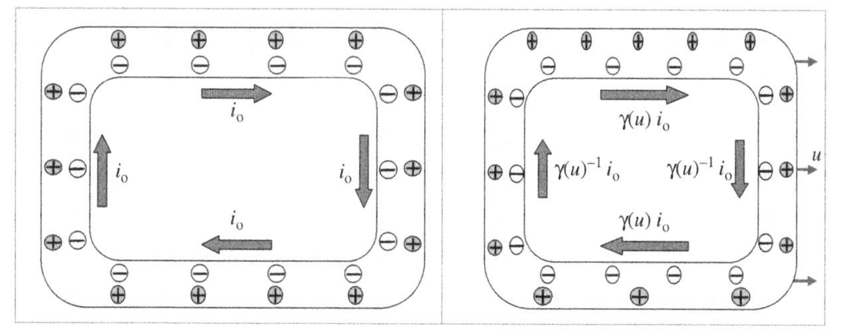

Figure 5.6. At left, a circuit at rest having null charge density and a stationary current i_o. At right, the same charge and current configuration as regarded in a frame where the circuit moves with velocity $\mathbf{u}$. (It has been assumed, but it is not relevant for density transformations, that current i_o is produced by the motion of positive charges).

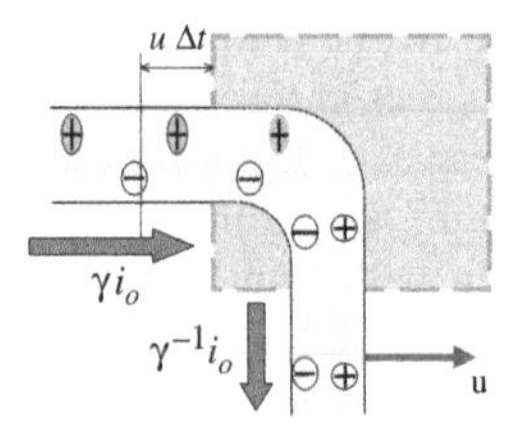

Figure 5.7. The incoming current is higher than the outgoing current. In this way the net charge entering the volume as a consequence of the circuit movement is taken into account.

coincides—as demanded by the continuity equation—with the charge carried by the circuit segment of length $u\ \Delta t$ that enters the volume in time Δt. In fact, in frame S the circuit upper branch has a charge density equal to $\rho = \gamma(u)\, u\, j_{ox}/c^2$; then, the charge carried by the incoming segment is $\Delta Q = \rho\, A\, u\, \Delta t = \gamma(u)\, j_{ox}\, A\, u^2\, \Delta t/c^2$, A being the circuit section. The two so-computed charges are equal, because $i_o = j_{ox}\, A$. The different values of the current in each branch are then necessary to take into account the charge transport due to the circuit movement.[5]

5.4. FIELD OF A UNIFORMLY MOVING CHARGE

The electric and magnetic fields due to a charge q that moves with a constant velocity $\mathbf{u}$ are easily obtained by transforming the known field values of the charge proper frame S_o. Since the magnetic field is null in the proper frame, because the charge is at rest, Eqs. (5.6) and (5.10b) state that

$$\mathbf{E}_{||} = \mathbf{E}_{o||} \qquad \mathbf{E}_{\perp} = \gamma(u)\, \mathbf{E}_{o\perp} \qquad \mathbf{B} = \gamma(u)\, c^{-2}\, \mathbf{u} \times \mathbf{E}_o$$

where field $\mathbf{E}_o$ is the Coulombian field of a charge at rest at the coordinate origin,

$$\mathbf{E}_o = \frac{q}{4\pi\varepsilon_o}\frac{\mathbf{r}_o}{r_o^{\,3}}$$

and $\mathbf{r}_o$ is the field point in the charge proper frame. $\mathbf{E}_o$ is a radial and isotropic field, inversely proportional to the squared distance between the field point and the charge q. Although it is not completely evident in the former expressions, $\mathbf{E}$ is also radial with respect to the charge position $\mathbf{r}_q = \mathbf{u}\, t$ in frame S. In fact, substituting the proper frame coordinates with those of frame S:

$$\mathbf{r}_{o||} = \gamma(u)(\mathbf{r}_{||} - \mathbf{u}\, t) \qquad \mathbf{r}_{o\perp} = \mathbf{r}_{\perp} \tag{5.13}$$

---

[5] Integrating the continuity equation (1B.3) in a time independent volume, like that of Figure 5.7, $0 = \int (\partial\rho/\partial t + \nabla\cdot\mathbf{j})\, \mathrm{d}^3 x = \mathrm{d}Q/\mathrm{d}t + \oint \mathbf{j}\cdot\mathbf{dS}$ is obtained. Then, the variation of charge Q inside the volume exclusively comes from the flux of $\mathbf{j}$ (charge cannot be created nor destroyed). See §7.6.

which are nothing but Lorentz transformations vectorially written, and calling $\mathbf{R} \equiv \mathbf{r} - \mathbf{u}\,t$ ($\mathbf{R}$ is the vector going from the charge to the field point), it results

$$\mathbf{E}(\mathbf{r}, t) = \frac{q}{4\pi\varepsilon_o} \frac{\gamma(u)\,\mathbf{R}}{\left(\gamma(u)^2\,|\mathbf{R}_{||}|^2 + |\mathbf{R}_{\perp}|^2\right)^{3/2}} \tag{5.14}$$

$\mathbf{E}(\mathbf{r}, t)$ has then the direction going from the source position to the field point $\mathbf{r}$. Although it is radial with respect to the charge, the field $\mathbf{E}(\mathbf{r}, t)$ is not isotropic in frame S. The expression in the denominator of $\mathbf{E}(\mathbf{r}, t)$ is not the distance between the source and the field point in S. So the modulus of $\mathbf{E}$ is not merely a function of the distance but depends on the direction. Let us call θ the angle between $\mathbf{R}$ and $\mathbf{u}$, then $|\mathbf{R}_{||}| = R\cos\theta$, $|\mathbf{R}_{\perp}| = R\sin\theta$; the modulus of $\mathbf{E}$ becomes

$$E(R, \theta) = \frac{q}{4\pi\varepsilon_o R^2} \frac{\gamma(u)^{-2}}{\left(1 - \dfrac{u^2}{c^2}\sin^2\theta\right)^{3/2}} \tag{5.15}$$

For a given distance R, E is minimum on the charge straight trajectory and maximum on the direction perpendicular to velocity $\mathbf{u}$.

The magnetic field in frame S is

$$\mathbf{B}(\mathbf{r}, t) = \gamma(u)\,c^{-2}\,\mathbf{u} \times \mathbf{E}_o = \frac{\mu_o}{4\pi} \frac{\gamma(u)\,q\,\mathbf{u} \times \mathbf{R}}{\left(\gamma(u)^2\,|\mathbf{R}_{||}|^2 + |\mathbf{R}_{\perp}|^2\right)^{3/2}} \tag{5.16}$$

so $\mathbf{B} = c^{-2}\,\mathbf{u} \times \mathbf{E}$. Field lines of $\mathbf{B}$ are circles around the charge straight trajectory. The modulus of $\mathbf{B}$ is

$$B(R, \theta) = \frac{\mu_o}{4\pi R^2} \frac{\gamma(u)^{-2} q\,u\sin\theta}{\left(1 - \dfrac{u^2}{c^2}\sin^2\theta\right)^{3/2}} \tag{5.17}$$

5.5. TRANSFORMATION OF POTENTIALS

Maxwell's laws (1B.1–2) are eight equations (since two of them are scalar and the other two are vector equations) describing six quantities: the components of $\mathbf{E}$ and $\mathbf{B}$. The four equations without sources can be used for introducing potentials ϕ and $\mathbf{A}$. In fact, by defining

$$\mathbf{B} \equiv \nabla \times \mathbf{A} \qquad \mathbf{E} \equiv -\nabla\phi - \frac{\partial \mathbf{A}}{\partial t} \tag{5.18}$$

then Eqs. (1B.1) are identically satisfied (note that $\nabla\cdot\{\nabla\times\ldots\} \equiv 0$, and $\nabla\times\{\nabla\ldots\} \equiv 0$). These four potentials are determined by the remaining four equations (1B.2) that include the sources.[6]

---

[6] Eqs. (5.18) do not completely define the potentials, because fields $\mathbf{E}$ and $\mathbf{B}$ do not change under a *gauge transformation* $\mathbf{A} \to \mathbf{A} - \nabla\xi$, $\phi \to \phi + \partial\xi/\partial t$, where $\xi(\mathbf{r}, t)$ is an arbitrary function. This "gauge freedom" can be used for providing the potentials with some suitable behavior.

In order to know the way the potentials transform, we shall express the invariance of parallel components of the fields (Eqs. (5.6)) in terms of potentials. If the relative velocity between reference systems is $\mathbf{V} = V\,\hat{\mathbf{x}}$, then

$$B_x = \frac{\partial A_z}{\partial y} - \frac{\partial A_y}{\partial z} = \frac{\partial A_z}{\partial y'} - \frac{\partial A_y}{\partial z'} = B'_x$$

$$\begin{aligned}
E_x = -\frac{\partial \phi}{\partial x} - \frac{\partial A_x}{\partial t} &= -\left(\frac{\partial t'}{\partial x}\frac{\partial}{\partial t'} + \frac{\partial x'}{\partial x}\frac{\partial}{\partial x'}\right)\phi - \left(\frac{\partial t'}{\partial t}\frac{\partial}{\partial t'} + \frac{\partial x'}{\partial t}\frac{\partial}{\partial x'}\right)A_x \\
&= -\left(-\gamma\frac{V}{c^2}\frac{\partial}{\partial t'} + \gamma\frac{\partial}{\partial x'}\right)\phi - \left(\gamma\frac{\partial}{\partial t'} - \gamma V\frac{\partial}{\partial x'}\right)A_x \\
&= -\frac{\partial}{\partial x'}\gamma(\phi - VA_x) - \frac{\partial}{\partial t'}\gamma\left(A_x - \frac{V}{c^2}\phi\right) = E'_x
\end{aligned}$$

These results show that the relation between fields and potentials will be the same in any reference system whenever the potentials transform according to

$$\phi' = \frac{\phi - VA_x}{\sqrt{1-\frac{V^2}{c^2}}}, \qquad A'_x = \frac{A_x - \frac{V}{c^2}\phi}{\sqrt{1-\frac{V^2}{c^2}}}, \qquad A'_y = A_y, \quad A'_z = A_z \tag{5.19}$$

It is easy to verify that this transformation also preserves the relation among fields and potentials perpendicular components.

5.6. FIELDS IN MATERIAL MEDIA

In electrostatics and magnetostatics in the presence of continuous media, polarization $\mathbf{P}$ and magnetization $\mathbf{M}$ of matter (electric and magnetic dipolar momentum per unit of volume) contribute to the potentials as a charge density $\rho_{\text{pol}} = -\nabla\cdot\mathbf{P}$, and a current density $\mathbf{j}_{\text{mag}} = \nabla\times\mathbf{M}$. For fields varying in time, conservation of ρ_{pol} requires a polarization current $\mathbf{j}_{\text{pol}} = \partial\mathbf{P}/\partial t$. In this way ρ_{pol} and $\mathbf{j}_{\text{pol}}$ fulfill a continuity equation similar to (1B.3). On the other hand, $\mathbf{j}_{\text{mag}}$ satisfies the continuity equation by itself, because it is identically stationary ($\nabla\cdot\mathbf{j}_{\text{mag}} = \nabla\cdot(\nabla\times\mathbf{M}) \equiv 0$). It is then useful to define the auxiliary fields $\mathbf{D}$ and $\mathbf{H}$ such that

$$\mathbf{D} \equiv \varepsilon_o\,\mathbf{E} + \mathbf{P} \qquad \mathbf{H} \equiv \mu_o^{-1}\,\mathbf{B} - \mathbf{M} \tag{5.20}$$

In fact, replacing these definitions in Maxwell's laws with sources (1B.2) it is obtained

$$\nabla\cdot\mathbf{D} = \varepsilon_o\nabla\cdot\mathbf{E} + \nabla\cdot\mathbf{P} = \rho - \rho_{\text{pol}} \equiv \rho_{\text{free}} \tag{5.21a}$$

$$\nabla \times \mathbf{H} - \frac{\partial \mathbf{D}}{\partial t} = \mu_o^{-1} \nabla \times \mathbf{B} - \nabla \times \mathbf{M} - \varepsilon_o \frac{\partial \mathbf{E}}{\partial t} - \frac{\partial \mathbf{P}}{\partial t}$$
$$= \mathbf{j} - \nabla \times \mathbf{M} - \frac{\partial \mathbf{P}}{\partial t} = \mathbf{j} - \mathbf{j}_{mag} - \mathbf{j}_{pol} \equiv \mathbf{j}_{free} \tag{5.21b}$$

In Eqs. (5.21) ρ_{free} and $\mathbf{j}_{free}$ consist of all distribution of charges and currents non-associated with polarization and magnetization of matter (for instance, charges and currents circulating in conductors). This "free" charge (non-bound to the matter structure) is separately conserved, as it results from Eqs. (5.21).

Relations (5.20–5.21) are valid in any inertial reference system, whenever $(\varepsilon_o^{-1}\,\mathbf{P},\ -\mu_o\,\mathbf{M})$ and $(\varepsilon_o^{-1}\ \mathbf{D},\ \mu_o\,\mathbf{H})$ transform as $(\mathbf{E},\ \mathbf{B})$. In particular:

$$\mathbf{P}'_{||} = \mathbf{P}_{||} \qquad \mathbf{P}'_{\perp} = \gamma(V)(\mathbf{P}_{\perp} - c^{-2}\,\mathbf{V} \times \mathbf{M}_{\perp})$$
$$\mathbf{M}'_{||} = \mathbf{M}_{||} \qquad \mathbf{M}'_{\perp} = \gamma(V)(\mathbf{M}_{\perp} + \mathbf{V} \times \mathbf{P}_{\perp}) \tag{5.22}$$

Analogously, ρ_{free} and $\mathbf{j}_{free}$ must transform like ρ and $\mathbf{j}$ in order that Eqs. (5.21) keep the same form in all inertial reference system. As a consequence, $\rho_{bound} = \rho_{pol}$ and $\mathbf{j}_{bound} = \mathbf{j}_{pol} + \mathbf{j}_{mag}$ also obey the transformation (5.11). As an example, Figure 5.8 shows two possible state of motions of a material slab having a uniform polarization $\mathbf{P}_o$ in its proper frame. The polarization generates a distribution of polarization charges on the slab surface. These polarization charges go with the material producing bound currents. If the slab velocity $\mathbf{u}$ is perpendicular to $\mathbf{P}_o$, the bound current $\rho_{pol}\,\mathbf{u}$ is a stationary current $\mathbf{j}_{mag}$ associated with the magnetization $\mathbf{M} = -\gamma(u)\,\mathbf{u} \times \mathbf{P}_o = -\mathbf{u} \times \mathbf{P}$ resulting from (5.22). Instead, if velocity $\mathbf{u}$ is parallel to $\mathbf{P}_o$, then $\mathbf{M} = 0$ and the charge movement is a polarization current. The behavior of $\mathbf{P}$ in each case is a consequence of length contraction, which affects both the distances between charges and the elementary volume $(\delta v = \gamma(u)^{-1}\,\delta v_o)$.

On the other hand, Figure 5.6 illustrates the existence of an electric dipolar momentum when matter moves with velocity $\mathbf{u}$ perpendicular to a magnetization $\mathbf{M}_o$.

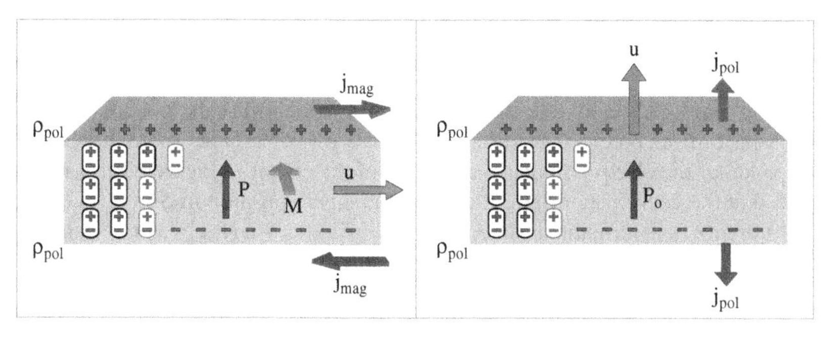

Figure 5.8. Bound currents in a polarized moving material medium.

Eqs. (5.21) joined to Eqs. (1B.1a–b) are used to describe the electromagnetic field in a continuous medium. In order to solve them, we have to know the *constitutive relations*, i.e., relations among **D**, **H**, **E**, and **B** that are dictated by the properties of the medium. Usually, these properties are formulated in the frame S_o where the medium is at rest. The simplest media are those whose polarization $\mathbf{P}_o$ and magnetization $\mathbf{M}_o$ are proportional to fields $\mathbf{E}_o$ and $\mathbf{B}_o$ respectively. The constitutive relations of these *linear media* are

linear media

$$\mathbf{D}_o = \varepsilon\,\mathbf{E}_o \qquad \mathbf{B}_o = \mu\,\mathbf{H}_o \tag{5.23}$$

where ε and μ are properties of the medium. We remark that relations (5.23) do not keep their form under reference system changes; they are only valid in the medium proper frame. Replacing the relations (5.23) in Eq. (5.21b), and considering a homogeneous medium (ε and μ do not depend on **r**), it results

$$\nabla \times \mathbf{B}_o - \mu\,\varepsilon \frac{\partial \mathbf{E}_o}{\partial t} = \mu\,\mathbf{j}_{\text{free}}$$

If we compare this equation with (1B.2b), and follow the procedure leading to Eq. (1B.5), we shall write a wave equation for the electromagnetic field in the medium proper frame, where the (non invariant) velocity of propagation will turn to be $(\mu\,\varepsilon)^{-1/2} < c$. This is the velocity that should be used in Huygens' construction of Figure 2.3 to obtain Snell and reflection laws. These laws are then valid only in the medium proper frame.

5.7. MOVING DIPOLES FIELDS

The potentials generated by a magnetic dipole $\mathbf{m}_o$ have their simplest form in the frame where the dipole is at rest (proper frame S_o):

$$\phi_o = 0 \qquad \mathbf{A}_o(\mathbf{r}_o) = \frac{\mu_o}{4\pi}\frac{\mathbf{m}_o \times \mathbf{r}_o}{r_o^{\,3}} \tag{5.24}$$

where $\mathbf{r}_o$ is the vector joining the dipole position with the field point. In a frame S where the dipole moves with velocity **u**, the potentials are (using the inverse transformation of (5.19)):

$$\phi = \gamma(u)\,\mathbf{u}\cdot\mathbf{A}_o \qquad \mathbf{A}_{||} = \gamma(u)\,\mathbf{A}_{o||} \qquad \mathbf{A}_\perp = \mathbf{A}_{o\perp}$$

Since $\mathbf{u}\cdot(\mathbf{m}_o \times \mathbf{r}_o) = \mathbf{r}_o\cdot(\mathbf{u}\times\mathbf{m}_o)$, it becomes evident that the potential ϕ has the aspect of the potential of an electric dipole $\mathbf{p} = c^{-2}\,\mathbf{u}\times\mathbf{m}_o$ (note that this is the value of the electric dipolar momentum of the circuit in Figure 5.6). If $\mathbf{m}_o$ is placed at the coordinate origin of S_o then the dipole position in S is $\mathbf{r}_p(t) = \mathbf{u}\,t$, and Lorentz transformations state that $\mathbf{r}_{o||} = \gamma(u)\,(\mathbf{r}_{||} - \mathbf{u}\,t)$, $\mathbf{r}_{o\perp} = \mathbf{r}_\perp$. Since **p** is transversal, the potential ϕ in frame S results

$$\phi(\mathbf{r},\,t) = \frac{1}{4\pi\varepsilon_o}\frac{\gamma(u)\,\mathbf{p}\cdot(\mathbf{r}-\mathbf{u}\,t)}{\left(\gamma(u)^2|\mathbf{r}_{||}-\mathbf{u}\,t|^2 + |\mathbf{r}_\perp|^2\right)^{3/2}} \tag{5.25}$$

The similarity between ϕ and the electrostatic potential of an electric dipole is complete when $u << c$ and factors $\gamma(u)$ approaches unity (in such a case, also $\mathbf{A}$ keeps the form of the magnetic potential belonging to a dipole $\mathbf{m}_o$).[7] Nevertheless, the electric field at each time is not equal to the electrostatic field of a dipole $\mathbf{p}$, because the temporal derivative of magnetic potential $\mathbf{A}$ is also involved in the computation of $\mathbf{E}$ (see Eq. (5.18)). In particular it is $\mathbf{E}_{||} = 0$, as it results from Eq. (5.6).

Let us now consider the transformation of the potential of an electric dipole $\mathbf{p}_o$ fixed to the coordinate origin in the proper frame S_o. The dipole potential is

$$\phi_o(\mathbf{r}_o) = \frac{1}{4\pi\varepsilon_o}\frac{\mathbf{p}_o \cdot \mathbf{r}_o}{r_o^{\,3}} \qquad \mathbf{A}_o = 0 \tag{5.26}$$

where $\mathbf{r}_o$ is again the vector joining the dipole position with the field point. Under a change to a frame S where the dipole moves with velocity $\mathbf{u}$, the potentials transform into

$$\phi = \gamma(u)\,\phi_o \quad \mathbf{A} = \gamma(u)\frac{\mathbf{u}}{c^2}\phi_o$$

Differing from the former problem, in this case the potential $\mathbf{A}$ has not the aspect of the magnetostatic potential of a magnetic dipole $\mathbf{m}$ placed at $\mathbf{r}_m(t) = \mathbf{u}\,t$. This is a consequence of the fact that the motion of a unique electric dipole does not generate a stationary current, as required in magnetostatics (instead, this requirement is effectively accomplished by the continuous distribution of electric dipoles in Figure 5.8 if it is $\mathbf{u} \perp \mathbf{P}$). To put it in another way, it is clear that field $\mathbf{B}$ differs from a magnetic dipole field because $\mathbf{B}_{||} = 0$ in transformations (5.6). Therefore $\mathbf{A}$ must be different from a magnetic dipole potential, since $\mathbf{B}$ exclusively derives from the behavior of $\mathbf{A}$. Nevertheless, the current due to the movement of dipole $\mathbf{p}_o$ could be described as the superposition of a stationary current and a non-stationary current, just with the aim of forcing the appearance of a magnetic dipole $\mathbf{m}$ in the expression of potential $\mathbf{A}$. For this, we notice that $\mathbf{u}\,(\mathbf{p}_o \cdot \mathbf{r}_o) = -(\mathbf{u} \times \mathbf{p}_o) \times \mathbf{r}_o + \mathbf{p}_o(\mathbf{r}_o \cdot \mathbf{u})$; thus we call $\mathbf{m} = -\mathbf{u} \times \mathbf{p}_o$ to obtain

$$\mathbf{A} = \frac{\mu_o}{4\pi}\gamma(u)\frac{\mathbf{m} \times [\gamma(u)\,(\mathbf{r}_{||} - \mathbf{u}\,t) + \mathbf{r}_\perp] + \mathbf{p}_o\,[\gamma(u)(\mathbf{r} - \mathbf{u}\,t) \cdot \mathbf{u}]}{\left(\gamma(u)^2|\mathbf{r}_{||} - \mathbf{u}\,t|^2 + |\mathbf{r}_\perp|^2\right)^{3/2}}$$

where the first term exhibits the characteristic form of a magnetic dipole potential.

On the other hand, the electric field $\mathbf{E}$ keeps the aspect of an electric dipole field (see transformations (5.6) and (5.10)) at the first order in u/c, in agreement with the fact that the contribution of $\partial\mathbf{A}/\partial t$ to the electric field is, in this case, of the order u^2/c^2.

[7] Energy U_{LS} in Eq. (4A.2) can be regarded as $U_{LS} = -\mathbf{m}_o \cdot (\mathbf{E} \times \mathbf{u}/c^2)/2 = -\mathbf{E} \cdot (\mathbf{u} \times \mathbf{m}_o)/2c^2 = -\mathbf{E} \cdot \mathbf{p}/2$. Except for the factor 2, which is analyzed in Complement 4A, it is the energy of orientation of an electric dipole $\mathbf{p}$ in an electric field $\mathbf{E}$.

5.8. LORENTZ FORCE TRANSFORMATION

Lorentz force $\mathbf{F} = q\,(\mathbf{E} + \mathbf{u} \times \mathbf{B})$—the force exerted by the electromagnetic field on a charge q moving with velocity $\mathbf{u}$—is not invariant or transforms like acceleration. The Lorentz force transformation is obtained from the transformations of fields (5.6) and (5.10), and velocities (3.42–3.43), together with the charge invariance. Let us begin with the x component, which corresponds to the direction of the relative velocity between frames:

$$\frac{F'_x}{q} = E'_x + u'_y B'_z - u'_z B'_y$$

$$= E_x + \frac{\gamma^{-1} u_y}{1 - \frac{u_x V}{c^2}} \gamma \left(B_z - \frac{V}{c^2} E_y \right) - \frac{\gamma^{-1} u_z}{1 - \frac{u_x V}{c^2}} \gamma \left(B_y + \frac{V}{c^2} E_z \right)$$

$$= \frac{E_x + u_y B_z - u_z B_y - \frac{V u_x}{c^2} E_x - \frac{V u_y}{c^2} E_y - \frac{V u_z}{c^2} E_z}{1 - \frac{u_x V}{c^2}}$$

which is summarized in the expression

$$\mathbf{F}'_{\|} = \frac{\mathbf{F}_{\|} - \frac{V}{c^2} \mathbf{F} \cdot \mathbf{u}}{1 - \frac{\mathbf{u} \cdot \mathbf{V}}{c^2}} \tag{5.27}$$

For components that are perpendicular to the relative movement, like y component, it is obtained

$$\frac{F'_y}{q} = E'_y + u'_z B'_x - u'_x B'_z$$

$$= \gamma\,(E_y - V B_z) + \frac{\gamma^{-1} u_z}{1 - \frac{u_x V}{c^2}} B_x - \frac{u_x - V}{1 - \frac{u_x V}{c^2}} \gamma \left(B_z - \frac{V}{c^2} E_y \right)$$

$$= \frac{\gamma^{-1}}{1 - \frac{u_x V}{c^2}} \left(\gamma^2 (1 - \frac{V^2}{c^2}) E_y + u_z B_x - \gamma^2 (1 - \frac{V^2}{c^2}) u_x B_z \right)$$

$$= \frac{\gamma^{-1}\,(E_y + u_z B_x - u_x B_z)}{1 - \frac{u_x V}{c^2}}$$

i.e.,

$$\mathbf{F}'_{\perp} = \frac{\gamma(V)^{-1} \mathbf{F}_{\perp}}{1 - \frac{\mathbf{u} \cdot \mathbf{V}}{c^2}} \tag{5.28}$$

5.9. ELECTROMAGNETIC FIELD INVARIANTS

The scalar product $\mathbf{E}\cdot\mathbf{B}$ and the difference of squared modules $\mathrm{E}^2 - c^2\mathrm{B}^2$ are invariant not only under spatial rotations, but under Lorentz transformations (5.6) and (5.10) as well. Let us consider the example of the electromagnetic plane wave described in §5.1; in this case, the condition (5.3) implies that both $\mathbf{E}\cdot\mathbf{B}$ and $\mathrm{E}^2 - c^2\,\mathrm{B}^2$ are null in *any* reference system. In other field configurations the invariants can have different values at different events; but their values at each event do not depend on the reference system.

Transformation (5.6) indicates that the proof of the invariance of $\mathbf{E}\cdot\mathbf{B} = \mathbf{E}_{||}\cdot\mathbf{B}_{||} + \mathbf{E}_{\perp}\cdot\mathbf{B}_{\perp}$ is just the proof of the invariance of $\mathbf{E}_{\perp}\cdot\mathbf{B}_{\perp}$. According to transformation (5.10a) it is

$$\begin{aligned}\mathbf{E}'_{\perp}\cdot\mathbf{B}'_{\perp} &= \gamma(V)^2\,(\mathbf{E}_{\perp} + \mathbf{V}\times\mathbf{B}_{\perp})\cdot(\mathbf{B}_{\perp} - c^{-2}\,\mathbf{V}\times\mathbf{E}_{\perp})\\ &= \gamma(V)^2\,[\mathbf{E}_{\perp}\cdot\mathbf{B}_{\perp} - c^{-2}\,(\mathbf{V}\times\mathbf{B}_{\perp})\cdot(\mathbf{V}\times\mathbf{E}_{\perp})]\end{aligned}$$

where we have used that $(\mathbf{V}\times\mathbf{E}_{\perp})\cdot\mathbf{E}_{\perp} = 0 = (\mathbf{V}\times\mathbf{B}_{\perp})\cdot\mathbf{B}_{\perp}$. We remark that vectors $\mathbf{V}\times\mathbf{E}_{\perp}$ and $\mathbf{V}\times\mathbf{B}_{\perp}$ belong to the plane orthogonal to $\mathbf{V}$, like $\mathbf{E}_{\perp}$ and $\mathbf{B}_{\perp}$ do. Moreover the angle between $\mathbf{V}\times\mathbf{E}_{\perp}$ and $\mathbf{V}\times\mathbf{B}_{\perp}$ is equal to that between $\mathbf{E}_{\perp}$ and $\mathbf{B}_{\perp}$. Therefore, it is $(\mathbf{V}\times\mathbf{E}_{\perp})\cdot(\mathbf{V}\times\mathbf{B}_{\perp}) = V^2\,\mathbf{E}_{\perp}\cdot\mathbf{B}_{\perp}$. By replacing this result in the former equation it is verified the invariance of $\mathbf{E}_{\perp}\cdot\mathbf{B}_{\perp}$. Then

$$\mathbf{E}'\cdot\mathbf{B}' = \mathbf{E}\cdot\mathbf{B} \tag{5.29}$$

Going to the second invariant, let us see how $\mathrm{E}^2 = \mathrm{E}_{||}{}^2 + \mathrm{E}_{\perp}{}^2$ transforms. Again, the invariance of $\mathbf{E}_{||}$ (transformation (5.6)) focuses our attention on the transformation of transversal component (5.10a):

$$\begin{aligned}\mathrm{E}'_{\perp}{}^2 = \mathbf{E}'_{\perp}\cdot\mathbf{E}'_{\perp} &= \gamma(V)^2\,(\mathbf{E}_{\perp} + \mathbf{V}\times\mathbf{B}_{\perp})\cdot(\mathbf{E}_{\perp} + \mathbf{V}\times\mathbf{B}_{\perp})\\ &= \gamma(V)^2\,[\mathbf{E}_{\perp}\cdot\mathbf{E}_{\perp} + 2\,(\mathbf{V}\times\mathbf{B}_{\perp})\cdot\mathbf{E}_{\perp} + (\mathbf{V}\times\mathbf{B}_{\perp})\cdot(\mathbf{V}\times\mathbf{B}_{\perp})]\\ &= \gamma(V)^2\,[\mathrm{E}_{\perp}{}^2 + 2\,(\mathbf{V}\times\mathbf{B}_{\perp})\cdot\mathbf{E}_{\perp} + V^2\,\mathrm{B}_{\perp}{}^2]\end{aligned}$$

Analogously, it is

$$c^2\mathrm{B}'_{\perp}{}^2 = \gamma(V)^2\,[c^2\mathrm{B}_{\perp}{}^2 - 2\,(\mathbf{V}\times\mathbf{E}_{\perp})\cdot\mathbf{B}_{\perp} + c^{-2}\,V^2\,\mathrm{E}_{\perp}{}^2]$$

Subtracting both results, and using the cyclic property of mixed product, $(\mathbf{V}\times\mathbf{E}_{\perp})\cdot\mathbf{B}_{\perp} = -(\mathbf{V}\times\mathbf{B}_{\perp})\cdot\mathbf{E}_{\perp}$, it is obtained

$$\mathrm{E}'_{\perp}{}^2 - c^2\,\mathrm{B}'_{\perp}{}^2 = \gamma(V)^2\,[(\mathrm{E}_{\perp}{}^2 - c^2\,\mathrm{B}_{\perp}{}^2) - c^{-2}\,V^2\,(\mathrm{E}_{\perp}{}^2 - c^2\,\mathrm{B}_{\perp}{}^2)] = \mathrm{E}_{\perp}{}^2 - c^2\mathrm{B}_{\perp}{}^2$$

Therefore

$$\mathrm{E}'^2 - c^2\,\mathrm{B}'^2 = \mathrm{E}^2 - c^2\,\mathrm{B}^2 \tag{5.30}$$

If $\mathbf{E} \cdot \mathbf{B} = 0$ at an event but $\mathrm{E}^2 - c^2\,\mathrm{B}^2 \neq 0$, then there are reference systems where one of the fields vanishes at the event. The sign of the invariant $\mathrm{E}^2 - c^2\,\mathrm{B}^2$ determines which is the field that can be cancelled. If $\mathrm{E}^2 - c^2\,\mathrm{B}^2 > 0$ then the magnetic field is null in the reference system moving with velocity $\mathbf{V} = c^2\,\mathbf{E} \times \mathbf{B}/\mathrm{E}^2$ (the velocity is perpendicular to the plane defined by $\mathbf{E}$ and $\mathbf{B}$; according to transformations (5.6) and (5.10a) it results $\mathbf{B}' = 0$). Analogously, if $\mathrm{E}^2 - c^2\,\mathrm{B}^2 < 0$ then the electric field is cancelled by means of a Lorentz transformation with velocity $\mathbf{V} = \mathbf{E} \times \mathbf{B}/\mathrm{B}^2$. In both cases it is guaranteed that $V < c$, because it is $V = c^2\,\mathrm{B}/\mathrm{E}$ in the first case and $V = \mathrm{E}/\mathrm{B}$ in the second. Inversely, whenever a frame exists where the field at an event is purely electric ($\mathbf{B} = 0$) or purely magnetic ($\mathbf{E} = 0$), then fields $\mathbf{E}$ and $\mathbf{B}$ will be orthogonal at the event in other reference systems (see, for instance, the case studied in §5.4).

If both invariants are non-null then there are frames where $\mathbf{E}$ and $\mathbf{B}$ are colinear. This condition can be reached by means of a Lorentz transformation with a velocity $\mathbf{V}$ such that

$$\frac{\mathbf{V}}{1 + V^2\,c^{-2}} = \frac{c^2\,\mathbf{E} \times \mathbf{B}}{E^2 + c^2\,B^2} \tag{5.31}$$

Whenever $\mathbf{E}$ and $\mathbf{B}$ share direction at an event (or one of them is null), then this feature is preserved by Lorentz transformations along the direction of fields.

6

Energy and Momentum

6.1. CONSERVATION LAWS

The laws of Newtonian dynamics are consistent with the notions of absolute space and time contained in Galileo transformations, since they keep their form under such transformations (Galilean Principle of relativity). In particular, the Second Law of Dynamics, $\mathbf{F} = m\,\mathbf{a}$, keeps its form in any inertial reference system because (i) the acceleration is a Galilean invariant, (ii) fundamental forces depend only on distances, so they are also invariant, and (iii) masses are assumed as properties of the particles that are independent of the reference system (see §1.6). The substitution of classical notions of space and time for the relativistic notion of space-time consequently obligates to modify the laws of Dynamics in order that they shall obey the Principle of relativity under Lorentz transformations. The concepts of acceleration and force do not seem to be a good starting point for this aim since, as shown in §3.13 and §5.8, these magnitudes transform in a complicated manner under Lorentz transformations.

A feature of Newtonian mechanics that was not considered in §1.6 is the Principle of action and reaction (Third Law of Dynamics) and its direct consequence: the conservation of the momentum of an isolated system of interacting particles. The Principle of action and reaction states that any interaction between two particles (including an interaction "at a distance") is such that the forces the particles mutually exert are, at each time, equal and opposite:

$$\mathbf{F}_{12} = -\mathbf{F}_{21} \tag{6.1}$$

In Eq. (6.1) $\mathbf{F}_{ij}$ is the force on the particle i exerted by particle j. Forces $\mathbf{F}_{12}$ and $\mathbf{F}_{21}$ constitute a *pair of action and reaction.*

By combining the Second and Third laws, a theorem of conservation is obtained. Let us consider two interacting particles m_1 and m_2. If no other forces are acting on the particles but the pair $\mathbf{F}_{12}$ and $\mathbf{F}_{21}$ they mutually exert (i.e., the particles system is isolated), then it is obtained that

$$0 = \mathbf{F}_{12} + \mathbf{F}_{21} = m_1\mathbf{a}_1 + m_2\mathbf{a}_2 = m_1\frac{\mathrm{d}\mathbf{u}_1}{\mathrm{d}t} + m_2\frac{\mathrm{d}\mathbf{u}_2}{\mathrm{d}t} = \frac{\mathrm{d}}{\mathrm{d}t}(m_1\mathbf{u}_1 + m_2\mathbf{u}_2)$$

According to this result, the magnitude $m_1\mathbf{u}_1 + m_2\mathbf{u}_2$ does not change during the interaction of two isolated particles (regardless of which is the nature of the interaction). This magnitude is said to be *conserved*. Due to the interest of this result, the product of mass and velocity of a particle is called classical *momentum*,

classical momentum

$$\mathbf{p} \equiv m\,\mathbf{u} \tag{6.2}$$

Thus the theorem says that the sum of momenta does not vary during an interaction. This result is straightforwardly generalized to an isolated system of several interacting particles, so leading to the conservation of *total momentum*

$$\mathbf{P} \equiv \sum_i \mathbf{p}_i \tag{6.3}$$

where the addition covers all the particles of the system. In other words, if the momentum of one of the particles varied at some time as a consequence of the interaction, this fact could only happen if this variation were *simultaneously* compensated by an opposite variation of the momentum of the rest of the system. If particles interact at a distance—as in the case of the gravitational interaction—the simultaneous compensation is only possible because the interaction force depends only on distance; thus any modification of the position of one of the particles has an *instantaneous* effect on the rest of the system.

The total momentum conservation of an isolated system of particles is verified in any inertial reference system, as it should be expected because the theorem has been proved starting from laws satisfying the Galilean Principle of relativity. In fact, if the Galilean addition theorem of velocities (1.7) is multiplied by the particle mass it will result in the Galilean transformation of momentum

$$\mathbf{p}' = m\,\mathbf{u}' = m\,\mathbf{u} - m\,\mathbf{V} = \mathbf{p} - m\,\mathbf{V} \tag{6.4}$$

then

$$\mathbf{P}' = \sum_i \mathbf{p}'_i = \sum_i (\mathbf{p}_i - m_i\mathbf{V}) = \sum_i \mathbf{p}_i - \mathbf{V}\sum_i m_i = \mathbf{P} - M\,\mathbf{V} \tag{6.5}$$

In Eq. (6.5) $M \equiv \sum m_i$ is the total mass of the particles system, a quantity that classical physics assumes as invariant (it has the same value in any frame) and conserved when the system is isolated. On the other hand, velocity $\mathbf{V}$ is the uniform velocity of frame S' relative to frame S. Therefore, the term $M\mathbf{V}$ does not change as time elapses. This means that if $\mathbf{P}$ is conserved then $\mathbf{P}'$ is conserved as well.

relativistic conservation laws must be local

A conservation law like the examined one, which allows that variations happened in a subsystem be simultaneously compensated by opposite variations in the rest of the system—which could be placed in other location—is not acceptable in Relativity. Due to the fact that the notion of simultaneity is not absolute in Relativity, a distant compensation that resulted simultaneous in a given frame would not be simultaneous in other frame moving with respect to the first one (see §3.8 and §4.6). The relativistic conservation laws have to be *local*, in the sense that conservation must be satisfied at each space-time event, instead of being the result of compensations among variations happening simultaneously at different places. The archetypal conservation law in Relativity is the continuity equation (1B.3) which expresses the charge conservation. The continuity equation is a differential equation (so it is local) stating that all charge variation at a place and a time can only come from the current (charge flux) arriving at the place at that time (see also §5.3). Therefore, if the charge is increasing at a given place and time, this rise is by no means ascribable to a simultaneous charge decreasing in a distant place.

Likewise, the archetypal interaction in Special Relativity is the electromagnetic interaction, where two distant charges can only influence each other through a field—the electromagnetic field— which propagates at a finite velocity. This means that any variation of the state of motion of one of the charges will not cause an immediate effect on the other, but after that the electromagnetic radiation produced by such acceleration reaches the position of the other charge. In this way, the Principle of action and reaction evaporates, since its statement implicitly assumes that interactions at a distance are instantaneous.[1] Relativity does not admit instantaneous interactions at a distance, and substitutes them by interactions that are mediated by fields carrying energy and momentum. Thus the particles momentum is not separately conserved, but the sum of field and particles momentum is (locally) conserved. A charge can transfer energy and momentum to the electromagnetic field at its vicinity (locally); the field carries the energy and momentum with finite velocity c, and can transfer energy and momentum to another charge at another place (locally as well). No instantaneous transfer exists between separated charges.

See §7.6 §7.7

The relativistic formulation of the laws of Dynamics will take as starting point the concept of momentum and its local conservation.

6.2. ENERGY AND MOMENTUM OF A PARTICLE

Let us consider an isolated system composed of several particles which do not interact "at a distance." In this way we shall avoid considering for the present the momentum associated to the fields mediating such interactions. The only possible interaction is local, and it is realized by means of collisions. Our aim is to define a relativistic notion of particle momentum $\mathbf{p}$ such that

(i) if $u << c$, $\mathbf{p}$ goes to the classical value $m\,\mathbf{u}$;
(ii) if the total momentum $\mathbf{P} = \sum \mathbf{p}_i$ is conserved in an inertial reference system, then it will be conserved in any other inertial reference system

[1] Indeed Lorentz forces between two moving charges are not, in general, equal and opposite, as it can be verified by using the fields obtained in §5.4.

(the conservation law satisfies the Principle of relativity under Lorentz transformations);

(iii) conservation of **P** is experimentally verified.

Item (ii) invites us to revisit the procedure followed in Eqs. (6.4–6.5), in the light of Lorentz velocity transformations. We shall then start from Eqs. (3.14) and (3.23–3.24); after multiplying them by the particle mass m (the same inertial mass of Newtonian mechanics), it is obtained

$$m\,\gamma(u')\,u'_x = \gamma(V)\,[m\,\gamma(u)\,u_x - m\,\gamma(u)\,V] \tag{6.6}$$

$$m\,\gamma(u')\,u'_y = m\,\gamma(u)\,u_y, \tag{6.7}$$

$$m\,\gamma(u')\,u'_z = m\,\gamma(u)\,u_z \tag{6.8}$$

If $u, V << c$, factors γ go to 1, and Eqs. (6.6–6.8) become the three components of the vector equation (6.4). This correspondence suggests trying with the following definition of relativistic momentum:

relativistic momentum

$$\mathbf{p} \equiv m\,\gamma(u)\,\mathbf{u} = \frac{m\,\mathbf{u}}{\sqrt{1-\frac{u^2}{c^2}}} \tag{6.9}$$

to write Eqs. (6.6–6.8) in the way

$$p'_x = \gamma(V)\,[p_x - m\gamma(u)\,V] \qquad p'_y = p_y \qquad p'_z = p_z \tag{6.10}$$

Then the total momentum for a system of several free particles transforms in the following manner:[2]

$$\begin{aligned} P'_x &= \sum_i p'_{i_x} = \gamma(V)\sum_i\left[p_{i_x} - m_i\gamma(u_i)V\right] = \gamma(V)[P_x - V\sum_i m_i\gamma(u_i)] \\ P'_y &= \sum_i p'_{iy} = \sum_i p_{i_y} = P_y \qquad P'_z = \sum_i p'_{i_z} = \sum_i p_{i_z} = P_z \end{aligned} \tag{6.11}$$

In relativistic transformation (6.9), the magnitude $\sum m_i\gamma(u_i)$ takes the place of the addition of masses $\sum m_i$ in classical transformation (6.5). In classical physics it is accepted that the total mass of an isolated system is conserved. In Eq. (6.5) this classical principle of mass conservation guarantees that $\mathbf{P}'$ is conserved whenever **P** is conserved; thus, the classical momentum conservation law is

[2] Since the notions of simultaneity in S and S' are not coincident, if **P** is the sum of "simultaneous" values of particle momenta, then this will not be true for $\mathbf{P}'$. However, since the system is isolated and particles can only interact by means of collisions, each $\mathbf{p}_i$ is conserved between successive collisions. Thus, for a two particle system it is irrelevant to add simultaneous values of $\mathbf{p}_1$ and $\mathbf{p}_2$; It is sufficient that both are previous or subsequent to the collision, which is a feature independent of the reference system. In general, the $\mathbf{p}_i$'s added in **P** are evaluated at a set of events spacelike separated.

consistent with the Galilean Principle of relativity. Instead, the relativistic transformation (6.11) expresses that the momentum conservation law accomplishes the Principle of relativity under Lorentz transformations only if the quantity $\sum m_i \gamma(u_i)$ is also conserved. Therefore, the relativistic notion of space-time, responsible for the form of velocity transformations (3.14) and (3.23–3.24), together with our confidence in the existence of a momentum conservation law, compels us to admit that the addition of masses is not conserved; what is conserved instead is a magnitude combining masses and velocities (certainly, $\sum m_i \gamma(u_i)$ approaches the total mass when all the velocities u_i are much smaller than c).

To become familiar with the magnitude[3] that substitutes the mass in **p** transformation, we shall examine the first two terms in the power series of the variable u^2/c^2:

$$m\,\gamma(u) = m + \frac{1}{2} m \frac{u^2}{c^2} + O(u^4/c^4) \tag{6.12}$$

Of course, the term that does not depend on the velocity is mass, while the contribution of the velocity at the lowest order results to be the classical kinetic energy divided by c^2. So, the magnitude that has to be conserved, in order that the **P** conservation shall be verified in all inertial reference system, is a combination of mass and kinetic energy. This quantity is called *relativistic energy* of the particle:

$$E \equiv m\,\gamma(u)\,c^2 = \frac{m\,c^2}{\sqrt{1 - \dfrac{u^2}{c^2}}} \tag{6.13}$$

relativistic energy

The relativistic *kinetic energy* of the particle is

$$T \equiv E - m\,c^2 \tag{6.14}$$

which goes to the classical kinetic energy when $u << c$ (cf. Eq. (6.12)). The magnitude mc^2 is called *rest energy*. By combining the definitions of momentum (6.9) and energy (6.13) we write

$$\mathbf{p} = c^{-2} E\,\mathbf{u} \tag{6.15}$$

The conservation of the total relativistic energy of an isolated system of particles

$$\mathcal{E} \equiv \sum_i E_i \tag{6.16}$$

makes compatible the relativistic momentum conservation with the Principle of relativity under Lorentz transformations. Naturally, also the conservation of

[3] Some authors call $m\gamma(u)$ the "relativistic mass."

the total relativistic energy of an isolated system of particles should be verified in any inertial frame. In fact, the former argumentation could start from the inverse transformation of (6.11) to conclude that the total energy in frame S' must be conserved in order that P_x and P'_x be conserved. The same conclusion is reached through the Lorentz transformation of the relativistic energy: multiplying Eq. (3.17) by mc^2 it is obtained

$$E' = \gamma(V)(E - V\, p_x) \tag{6.17}$$

The transformation of total energy results from adding (6.17) for all the particles. The linearity of transformation (6.17) guarantees that the total energy and total momentum transform in the same way as the energy and momentum of a unique particle do:

$$\mathcal{E}' = \sum_i E_i' = \gamma(V)\left(\sum_i E_i - V \sum_i p_{i_x}\right) = \gamma(V)\,(\mathcal{E} - V\,P_x) \tag{6.18}$$

In transformation (6.18) it is evident that the conservation of both $\mathcal{E}$ and $\mathcal{E}'$ requires the conservation of P_x. Thus, the conservation laws for relativistic energy and momentum of an isolated system are deeply intertwined.[4]

What follows is a summary of E and $\mathbf{p}$ transformations, as they result from Eqs. (6.17) and (6.10), and the respective inverse transformations (which are obtained by replacing V by $-V$):

energy and momentum transformations

$$E' = \frac{E - V\,p_x}{\sqrt{1 - \frac{V^2}{c^2}}} \qquad E = \frac{E' + V p'_x}{\sqrt{1 - \frac{V^2}{c^2}}} \tag{6.19a–b}$$

$$p'_x = \frac{p_x - Vc^{-2}\,E}{\sqrt{1 - \frac{V^2}{c^2}}} \qquad p_x = \frac{p'_x + Vc^{-2}\,E'}{\sqrt{1 - \frac{V^2}{c^2}}} \tag{6.20a–b}$$

$$p'_y = p_y \qquad p'_z = p_z \tag{6.21a–b}$$

If the interaction among particles are mediated by fields, then the energy and momentum of the fields are also involved in conservation laws. The conservation of the relativistic energy of an isolated physics system—in the place of

[4] Instead, there is no such interdependence between momentum conservation and mass conservation in classical physics. Although **P** conservation in different inertial frames requires mass conservation, what makes the consistency between classical mass conservation and Galilean principle of relativity is the mere mass invariance (mass has the same value in all the frames; so, if mass is conserved in a frame, then it will be conserved in any frame).

the classical principle of mass conservation—entails the possibility of converting mass in other types of energy and vice versa, what constitutes the more dramatic result of Special Relativity. Several sections of this chapter will describe the nature of the experiments that fully confirm the conservation of relativistic energy and momentum.

6.3. ENERGY-MOMENTUM INVARIANT. FORCE

Transformations (6.19–6.21) exhibit a remarkable similarity with Lorentz transformations of coordinates (Eqs. (3.37–3.39)). Clearly, the ones become the others when $\mathbf{p}$ is substituted for $\mathbf{r}$, and $c^{-2}E$ is replaced by t. The reason for this similarity can be easily discovered by writing $\mathbf{p}$ and E with the help of the proper time along the particle world line $\mathrm{d}\tau = c^{-1}\mathrm{d}s = (1 - u^2/c^2)^{1/2}\mathrm{d}t$ (see Eq. 4.7)):

$$\mathbf{p} = m\,\gamma(u)\,\mathbf{u} = \frac{m}{\sqrt{1-\frac{u^2}{c^2}}}\frac{\mathrm{d}\mathbf{r}}{\mathrm{d}t} = m\frac{d\mathbf{r}}{\mathrm{d}\tau} \tag{6.22}$$

$$E = m\,\gamma(u)\,c^2 = \frac{mc^2}{\sqrt{1-\frac{u^2}{c^2}}} = m\,c^2\frac{\mathrm{d}t}{\mathrm{d}\tau} \tag{6.23}$$

Since m and $d\tau$ are invariant (they do not change under Lorentz transformations), then the $\mathbf{p}$ and E transformations are governed by the transformations of $\mathrm{d}\mathbf{r}$ and $\mathrm{d}t$, what explains the similarity between coordinates transformations and energy–momentum transformations.

The rapidity Θ_u (§4.7) offers another way to write relations among E, $\mathbf{p}$, m, and the particle velocity. According to Eq. (4.15) it is $\mathrm{d}t = \mathrm{d}\tau\cosh\Theta_u$ and $|\mathrm{d}\mathbf{r}| = c\mathrm{d}\tau\sinh\Theta_u$, then

$$p \equiv |\mathbf{p}| = m\,c\sinh\Theta_u \qquad E = m\,c^2\cosh\Theta_u \tag{6.24}$$

Since energy–momentum transformations are entirely similar to those for the coordinates, it should be expected that a combination of E and p resembling the interval will remain invariant under Lorentz transformations. So long as $c^{-1}E$ transforms like $c\ \mathrm{d}t$, and p does it as $|\mathrm{d}\mathbf{r}|$, then the combination $c^{-2}E^2 - p^2$ has to be a Lorentzian invariant. This energy–momentum invariant can be evaluated taking into account Eqs. (6.22–6.23) or Eq. (6.24):

energy–momentum invariant

$$\begin{aligned} E^2 - p^2c^2 &= m^2c^4\gamma(u)^2 - m^2c^2\gamma(u)^2u^2 \\ &= m^2c^4\gamma(u)^2\,(1-\frac{u^2}{c^2}) = m^2c^4 \end{aligned} \tag{6.25}$$

So, the energy–momentum invariant gives the measure of the particle mass.

By means of Eq. (6.25), we can relate the variations of the particle energy and momentum:

$$0 = \mathrm{d}(E^2 - p^2c^2) = \mathrm{d}(E^2 - \mathbf{p}\cdot\mathbf{p}c^2) = 2E\,\mathrm{d}E - 2c^2\mathbf{p}\cdot\mathrm{d}\mathbf{p}$$

and using Eq. (6.15),

$$\mathrm{d}E = \mathbf{u}\cdot\mathrm{d}\mathbf{p} = \frac{\mathrm{d}\mathbf{p}}{\mathrm{d}t}\cdot\mathbf{u}\,\mathrm{d}t = \frac{\mathrm{d}\mathbf{p}}{\mathrm{d}t}\cdot\mathrm{d}\mathbf{r} \tag{6.26}$$

force in Relativity

In order to get the work–energy relation, the force acting on the particle has to be associated, according to Eq. (6.26), with the variation of momentum $\mathbf{p}$ with respect to the coordinate time t:

$$\mathbf{F} = \frac{\mathrm{d}\mathbf{p}}{\mathrm{d}t} \tag{6.27}$$

Equation (6.27) would then accomplish the role of the Second Law of Dynamics (§1.6). The consistency of this relation demands that force transforms in the same way as $\mathrm{d}\mathbf{p}/\mathrm{d}t$; thus the relation $\mathbf{F} = \mathrm{d}\mathbf{p}/\mathrm{d}t$ would be valid in all inertial reference system. In order to know how the force transforms it is necessary to have a law that, completing the relation (6.27), expresses the form of the interaction force $\mathbf{F}$ under consideration. The paradigmatic interaction in Special Relativity is the electromagnetic interaction, whose force is the Lorentz force $\mathbf{F} = q(\mathbf{E} + \mathbf{u}\times\mathbf{B})$. The transformation of Lorentz force was obtained in Eqs. (5.27) and (5.28), and it is easy to verify that it coincides with $\mathrm{d}\mathbf{p}/\mathrm{d}t$ transformation (differentiating Eqs. (6.20a) and (6.21a), and dividing them by Eq. (3.41); then using Eq. (6.26) to substitute $\mathrm{d}E/\mathrm{d}t = \mathbf{F}\cdot\mathbf{u}$).

By integrating Eq. (6.26) the relation between the work of the force and the variation of the particle energy is obtained:

$$E_2 - E_1 = \int_{\mathbf{r}_1}^{\mathbf{r}_2} \mathbf{F}\cdot\mathrm{d}\mathbf{r} \tag{6.28}$$

speed of light cannot be reached

Since the relativistic energy diverges when u goes to c, it would be necessary an infinite work in order that a particle could reach the speed of light. The speed of light is then an upper bound for the particle velocity.

To make a comparison with the Newtonian law of proportionality between force and acceleration, it is necessary to express Eq. (6.27) as a function of the derivative of $\mathbf{u}$:

$$\mathbf{F} = \frac{\mathrm{d}\mathbf{p}}{\mathrm{d}t} = m\,\frac{\mathrm{d}}{\mathrm{d}t}(\gamma(u)\,\mathbf{u}) = m\gamma(u)\frac{\mathrm{d}\mathbf{u}}{\mathrm{d}t} + m\,\mathbf{u}\frac{\mathrm{d}\gamma(u)}{\mathrm{d}t}$$

where

$$\frac{\mathrm{d}\gamma(u)}{\mathrm{d}t} = \frac{\mathrm{d}}{\mathrm{d}t}\left(1 - \frac{u^2}{c^2}\right)^{-1/2} = \frac{1}{2c^2}\left(1 - \frac{u^2}{c^2}\right)^{-3/2}\frac{\mathrm{d}u^2}{\mathrm{d}t}$$

$$= \frac{1}{2c^2}\gamma(u)^3\frac{\mathrm{d}(\mathbf{u}\cdot\mathbf{u})}{\mathrm{d}t} = c^{-2}\gamma(u)^3\mathbf{u}\cdot\frac{\mathrm{d}\mathbf{u}}{\mathrm{d}t}$$

Acceleration $\boldsymbol{a} = \mathrm{d}\mathbf{u}/\mathrm{d}t$ can be decomposed along directions being transversal and longitudinal to $\mathbf{u}$, which allows us to write $\mathbf{u}\cdot \mathrm{d}\mathbf{u}/\mathrm{d}t = u\, a_{\text{longitudinal}}$, and $\mathbf{u}\, a_{\text{longitudinal}} = u\, \boldsymbol{a}_{\text{longitudinal}}$. Thus

$$\begin{aligned}\mathbf{F} = \frac{\mathrm{d}\mathbf{p}}{\mathrm{d}t} &= m\gamma(u)(\boldsymbol{a}_{\text{longitudinal}} + \boldsymbol{a}_{\text{transversal}}) + m\,\frac{u^2}{c^2}\gamma(u)^3 \boldsymbol{a}_{\text{longitudinal}} \\ &= m\gamma(u)^3\,[\gamma(u)^{-2} + \frac{u^2}{c^2}]\,\boldsymbol{a}_{\text{longitudinal}} + m\gamma(u)\,\boldsymbol{a}_{\text{transversal}} \\ &= m\gamma(u)^3\,\boldsymbol{a}_{\text{longitudinal}} + m\gamma(u)\,\boldsymbol{a}_{\text{transversal}}\end{aligned} \tag{6.29}$$

relation between force and acceleration

Equation (6.29) means that the force direction differs from the acceleration direction. If $u << c$ then $\gamma(u)$ approximates to 1, so the Newtonian relation between force and acceleration is recovered. In the particle proper frame ($\mathbf{u}_{\text{o}} = 0$) it is

$$\mathbf{F}_{\text{o}} = m\ \mathbf{a}_o \tag{6.30}$$

According to Eqs. (3.46) and (3.47), Eq. (6.29) can be rewritten in terms of the acceleration in the particle proper frame as

$$\mathbf{F} = m\boldsymbol{a}_{\text{o longitudinal}} + m\gamma(u)^{-1}\,\boldsymbol{a}_{\text{o transversal}} \tag{6.31}$$

Therefore

$$\mathbf{F}_{\text{longitudinal}} = \mathbf{F}_{\text{o longitudinal}} \tag{6.32}$$

$$\mathbf{F}_{\text{transversal}} = \gamma(u)^{-1}\,\mathbf{F}_{\text{o transversal}} \tag{6.33}$$

Equations (6.32–6.33) can be compared with Eqs. (5.27–5.28). We can suppose in (5.27) and (5.28) that S is the proper frame S_{o}; then $\mathbf{u} = 0$, V is the particle velocity in the other frame, and transformations (5.27–5.28) turn out to be identical to (6.32–6.33). Transformations of this kind reveal that forces in Relativity unavoidably depend on the particle velocity. Even if the force resulted independent of the velocity in a peculiar frame, it would depend on the velocity in any other frame as a consequence of the transformation form.

6.4. CHARGE MOVEMENT IN UNIFORM FIELDS

charge in a uniform magnetostatic field

Let be a charge q moving in the presence of a uniform magnetostatic field $\mathbf{B} = \mathrm{B}\hat{\mathbf{z}}$. According to Eq. (6.27), the charge movement is governed by the equation

$$\frac{\mathrm{d}\mathbf{p}}{\mathrm{d}t} = q\,\mathbf{u}\times\mathbf{B}$$

Since the magnetic force is perpendicular to $\mathbf{p} = m\gamma(u)\mathbf{u}$, then $p = |\mathbf{p}|$ remains constant. Decomposing $\mathbf{p}$ along directions parallel and perpendicular to $\mathbf{B}$ it results

$$\frac{\mathrm{d}\mathbf{p}_\perp}{\mathrm{d}t} = q\mathbf{u} \times \mathbf{B} \qquad \frac{\mathrm{d}p_z}{\mathrm{d}t} = 0,$$

which means that p_z and $p_\perp = |\mathbf{p}_\perp|$ are separately conserved. Therefore $\mathbf{p}_\perp$ can only change direction, without changing modulus. The change of direction in a time $\mathrm{d}t$ is measured by the angle $\mathrm{d}\theta = |\mathrm{d}\mathbf{p}_\perp|/p_\perp$:

$$\mathrm{d}\theta = \frac{|\mathrm{d}\mathbf{p}_\perp|}{p_\perp} = q\frac{u_\perp \mathrm{B}}{p_\perp}\,\mathrm{d}t = \frac{q\mathrm{B}}{m\,\gamma(u)}\,\mathrm{d}t$$

Since B does not depend on position or time, and p is conserved, then the change of direction of $\mathbf{p}_\perp$ is uniform. Therefore, the motion projected on a plane perpendicular to $\mathbf{B}$ consists in a uniform circular movement. So long as the angle spanned by the charge from the center of the circular trajectory is equal to the change of $\mathbf{p}_\perp$ direction, it results that the velocity of the circular movement is

$$\omega = \frac{\mathrm{d}\theta}{\mathrm{d}t} = \frac{q\,\mathrm{B}}{m\gamma(u)} \tag{6.34}$$

On the other hand, conservation of p_z and E (note that the magnetic force does not make any work) means that the charge movement along $\mathbf{B}$ direction has a constant velocity $u_z = p_z c^2/E$.

cyclotron

The circular movement of a charge in a uniform magnetic field is exploited in designing the *cyclotron*, which is a particle accelerator where the charge is accelerated by an electric field acting only in a fringe intersecting two diametrically opposite segments of the circular trajectory. In order that the charge be accelerated in each segment, but never slowed down, the electric field has to be in phase with the circular motion. Thus the electric field has to invert direction each half turn of the charge; i.e., the electric field oscillates with an angular frequency ω equal to the charge angular velocity. According to classical physics the rise of linear velocity does not modify the angular velocity ω, since classical value $\omega = qB/m$ does not depend on u (the sole consequence of the rise of u would be the increase of the trajectory radius, as $u = \omega R$). However, when the charge velocity approaches the speed of light, the factor $\gamma(u)^{-1}$ in Eq. (6.34)—coming from the relativistic form of momentum—will play its role, and the frequency of the electric field (*synchrocyclotron*) or the intensity of the magnetic field (*synchrotron*) must be adjusted as a function of the charge velocity to avoid a phase shift between the electric field and the charge motion. Certainly, the exactitude of result (6.34) is experimentally verified in such particle accelerators.

charge in a uniform electrostatic field

Let us now consider the movement of a charge q in a uniform electrostatic field $\mathbf{E} = \mathrm{E}\,\hat{\mathbf{y}}$. The corresponding equation of motion is

$$\frac{\mathrm{d}\mathbf{p}}{\mathrm{d}t} = q\,\mathbf{E}$$

which can be integrated to obtain $\mathbf{p}(t) = \mathbf{p}(0) + q\mathbf{E}t$. We can choose $t = 0$ at the time when $\mathbf{p}(0) \cdot \mathbf{E} = 0$; so, after a proper choice of Cartesian axes it results

$$\mathbf{p}(t) = p_x \hat{\mathbf{x}} + q\mathrm{E}t\,\hat{\mathbf{y}}$$

where p_x is a conserved magnitude. Differing from what happens in classical physics, conservation of p_x does not means that $u_x = p_x c^2/E$ (see Eq. (6.15)) remains constant. In fact, due to the work of the electric field, the energy E varies with time (see Eq. (6.28)). In other words, energy $E = (p^2c^2 + m^2c^4)^{1/2}$ changes because p_y changes; this change of energy affects all velocity components, including those associated with the conserved components of $\mathbf{p}$. A further important difference is that conservation of p_x does not guarantee the conservation of p_x' in other frame S', as another consequence of the lack of conservation of energy E (see Eq. (6.20a)).

The charge deflects, because the orientation of $\mathbf{p}$ varies with time. The angle between the direction of $\mathbf{p}(t)$ and the x axis is equal to the quotient between the components of $\mathbf{p}(t)$:

$$\frac{p_y(t)}{p_x} = \frac{q\mathrm{E}t}{m\gamma(u(0))u(0)} \tag{6.35}$$

The deflection of a beam of charged particles in the presence of an electric field can be observed in the laboratory and is used to determine the ratio between charge and mass of the particles. The classical result does not include the factor $\gamma(u)$ in Eq. (6.35); since u_x is classically constant, the classical deflection can be written as $p_y/p_x = q\mathrm{E}\,t/(m\,u_x) = (q/m)\,\mathrm{E}\,\Delta x/u_x^2$. The velocity u_x of charges in the beam can be tuned by means of a velocity selector.[5] So, in the classical result the tangent of the angle between the motion direction and the x axis is proportional to the displacement Δx. In 1908, A.H. Bucherer examined the deflection of high-energy β rays (electrons) and obtained a conclusive experimental confirmation that the deflection for high energies strays away from the classical result and adjusts to the values predicted by relativistic dynamics.

We shall complete the description of the movement by writing the velocity $\mathbf{u}(t) = \mathbf{p}(t)\; c^2/E(t)$, where

$$E(t) = \sqrt{p(t)^2c^2 + m^2c^4} = \sqrt{p_x^2c^2 + m^2c^4 + q^2c^2E^2t^2} = E(0)\sqrt{1 + \left(\frac{q\,c\,\mathrm{E}\,t}{E(0)}\right)^2}$$

then

$$u_x(t) = \frac{p_x\,c^2}{E(t)} = \frac{u_x(0)}{\sqrt{1 + \left(\dfrac{q\,\mathrm{E}\,t}{m\gamma(u(0))c}\right)^2}}$$

[5] In a velocity selector, the charge velocity is controlled by applying crossed (mutually orthogonal) fields $\mathbf{E}$ and $\mathbf{B}$. The selector is a device that allows the passing of those charges that, traveling perpendicularly to the fields, are not deviated. Thus the velocity of these charges must be such that the Lorentz force $\mathbf{F} = q\,(\mathbf{E} + \mathbf{u} \times \mathbf{B})$ vanishes; then $u = \mathrm{E}/\mathrm{B}$.

$$u_y(t) = \frac{p_y(t)\,c^2}{E(t)} = \frac{\dfrac{q\mathrm{E}\,t}{m\gamma(u(0))}}{\sqrt{1+\left(\dfrac{q\mathrm{E}t}{m\gamma(u(0))c}\right)^2}}$$

Integrating the equation for the x component it is obtained

$$x(t) = x(0) + u_x(0)\,\frac{m\gamma(u(0))c}{qE}\,\mathrm{arcsinh}\left[\frac{q\mathrm{E}t}{m\gamma(u(0))c}\right] \tag{6.36}$$

On the other hand the velocity $u_y(t)$ has the form corresponding to a uniformly accelerated movement with proper acceleration $q\mathrm{E}/[m\gamma(u(0))]$ (see Eq. (3.48)). therefore the movement $y(t)$ is the hyperbolic motion described in §3.13:

$$y(t) = y(0) + \frac{m\gamma(u(0))c^2}{qE}\left[\sqrt{1+\left(\frac{q\mathrm{E}\,t}{m\gamma(u(0))c}\right)^2} - 1\right] \tag{6.37}$$

The charge movements studied in this section can be transformed to a frame S' moving with relative velocity $\mathbf{V} = V\hat{\mathbf{x}}$, to obtain the charge movements in uniform "crossed" fields $\mathbf{E}' = \mathrm{E}'\,\hat{\mathbf{y}}$, $\mathbf{B}' = \mathrm{B}'\,\hat{\mathbf{z}}$. In fact, since we have considered the cases of purely magnetostatic or electrostatic fields in S, then the fields will be orthogonal in S' (note the invariance of $\mathbf{E}\cdot\mathbf{B}$), as it follows from transformations (5.10). After the transformation is done, the two studied cases are still distinguishable by means of the value of the invariant $\mathrm{E}^2 - c^2\,\mathrm{B}^2$, which is negative in the case of a purely magnetostatic field (or the respective Lorentz transformed fields) but is positive for a purely electrostatic field or the transformed one.

6.5. CENTER-OF-MOMENTUM FRAME

In classical physics the total momentum of a system of particles can be written in terms of the velocity of a representative point of the system, which is called *center of mass*. In fact, $\mathbf{P} = \sum m_i\,\mathbf{u}_i(t) = (\mathrm{d}/\mathrm{d}t)\,[\sum m_i\,\mathbf{r}_i\,(t)]$; so, defining the center of mass as the point whose position is given by the vector $\mathbf{R}_{\mathrm{cm}}(t) = \sum m_i\,\mathbf{r}_i(t)/\sum m_i$, then $\mathbf{P} = (\sum m_i)\,\mathrm{d}\mathbf{R}_{\mathrm{cm}}/\mathrm{d}t = \mathrm{M}\,\mathbf{U}_{\mathrm{cm}}$. The components of vector $\mathbf{R}_{\mathrm{cm}}$ change under Galileo transformations as the components of the position vector of a particle do; so, vector $\mathbf{R}_{\mathrm{cm}}$ certainly identifies a point in space at each time, independently of the reference system where its components are evaluated. If $\mathbf{P}$ is conserved then the center of mass velocity will be constant. In the center of mass frame, the center of mass velocity is null; consequently the total momentum of the system of particles is null as well.

As it will be shown in §6.7, in Relativity it is not possible to define a representative point of the system of particles—a center of mass or energy—in a way independent of the frame. This is not an obstacle to find a reference system where the total momentum of a compound system vanishes. According to Eqs. (6.11) and (6.18), **P** and ε transform like **p** and E:

$$P_x = \gamma(V)\,(P'_x + V\,c^{-2}\mathcal{E}') \qquad P_y = P'_y \qquad P_z = P'_z$$
$$\mathcal{E} = \gamma(V)\,(\mathcal{E}' + V\,P'_x)$$

Let S' be the frame S_c—which will be called *center-of-momentum frame*—where the total momentum vanishes ($\mathbf{P}_c = 0$); $\mathbf{U}_c$ is the velocity of S_c relative to S. Then, by replacing in the former transformation it results

$$\mathbf{P} = c^{-2}\mathcal{E}_c\ \gamma(U_c)\ \mathbf{U}_c = c^{-2}\mathcal{E}\,\mathbf{U}_c \tag{6.38}$$

The relation among total energy and momentum of the physical system, **P** and $\mathcal{E}$, respectively, and velocity $\mathbf{U}_c$ is the same relation (6.15) characteristic of a unique particle. This is not surprising, since it is a consequence of the fact that the energy and momentum of the compound physical system transform like those of a unique particle (the same remark is valid in classical physics for the relation $\mathbf{P} = M\,\mathbf{U}_{cm}$, which results from taking $\mathbf{P}'$ equal to zero in Eq. (6.5)). In order that $U_c = P\,c^2/\mathcal{E}$ be smaller than c, the invariant $\mathcal{E}^2 - P^2c^2$ must be positive, as it is for a particle (this condition is guaranteed in a physical system that is composed by several free particles[6]).

Comparing Eqs. (6.38) with (6.9) it is evident that, whenever a compound physical system is regarded as a whole, without taking into account its internal structure, then the role of mass will be played by $c^{-2}\mathcal{E}_c$. Like it happens with a sole particle, the invariant $\mathcal{E}^2 - P^2c^2 = \mathcal{E}_c^{\,2}$ (evaluate it in S_c) gives the measure of the squared mass of a compound system (multiplied by c^4). This means that the internal energy of a compound system contributes to its mass. In the case of a set of free particles, which only interacts through collisions, the mass of the system is larger than the sum of masses of the constituent particles because $c^{-2}\mathcal{E}_c = \sum(m_i + c^{-2}T_{i_c})$. The internal kinetic energy contributes to the mass of a compound system. For instance, the temperature of an ideal gas reflects the mean kinetic energy of gas particles in frame S_c; so, the rise of temperature produces an increase of the gas net mass.[7]

internal energy contributes to mass

The mass of a compound system is not then the result of adding the masses of its constituents, but receives the contributions of the internal energies found in the system. This conclusion is not limited to the internal kinetic energy,

[6] $(E_1+E_2)^2 - (\mathbf{p}_1+\mathbf{p}_2)^2c^2 = (E_1{}^2 - \mathbf{p}_1{}^2c^2) + (E_2{}^2 - \mathbf{p}_2{}^2c^2) + 2(E_1E_2 - \mathbf{p}_1\cdot\mathbf{p}_2\ c^2)$. Each term is positive, because each particle satisfies $E_i > |\mathbf{p}_i|c$ (i.e., $c > |\mathbf{u}_i|$). In particular the two first terms correspond to $m_1{}^2c^4 + m_2{}^2c^4$.

[7] Of course, the velocities of the particles should be comparable to c in order that the internal kinetic energy significantly contributes to the gas net mass.

but it applies to all kind of internal energy whatever its nature is. The mass of a body increases when it is deformed: the (positive) work done remains stored in the body as an internal energy, both as elastic energy associated with the deformation and internal kinetic energy with the consequent increase of temperature. In a *completely inelastic* collision, the colliding particles form a unique particle after the collision. This resulting particle is at rest in S_c; its mass is equal to $c^{-2}\mathcal{E}_c$. Since the relativistic energy is conserved, the value of $c^{-2}\mathcal{E}_c$ after the collision is equal to the relativistic energy before the collision: $c^{-2}\mathcal{E}_c = \sum E_{i_c} = \sum(m_i + c^{-2}T_{i_c})$. Therefore the kinetic energies of the colliding particles, measured in frame S_c, remains stored as internal energy of the compound system, so contributing to the mass of the resulting particle.

completely inelastic collision

6.6. PHENOMENA DERIVED FROM MASS-ENERGY EQUIVALENCE

In 1905, Einstein wrote *"The mass of a body is a measure of its energy-content"* (1905c). In a daring interpretation of the energy mc^2 of a body at rest, Einstein also conjectured that bodies emitting some type of energy, like radium salts, must do it at the expense of their masses. Einstein (1910) was conscious that "for the moment there is no hope whatsoever for the experimental verification of mass-energy equivalence," since the typical energies of those emissions would involve mass changes too small to be detectable. Only in the 1930s, when a better knowledge of the nuclear structure was reached and it was possible to measure the masses of the nuclear constituents with enough precision, Einstein's prediction was confirmed. In this section we shall refer to several features of the phenomenology associated with the relativistic equivalence between mass and energy, in particular mass conversion in other types of energies and vice versa.

First, we shall mention the *binding energy* contribution to the atomic nucleus mass. If several particles are bound in a stable configuration, through some interaction mechanism, then it will be necessary to deliver some energy (to do some positive work on the system) to separate and leave the particles at rest and free. This implies that the mass of the bound configuration—the initial energy of the compound system measured in S_c—is lower than the sum of the masses of the free constituents at rest—which is equal to the final energy in S_c—because the last one includes the positive work delivered to the system for separating its constituents. Thus the mass of a stable atomic nucleus is lower than the sum of masses of its constituent *nucleons* (protons and neutrons), because the nucleus mass includes the contribution of the negative binding energy associated with the nuclear interactions keeping the configuration stable. This *mass defect* predicted by Relativity is experimentally verified by means of precise measurements of the masses of nuclei and their constituents performed with mass spectrometers. For instance, the helium nucleus, made up of two

mass defect

protons and two neutrons (^{4}He or α particle), can be regarded as constituted by two deuterons.[8] The ^{4}He mass and its rest energy are[9]

$$m_{^4\mathrm{He}} = 6.644656 \times 10^{-27}\,\mathrm{kg} \qquad m_{^4\mathrm{He}}c^2 = 3727.379\,\mathrm{MeV}$$

while the deuteron mass and its rest energy are

$$m_\mathrm{D} = 3.343583 \times 10^{-27}\,\mathrm{kg} \qquad m_\mathrm{D}c^2 = 1875.613\,\mathrm{MeV}.$$

Therefore the mass defect of ^{4}He, when regarded as constituted by two deuterons, is

$$m_{^4\mathrm{He}} - 2\,m_\mathrm{D} = (6.644656 - 2 \times 3.343583) \times 10^{-27}\,\mathrm{kg} = -4.251 \times 10^{-29}\,\mathrm{kg}$$

which corresponds to a binding energy equal to $(3727.379 - 2 \times 1875.613)$ MeV $= -23.847$ MeV.

Furthermore, the proton and neutron[10] masses are

$$m_\mathrm{p} = 1.672621 \times 10^{-27}\,\mathrm{kg} \qquad m_\mathrm{p}c^2 = 938.272\,\mathrm{MeV}$$
$$m_\mathrm{n} = 1.674927 \times 10^{-27}\,\mathrm{kg} \qquad m_\mathrm{n}c^2 = 939.565\,\mathrm{MeV}$$

so deuteron mass defect, when regarded as constituted by one proton and one neutron, is

$$m_\mathrm{D} - (m_\mathrm{p} + m_\mathrm{n}) = (3.343583 - 1.672621 - 1.674927) \times 10^{-27}\,\mathrm{kg}$$
$$= -3.965 \times 10^{-30}\,\mathrm{kg}$$

which corresponds to a binding energy equal to $(1875.613 - 938.272 - 939.565)$ MeV $= -2.224$ MeV.

The magnitude of the binding energy per nucleon increases with the number of nucleons in the nucleus up to be maximum for nuclei having around 60 nucleons, and then decreases slowly. Therefore, two types of nuclear reactions exist able to convert mass into another type of energy: fusion of light nuclei and fission of heavy nuclei.

nuclear fusion

In a *nuclear fusion*, two light nuclei—such as hydrogen, deuterium, helium isotopes, lithium, etc.—join to make up a unique nucleus together with byproducts of the reaction.[11] The total mass of the reaction products is lower

[8] Deuteron is deuterium nucleus, which is the hydrogen isotope whose nucleus is made up of one proton and one neutron.

[9] 1 eV (electron-volt) is the energy gained by an electron when goes through a potential difference of 1 V, and is equal to $1.602176462 \times 10^{-19}$ J. 1 MeV is equal to 10^6 eV.

[10] The neutron mass cannot be measured in a mass spectrometer since neutron has not electric charge. Indeed it is inferred from the experiments by applying conservation laws.

[11] No reaction can yield, as the unique result, one particle whose mass is smaller than the sum of the reactants masses, because the energy and momentum conservation laws would be violated (in the center-of-momentum frame the final energy would be equal to the mass of the resulting particle multiplied by c^2, so it would be smaller than the initial energy).

than the total mass of the light reactants because the magnitude of the (negative) binding energy of the resulting nucleus is larger than the sum of the binding energies of the reactants. Since the relativistic energy is conserved, the disappearing mass (multiplied by c^2) is converted into the (external) kinetic energy of the resulting nucleus and byproducts.[12] For instance, if two deuterons have enough kinetic energy for surmounting the electric repulsion and approaching up to the range of the nuclear interaction, then the following fusion reactions can happen

$$\mathrm{D} + \mathrm{D} \longrightarrow n + {}^3\mathrm{He}$$

$$\mathrm{D} + \mathrm{D} \longrightarrow p + {}^3\mathrm{H}$$

In the first case, the products of the fusion are a neutron and a helium 3 nucleus (the helium isotope made up of two protons and one neutron); in this reaction, a mass equivalent to an energy of 3.2 MeV is converted into kinetic energy of the reaction products.[13] In the second case, the products are one proton and one tritium nucleus (the hydrogen isotope made up of one proton and two neutrons), and the converted mass is equivalent to around 4 MeV. Each "released" MeV corresponds to the conversion of a mass of 1.782662×10^{30} kg. Although the converted mass can seem non-significant, the released energy is very large if compared with the typical energy of chemical reactions which is of around 10 eV per atom.

In the second one of these fusion reactions, both the reactants and the products have electric charge; this feature allows to determine their masses by means of a mass spectrometer. On the other hand, the mass of anyone of them could also be inferred by applying the energy and momentum conservation laws, together with the measurement of the remaining dynamical variables entering the laws. The agreement between both types of determination is excellent, which confirms the soundness of relativistic mechanics.

Fusion reactions as the ones here described, which are characteristic of an hydrogen bomb, are fundamental to understand the origin of the energy released from the stars: inside the stars the temperature is high enough (several millions of degrees) to make the hydrogen isotopes have the indispensable kinetic energy for fusing them. Moreover, the products of these reactions take part in other fusions making up heavier elements. The energy released

---

[12] Sometimes the resulting nucleus momentarily stores an (internal) quantum excitation energy, which it subsequently delivers by *decaying* to a fundamental quantum state in the form of kinetic energies of the decay products and/or through the emission of electromagnetic radiation (for instance, through a gamma decay to the nucleus fundamental state).

[13] For charged interacting particles, the initial and final energies of the Coulombian interaction are negligible whenever the initial and final distance among the particles are big enough. If there is electromagnetic radiation emitted, its energy and momentum will enter the balance through zero mass particles named *photons* (see §6.9).

by these thermonuclear reactions is eventually converted into electromagnetic radiation and kinetic energy of particles such as protons, neutrons, electrons, and neutrinos.

nuclear fission

In a *nuclear fission*, instead, it is exploited the fact that the binding energy per nucleon decreases for heavy nuclei. Thus it is also possible to convert mass into other type of energy by dividing a heavy nucleus because the products of the division will add a (negative) binding energy of larger absolute value. The nuclear fission of a heavy nucleus like uranium 235 can be induced by bombarding it with a neutron, which, having zero charge, does not need a high energy to approach the uranium nucleus; this fission releases an energy of around 200 MeV. The controlled nuclear fission is the source of the energy generated in nuclear power stations.

spontaneous decay

Radioactive nuclei are unstable compound systems which spontaneously decay in several ways: by fission, releasing a ^{4}He nucleus (α decay), or by emitting an electron and an antineutrino or a positron and a neutrino (β^- and β^+ decays), etc. As a consequence of the decay, there will be two or more particles in the final state. In the center-of-momentum frame the initial energy is equal to the particle rest energy $M\,c^2$ before decaying. Since energy is conserved, the sum of the masses of the decay products must be equal or lower than the decaying particle mass M; it will be equal when the products remain at rest in the center-of-momentum frame, but it will be lower when the products have kinetic energy or electromagnetic energy is radiated. For instance, when a neutron undergoes a β^- decay to a proton, an electron, and an antineutrino[14]

$$n \longrightarrow p + e^- + \bar{\nu}_e$$

the sum of the masses of proton, electron, and antineutrino is lower than the neutron mass. Since the electron mass and rest energy are

$$m_e = 9.10938 \times 10^{-31}\,\text{kg} \qquad m_e\,c^2 = 0.510999\,\text{MeV}$$

and the neutrino mass is practically zero ($m_{\nu e}\,c^2 < 2.3\,\text{eV}$), then a mass equivalent of around 0.783 MeV is converted into kinetic energy.[15]

---

[14] While energy and momentum conservation laws are universally fulfilled by any isolated system, because they result only from the properties of space-time, there exist besides other conservation laws that are specific of the interactions—conservation of electric charge, *baryon* number, *lepton* number, etc.—which also constraint the admissible final states. Neutron and proton are baryons; they have baryon number 1 and lepton number 0. Electron and neutrino are leptons; they have lepton number 1 and baryon number 0. The respective anti-particles have numbers of opposite sign.

[15] The neutrino existence was proposed by W. Pauli in 1930, to heal the apparent violation of conservation laws in beta decay. Pauli suspected that there was a particle taking part in the decay, whose detection was difficult. The first direct experimental evidence of neutrino existence happened in 1956.

Complement 6A: *Using conservation laws*

Problem 1. The reaction $\pi^+ + n \rightarrow \kappa^+ + \Lambda^\circ$, where the collision between a pion and a neutron creates a kaon and a *lambda* particle, requires a reactant energy exceeding a threshold energy. In fact: $m_{\pi+}c^2 = 140\,\text{MeV}$, $m_n c^2 = 940\,\text{MeV}$, $m_{\kappa+}c^2 = 494\,\text{MeV}$ and $m_{\Lambda^\circ}c^2 = 1116\,\text{MeV}$, so the sum of the product masses is higher than the sum of the reactant masses. In the frame where the neutron is at rest, calculate the minimum pion energy needed for the reaction to occur.

Solution: If the reaction happens with the minimum energy, then the products will remain at rest in the center-of-momentum frame. In such a case, in a different frame the products have the velocity U_c of the center-of-momentum frame; thus their energies are $E_i = m_i \gamma(U_c)\, c^2$. Therefore

$$\mathcal{E}_{\text{minimum}} = (E_\pi + E_n)_{\text{minimum}} = \gamma(U_c)(m_\kappa + m_\Lambda)c^2$$

In the frame where the neutron is at rest it is $U_c = Pc^2/\mathcal{E} = p_\pi c^2/(E_\pi + m_n c^2)$, and the former equation for $E_{\pi\ \text{minimum}}$ results to be

$$\begin{aligned}(m_\kappa + m_\Lambda)^2 c^4 =& (1 - U_c^2 c^{-2})(E_\pi + m_n c^2)^2 = (E_\pi + m_n c^2)^2 - p_\pi^2 c^2 \\ =& E_\pi^2 - p_\pi^2 c^2 + 2 m_n c^2 E_\pi + m_n^2 c^4 = m_\pi^2 c^4 + 2 m_n c^2 (T_\pi + m_\pi c^2) + m_n^2 c^4 \\ =& (m_\pi + m_n)^2 c^4 + 2 m_n c^2 T_\pi\end{aligned}$$

therefore

$$T_{\pi\ \text{minimum}} = \frac{(m_\kappa + m_\Lambda)^2 - (m_\pi + m_n)^2}{2\, m_n} c^2$$

Problem 2. In the frame where the neutron is at rest, the kaon of the former reaction is detected with energy E_κ and momentum $\mathbf{p}_\kappa$ perpendicular to $\mathbf{p}_\pi$. Which was the pion energy?

Solution: Being $\mathbf{p}_n = 0$, the conservation laws reduce to $\mathbf{p}_\pi = \mathbf{p}_\kappa + \mathbf{p}_\Lambda$, $E_\pi + m_n c^2 = E_\kappa + E_\Lambda$. In this kind of problems it is useful to eliminate from the equations the particle whose dynamics does not matter (we mean the *lambda* particle). Therefore, we write the energy–momentum invariant of that particle and then use the conservation laws:

$$\begin{aligned}m_\Lambda^2 c^4 =& E_\Lambda^2 - p_\Lambda^2 c^2 = (E_\pi + m_n c^2 - E_\kappa)^2 - (\mathbf{p}_\pi - \mathbf{p}_\kappa)\cdot(\mathbf{p}_\pi - \mathbf{p}_\kappa)c^2 \\ =& (E_\pi^2 - p_\pi^2 c^2) + (E_\kappa^2 - p_\kappa^2 c^2) + m_n c^2 (m_n c^2 - 2 E_\kappa) \\ & + 2 E_\pi (m_n c^2 - E_\kappa) + 2\, \mathbf{p}_\kappa \cdot \mathbf{p}_\pi\, c^2\end{aligned}$$

The two first terms are the invariants of the respective particles; the last one is zero by hypothesis. Finally E_π is solved:

$$E_\pi = \frac{(m_\Lambda^2 - m_\pi^2 - m_\kappa^2 - m_n^2)\, c^4 + 2 m_n c^2 E_\kappa}{2\,(m_n c^2 - E_\kappa)}$$

While mass can be converted into kinetic energy or electromagnetic radiation, there also exists the inverse phenomenon where part of the kinetic energy of the incoming particles is converted into mass of the products. For instance, if two protons collide with high enough kinetic energy, then a neutron-antineutron pair can be *created*

particle creation

$$p+p \longrightarrow p+p+n+\bar{n}$$

In order that this kind of reaction can happen, the colliding protons must have enough kinetic energy to transform it into the pair of neutrons to be created. Considering the reaction in the center-of-momentum frame S_c, where the total momentum is zero, it becomes clear that the *threshold energy* $\mathcal{E}_{c\ \text{threshold}}$—the minimum initial energy—required to produce the reaction is such that the products of the reaction remain at rest in the final state; in fact, this final state satisfies the momentum conservation. A larger value of initial energy will cause the increase of the kinetic energies of the products:

$$\mathcal{E}_{c\ \text{initial}} \geqslant \mathcal{E}_{c\ \text{threshold}} = \sum_{\text{products}} m_i\ c^2$$

In the former example, it is $\mathcal{E}_{c\ \text{initial}} = 2\ m_p c^2 + T_{c\ \text{initial}}$, where $T_{c\ \text{initial}}$ is the initial total kinetic energy in the center-of-momentum frame. Therefore, in order that the creation of the pair neutron–antineutron be possible, it is necessary that $T_{c\ \text{initial}} \geq 2\ m_n c^2 = 1879.13\,\text{MeV}$. So great kinetic energies are reached in particle accelerators, where the creation of particle pairs is a quotidian fact certifying the validity of relativistic dynamics.

According to transformation (6.18), the threshold energy in a frame different from the center-of-momentum frame is $\mathcal{E}_{\text{threshold}} = \gamma(U_c)\,\mathcal{E}_{c\ \text{threshold}}$, and is always greater than the threshold energy in the center-of-momentum frame. The reason for this property is traced to the fact that, in general, it is not possible to convert all the kinetic energy into mass, because a fraction of kinetic energy must be preserved to fulfill the momentum conservation. The value $\mathcal{E}_{\text{threshold}}$ can be directly computed in the frame S, by imposing that all the products of the reaction have the same velocity U_c (see Complement 6A).

6.7. CENTER OF INERTIA

The velocity $\mathbf{U}_c$ in Eq. (6.38) can be associated with the movement of a representative point of the system that will be called *center of inertia*. Let us consider a system of free particles in a frame S, and define

$$\mathbf{R}_S(t) \equiv \frac{\sum_i E_i(t)\,\mathbf{r}_i(t)}{\sum_i E_i(t)} \tag{6.39}$$

center of inertia in frame S

The velocity of the point whose position vector is $\mathbf{R}_S$ is obtained by differentiating with respect to t. If the system is isolated then the total energy will be constant, thus it is only necessary to differentiate the numerator in Eq. (6.39):

$$\frac{\mathrm{d}\mathbf{R}_S}{\mathrm{d}t} = \frac{\sum_i \left(E_i \frac{\mathrm{d}\mathbf{r}_i}{\mathrm{d}t} + \frac{\mathrm{d}E_i}{\mathrm{d}t}\mathbf{r}_i \right)}{\sum_i E_i}$$

Since the sole admissible interactions are local (collisions), then two particles can only exchange energy at the time when they share the same position. Thus, when the i particle collides with the j particle it is valid that: $\Delta E_i \mathbf{r}_i + \Delta E_j \mathbf{r}_j = (\Delta E_i + \Delta E_j)\mathbf{r}_i$. But $\Delta E_i + \Delta E_j$ must be zero in order to conserve the energy; then $\sum (\mathrm{d}E_i/\mathrm{d}t)\mathbf{r}_i = 0$. Therefore the velocity of the center of inertia is

$$\frac{\mathrm{d}\mathbf{R}_S}{\mathrm{d}t} = \frac{\sum_i E_i \frac{\mathrm{d}\mathbf{r}_i}{\mathrm{d}t}}{\sum_i E_i} = \frac{\sum_i E_i \mathbf{u}_i}{\sum_i E_i} = \frac{\sum_i c^2 \mathbf{p}_i}{\sum_i E_i} = \frac{c^2\,\mathbf{P}}{\varepsilon} = \mathbf{U}_\mathrm{c} \tag{6.40}$$

The center-of-inertia definition involves the simultaneous positions of all the particles. Since the simultaneity is not absolute in Relativity, the simultaneous events entering the computation of the center of inertia in a given frame are not recognized as simultaneous in another frame. As a consequence, the notion of center of inertia depends on the reference system: the centers of inertia relative to different frames trace, in general, different world lines. Nevertheless, these different world lines have to be parallel because the center of inertia—whatever the frame where it is computed—has to move with the velocity $\mathbf{U}_\mathrm{c}$ of the center-of-momentum frame. To illustrate these features of the notion of center of inertia, let us consider two equal particles in the center-of-momentum frame S_c where their velocities are equal and opposite. By choosing the y axis along the direction of the motion, and a suitable coordinate origin, the equations describing the particle movements in frame S_c turn out to be

$$x_1(t) = a \qquad\qquad x_2(t) = -a$$

$$y_1(t) = u\,t \qquad\qquad y_2(t) = -u\,t$$

Since the energies of both particles are equal, the center of inertia is at the point $\mathbf{R}_\mathrm{c}$=$(\mathbf{r}_1 + \mathbf{r}_2)/2$, which coincides with the coordinate origin at any time t. Let us regard the situation in a frame S' moving with respect to S_c with velocity V along the x axis. By using Lorentz transformations it is

$$a = \gamma(V)(x_1{}' + Vt_1{}') \quad \Rightarrow \quad x_1{}'(t') = a\,\gamma(V)^{-1} - V\,t'$$

$$t_1{}' = \gamma(V)(t_1 - Vc^{-2}x_1) = \gamma(V)(t_1 - Vc^{-2}a) \quad \Rightarrow \quad t_1 = \gamma(V)^{-1}\,t_1{}' + Vc^{-2}a$$

$$y_1{}' = y_1 = u\,t_1 \quad \Rightarrow \quad y_1{}'(t') = u\gamma(V)^{-1}\,t' + u\,Vc^{-2}a$$

The movement of particle 2 relative to S' is obtained from the previous equations by replacing a with $-a$ and u with $-u$:

$$x_2'(t') = -a\,\gamma(V)^{-1} - V\,t'$$
$$y_2'(t') = -u\,\gamma(V)^{-1}\,t' + uVc^{-2}a$$

The frame S' was chosen in such a way that the particle energies are still equal in $S' : E' = \gamma(V)(E - p_x) = \gamma(V)E$ for both particles. Thus, it remains valid that $\mathbf{R}'_{S'}=(\mathbf{r}_1' + \mathbf{r}_2')/2$. Then

$$X'_{S'}(t') = -V\,t'$$
$$Y'_{S'}(t') = uVc^{-2}a$$

While in the center-of-momentum frame the center of inertia is at rest at the coordinate origin, $x_c(t) = 0 = y_c(t)$, in frame S' the center of inertia $\mathbf{R}'_{S'}$ moves with velocity $-V\hat{\mathbf{x}}$ (as it was expected), but its world line is displaced along the direction of the y coordinate. This displacement seems to be essentially related to the existence of a non-null intrinsic angular momentum ($a \neq 0$) in the examined system of particles. The successive positions of the center of inertia $\mathbf{R}'_{S'}$ in S' do not then result from applying a Lorentz transformation to the position of the center of inertia $\mathbf{R}_c$ in S_c. In particular, $Y_c = 0$, while $Y'_{S'} \neq 0$. This dependence of the notion of center of inertia on the reference system employed could be already noticed in definition (6.39), where it is evident that the components of vector $\mathbf{R}_S$ do not transform as particle coordinates.

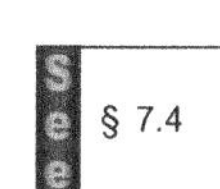

6.8. ELASTIC COLLISIONS

A collision in Relativity is said to be *elastic* if the identities of the particles do not undergo changes as a result of the collision. This means that the physical system consists of the same set of particles before and after the collision; so their masses do not undergo changes at all.[16] Since the rest energies $m_i c^2$ are not modified by the collision, then the conservation of the total relativistic energy implies the conservation of the total kinetic energy (see the definition of kinetic energy in Eq. (6.14)).

[16] If the particle has an internal structure, the internal energy associated with this structure should not change in order that the particle mass be the same before and after the collision.

Let us consider two particles with masses m_1 and m_2 elastically colliding. The magnitude $\mathcal{E}^2 - P^2c^2$ is not only invariant but a conserved magnitude which has the value

$$\begin{aligned}\mathcal{E}^2 - P^2c^2 &= (E_1 + E_2)^2 - (\mathbf{p}_1 + \mathbf{p}_2)\cdot(\mathbf{p}_1 + \mathbf{p}_2)\,c^2 \\ &= ({E_1}^2 - {p_1}^2c^2) + ({E_2}^2 - {p_2}^2c^2) + 2\,(E_1E_2 - \mathbf{p}_1\cdot\mathbf{p}_2\,c^2) \\ &= {m_1}^2c^4 + {m_2}^2c^4 + 2\,(E_1E_2 - \mathbf{p}_1\cdot\mathbf{p}_2\,c^2)\end{aligned} \tag{6.41}$$

In the last member of Eq. (6.41) the two first terms are themselves invariant. Therefore it is concluded that the last term, $E_1E_2 - \mathbf{p}_1 \cdot \mathbf{p}_2c^2$, must also be invariant. This conclusion is valid whatever the type of collision between two particles is. In the case of an elastic collision, the identities of the particles do not undergo changes; then the term ${m_1}^2c^4 + {m_2}^2c^4$ in the conserved magnitude (6.41) will be the same before and after the collision. Therefore, if the collision is elastic, then the magnitude $E_1E_2 - \mathbf{p}_1\cdot\mathbf{p}_2c^2$ is not only invariant but conserved. In a frame where one of the two particles is at rest, the invariant reduces to $E_1E_2 = m_1\ m_2c^4(1 - u^2/c^2)^{-1/2}$, where u is the velocity of the other particle. It is concluded that, in an elastic collision, the modulus of the relative velocity between the particles does not change.

Our aim is to establish how the conservation of total energy and momentum, $\mathbf{P}$ and $\mathcal{E}$, constrains the collision outcome. We shall assume that the values of $\mathbf{P}$ and $\mathcal{E}$ are known. The definitions of $\mathbf{P}$ and $\mathcal{E}$

$$\mathbf{P} = \mathbf{p}_1 + \mathbf{p}_2 \tag{6.42}$$

$$\mathcal{E} = E_1 + E_2 \tag{6.43}$$

are then equations to obtain information about the individual magnitudes $\mathbf{p}_1$ and $\mathbf{p}_2$ (the energies E_1 and E_2 are functions of momenta and masses) both before and after the collision. If the values of $\mathbf{P}$ and $\mathcal{E}$ come from the knowledge of the state of the system before the collision, then Eqs. (6.42–6.43) will be regarded as equations for the final values of the individual magnitudes. While the vector equation (6.42) gives the value of $\mathbf{p}_2$ in case that $\mathbf{p}_1$ were already known, Eq. (6.43) adds some information about $\mathbf{p}_1$. Since $\mathbf{p}_1$ is a vector, a sole equation is not enough to determine its modulus and direction. The conservation laws impose relations among modules and directions. But the precise values of $\mathbf{p}_1$ and $\mathbf{p}_2$ cannot be completely determined by the conservation laws: they are subjected to interaction details.

Our purpose is to eliminate magnitudes E_1, $\mathbf{p}_2$, and E_2 in Eqs. (6.42–6.43) to obtain an equation for $\mathbf{p}_1$. The general technique for eliminating magnitudes associated with one of the particles consists in writing its energy–momentum invariant and replacing the involved magnitudes by means of the conservation laws. In this case we shall eliminate $\mathbf{p}_2$ and E_2:

$$\mathbf{p}_2 = \mathbf{P} - \mathbf{p}_1 \qquad E_2 = \mathcal{E} - E_1$$

Therefore

$$m_2^2 c^4 = E_2^2 - p_2^2 c^2 = (\mathcal{E} - E_1)^2 - (\mathbf{P} - \mathbf{p}_1) \cdot (\mathbf{P} - \mathbf{p}_1) c^2$$
$$= \mathcal{E}^2 - 2\mathcal{E}E_1 + E_1^2 - p_1^2 c^2 - P^2 c^2 + 2\mathbf{P} \cdot \mathbf{p}_1 c^2$$

Finally we make use of the invariant energy–momentum of the particle 1 to eliminate E_1 in favor of $\mathbf{p}_1$:

$$m_2^2 c^4 = \mathcal{E}^2 - 2\mathcal{E}\sqrt{p_1^2 c^2 + m_1^2 c^4} + m_1^2 c^4 - P^2 c^2 + 2\mathbf{P} \cdot \mathbf{p}_1 c^2$$

Solving the term with the squared root, it is obtained

$$4\mathcal{E}^2 \left(p_1^2 c^2 + m_1^2 c^4\right) = \left[\mathcal{E}^2 - P^2 c^2 - m_2^2 c^4 + m_1^2 c^4 + 2\mathbf{P} \cdot \mathbf{p}_1 c^2\right]^2 \qquad (6.44)$$

Equation (6.44) is valid whenever a physical system having total energy $\mathcal{E}$ and total momentum $\mathbf{P}$ can be regarded as constituted by two sub-systems having masses m_1 and m_2. In the particular case of an elastic collision between two particles, Eq. (6.44) is valid before and after the collision, since the particles composing the system keep their identities.

We shall apply Eq. (6.44) to the final state of an elastic collision between two particles. Although Eq. (6.44) is valid in any frame, there are two types of reference systems of special interest: the frames where one of the particles is initially at rest and the center-of-momentum frame. We shall begin by using the reference system where m_1—the particle whose final momentum is the unknown quantity in Eq. (6.44)—is initially at rest. Thus the (conserved) values of $\mathcal{E}$ and $\mathbf{P}$ in this particular frame are

$$\mathbf{P} = \mathbf{p}_{2\ \text{initial}} \qquad \mathcal{E} = E_{1\ \text{initial}} + E_{2\ \text{initial}} = m_1 c^2 + \sqrt{P^2 c^2 + m_2^2 c^4}$$

By solving the radicand, it results

$$\begin{aligned} \mathcal{E}^2 - P^2 c^2 &= m_2^2 c^4 + m_1^2 c^4 + 2 m_1 c^2 E_{2\ \text{initial}} \\ &= m_2^2 c^4 - m_1^2 c^4 + 2 m_1 c^2 \mathcal{E} \end{aligned} \qquad (6.45)$$

so the substitution in Eq. (6.44) yields

$$\mathcal{E}^2 \left(p_1^2 c^2 + m_1^2 c^4\right) = \left[m_1 c^2 \mathcal{E} + \mathbf{P} \cdot \mathbf{p}_1 c^2\right]^2$$

i.e.,

$$\mathcal{E}^2 p_1^2 = 2\mathcal{E} m_1 c^2 \mathbf{P} \cdot \mathbf{p}_1 + c^2 (\mathbf{P} \cdot \mathbf{p}_1)^2 \qquad (6.46)$$

$\mathcal{E}$ and $\mathbf{P}$ conservation constrains the final value of $\mathbf{p}_1$

Equation (6.46) does not give the value of $\mathbf{p}_1$ but expresses a relation between its modulus and its direction with respect to $\mathbf{P}$, which is determined by the conservation laws for $\mathcal{E}$ and $\mathbf{P}$. The classical limit of Eq. (6.46) corresponds to the case $Pc << \mathcal{E}$, $p_1 c << E_1$. Thus, the last term in Eq. (6.46) can be neglected, and the energy can be replaced by $\varepsilon \cong (m_1 + m_2)c^2$. In this way the conservation of classical kinetic energy is obtained:

$$(m_1 + m_2)\, p_1{}^2 = 2\, m_1\, \mathbf{P} \cdot \mathbf{p}_1$$

$$\Rightarrow \frac{p_1{}^2}{2m_1} + \frac{p_2{}^2}{2m_2} = \frac{p_1{}^2}{2m_1} + \frac{(\mathbf{P} - \mathbf{p}_1)^2}{2m_2} = \frac{P^2}{2m_2} = T_{\text{initial}}$$

Equation (6.46) could also be reached starting from considerations performed in the center-of-momentum frame S_c. In S_c the momenta $\mathbf{p}_{1c}$ and $\mathbf{p}_{2c}$ are equal and opposite before and after the collision, since $\mathbf{P}_c$ must remain null all the time:

$$\mathbf{p}_{1_c} = -\mathbf{p}_{2_c}$$

Moreover, since the energy is conserved, the collision cannot modify the (equal) modules of these vectors. Therefore, the only admissible change in S_c, as a consequence of the collision, is a rotation of vectors $\mathbf{p}_{1c}$ and $\mathbf{p}_{2c}$ by an angle χ (see Figure 6.1). Conservation theorems do not determine the angle χ, but χ depends on the interaction details. Since particle m_1 is initially at rest in frame S, then its velocity in frame S_c is $\mathbf{u}_{1c} = -\mathbf{U}_c$, $\mathbf{U}_c = \mathbf{P}\, c^2/\varepsilon$ being the velocity of S_c relative to S (see Eq. (6.38)). Therefore

$$\mathbf{p}_{1_c\ \text{initial}} = -m_1\, \gamma(U_c)\, \mathbf{U}_c \qquad E_{1_c\ \text{initial}} = m_1\, \gamma(U_c)\, c^2$$

After the collision, vector $\mathbf{p}_{1c}$ has the same modulus but a different direction:

$$\mathbf{p}_{1_c} = m_1\, \gamma(U_c)\, U_c\, (\cos\chi\; \hat{\mathbf{x}} + \sin\chi\; \hat{\mathbf{y}}) \qquad E_{1_c} = m_1\, \gamma(U_c)\, c^2 \tag{6.47}$$

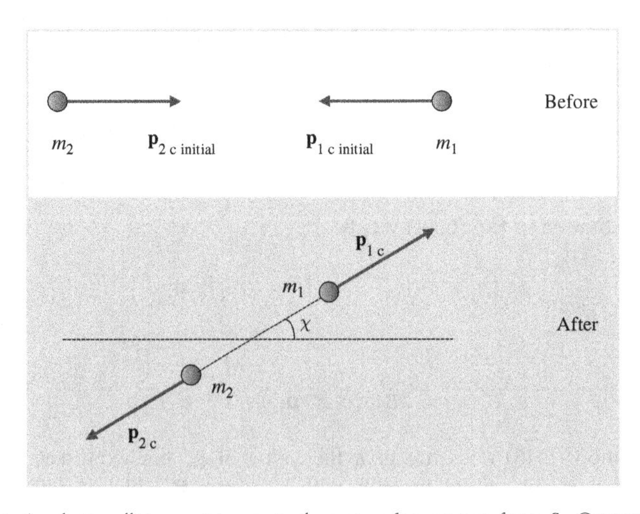

Figure 6.1. An elastic collision as it is seen in the center-of-momentum frame S_c. Conservation laws just allow that the (equal and opposite) momenta change their direction without changing modulus.

where $\hat{\mathbf{x}}$ has the direction of $\mathbf{U}_c$ (i.e., the direction of $\mathbf{P}$). In the reference system S it is

$$p_{1x} = \gamma(U_c)\,(p_{1_c}\cos\chi + U_c\,c^{-2}\,E_{1_c}) = m_1\,\gamma(U_c)^2\,U_c(\cos\chi + 1)$$
$$p_{1y} = p_{1_c}\sin\chi = m_1\,\gamma(U_c)\,U_c\sin\chi \tag{6.48}$$

So, χ can be eliminated by solving the trigonometric functions and using the identity $\cos^2\chi + \sin^2\chi = 1$. In this way a relation between the components of $\mathbf{p}_1$ is obtained:

$$\left[\frac{p_{1x}}{m_1\,\gamma(U_c)^2\,U_c} - 1\right]^2 + \left[\frac{p_{1y}}{m_1\,\gamma(U_c)\,U_c}\right]^2 = 1$$

$\mathbf{p}_{1\,\text{final}}$ *is inscribed in an ellipse*

which is the equation of an ellipse going through the origin (see Figure 6.2):[17]

$$\frac{(p_{1x} - a)^2}{a^2} + \frac{{p_{1y}}^2}{b^2} = 1 \tag{6.49}$$

where

$$a = m_1\,\gamma(U_c)^2\,U_c = m_1\frac{\dfrac{Pc^2}{\mathcal{E}}}{1 - \dfrac{P^2c^2}{\mathcal{E}^2}} = \frac{m_1\mathcal{E}\,P\,c^2}{\mathcal{E}^2 - P^2c^2}$$

$$b = m_1\,\gamma(U_c)\,U_c = \frac{m_1\,P\,c^2}{\sqrt{\mathcal{E}^2 - P^2c^2}} \tag{6.50}$$

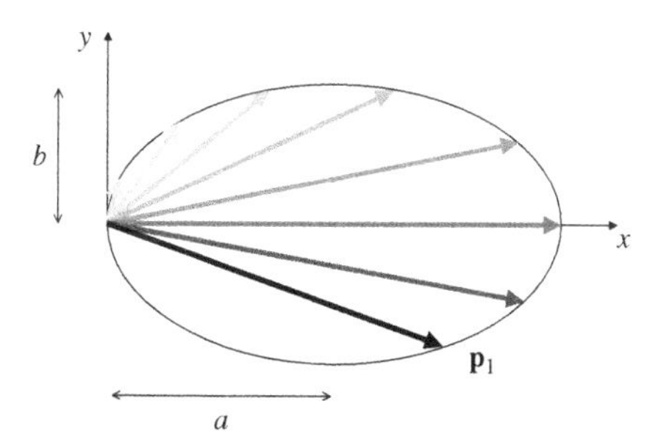

Figure 6.2. In an elastic collision, the momentum $\mathbf{p}_1$ acquired by the (initially at rest) target particle is inscribed in an ellipse defined by the projectile initial velocity and the masses of both particles.

[17] The solution $p_{1x} = 0 = p_{1y}$ (i.e., $\chi = \pi$) corresponds to missing the shot: projectile m_2 does not hit the target m_1; final and initial states are equal (remember that equations are accomplished by both the initial and final states).

The substitution of these a and b values in Eq. (6.51) leads to obtain the result (6.46) again.

We shall make a graphic representation of the result (6.49) (which is equivalent to the relation (6.46)). According to Eq. (6.49), conservation theorems constrain the final momentum of the target particle in such a way that $\mathbf{p}_1$ results to be a vector inscribed in an ellipse. Equation (6.50) exhibits the ellipse semi-axes; notice that $a/b = \gamma(U_c) > 1$. Taking into account Eq. (6.45), it is

$$\begin{array}{ll} a = P/2 & \text{if} \quad m_1 = m_2 \\ a > P/2 & \text{if} \quad m_1 > m_2 \\ a < P/2 & \text{if} \quad m_1 < m_2 \end{array}$$

Figure 6.3 shows these alternatives. Vector **P** has been included for comparing with the length of the major axis; besides, Figure 6.3 displays the projectile final momentum $\mathbf{p}_2 = \mathbf{P} - \mathbf{p}_1$. If the target mass m_1 is larger than the projectile mass m_2 , then the projectile can move back or forward after the collision, depending on whether the collision is or not head-on enough (see Figure 6.4). Instead, if $m_1 < m_2$ the projectile always moves forward after the collision. If $m_1 = m_2$ and the collision is exact head-on, then the projectile stops because the target takes all its momentum. In the three cases the maximum momentum that can be acquired by the target m_1 after the collision is equal to the ellipse major axis, which happens in a head-on collision:

$$\mathbf{p}_{1_{max}} = 2a = 2m_1\, \gamma(U_c)^2\, \mathbf{U}_c$$

In this case, the final energy of particle m_1 is

$$E_{1\,max} = \sqrt{p_{1\,max}{}^2 + m_1{}^2 c^4} = m_1 c^2 \sqrt{\frac{4\, U_c{}^2\, c^{-2}}{(1 - U_c{}^2\, c^{-2})^2} + 1} = m_1 c^2 \frac{1 + U_c{}^2\, c^{-2}}{1 - U_c{}^2\, c^{-2}}$$

$$= m_1 c^2 \frac{\mathcal{E}^2 + P^2 c^2}{\mathcal{E}^2 - P^2 c^2} = m_1 c^2 + \frac{2\, m_1 c^2\, P^2\, c^2}{\mathcal{E}^2 - P^2 c^2}$$

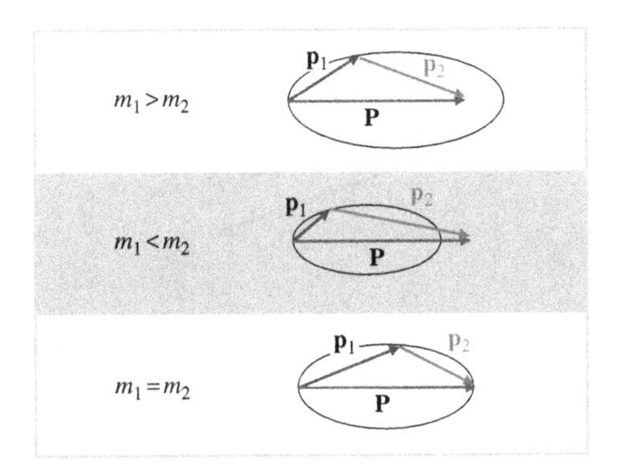

Figure 6.3. The relation between the ellipse major semi-axis and the projectile initial momentum **P** depends on the comparison between projectile mass m_2 and target mass m_1.

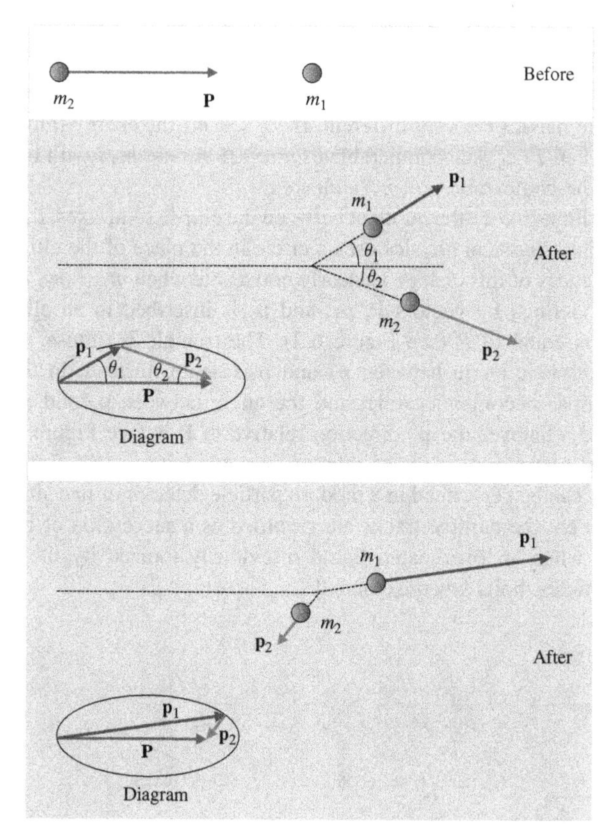

Figure 6.4. If the target mass is larger than the projectile mass, then the projectile can move back or forward after the collision.

The second term in the last member is the maximum kinetic energy that can be acquired by the target. Using Eq. (6.45), the maximum kinetic energy transfer from the projectile to the target results

$$\frac{T_{1\,\max}}{T_{2\ \text{initial}}} = \frac{4\dfrac{m_2}{m_1} + 2\dfrac{T_{2\ \text{initial}}}{m_1 c^2}}{\left(1+\dfrac{m_2}{m_1}\right)^2 + 2\dfrac{T_{2\ \text{initial}}}{m_1 c^2}} \tag{6.51}$$

maximum kinetic energy transfer

Taking the limit $c \to \infty$ (i.e., neglecting the initial kinetic energy) the well-known classical result is recovered:

$$\frac{T_{1\,\max}}{T_{2\ \text{initial}}} \underset{c\to\infty}{\longrightarrow} \frac{4\dfrac{m_2}{m_1}}{\left(1+\dfrac{m_2}{m_1}\right)^2} \tag{6.52}$$

In the classical case (6.52) the maximum kinetic energy transfer depends just on the ratio of masses (in particular, if $m_2 << m_1$ or $m_1 << m_2$ the energy transfer is poor). Instead, the relativistic energy transfer can be significative even if the particle masses are very different. If $m_2 << m_1$ the energy transfer will be important when $T_{2\ \text{initial}}$ be comparable to $m_1 c^2$. If $m_1 << m_2$ it will be necessary that $T_{2\ \text{initial}}$ be comparable to $(m_2/m_1) m_2 c^2$.

the angle between particle directions differs from the classical one

In the limit $c \to \infty$ the quotient between the ellipse semi-axes, $a/b = \gamma(U_c)$, goes to 1. Thus classical physics uses a circle in the place of the ellipse. One of the consequences of this change is clearly noticeable when $m_1 = m_2$. In this case, the triangle defined by vectors **P**, $\mathbf{p}_1$, and $\mathbf{p}_2$ is inscribed in an ellipse whose major axis is equal to P (see Figure 6.3). The triangle is obtuse, so showing that the relativistic angle between $\mathbf{p}_1$ and $\mathbf{p}_2$is acute. Instead, in the classical limit the ellipse becomes a circle, and the angle between $\mathbf{p}_1$ and $\mathbf{p}_2$ becomes a right angle whatever the $\mathbf{p}_1$ direction relative to **P** is (see Figure 6.5).[18] The experimental corroboration of the relativistic angle between the final directions of $\mathbf{p}_1$ and $\mathbf{p}_2$ can be performed in a modern particle detector or in a simple *bubble chamber*, where the particle tracks are recorded as a succession of bubbles that are formed when an overheated liquid, previously ionized by the passing of charged particles, boils when expanded.

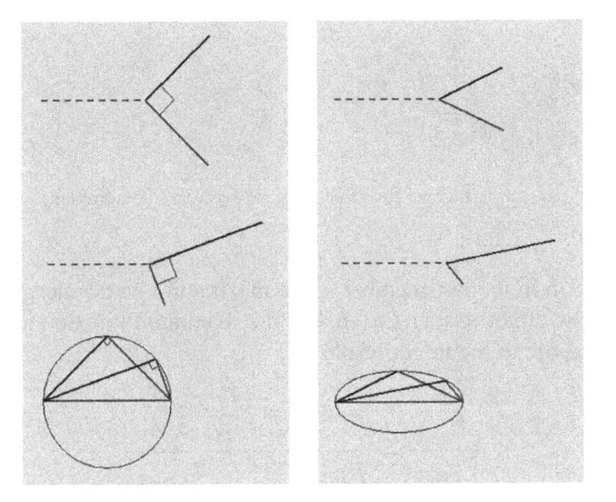

Figure 6.5. On the left, the classical elastic collision between equal particles in the frame where the target is initially at rest. On the right, the relativistic result for the same situation: after the collision the angle between the directions of motion is lower than 90°.

[18] If the particles have the same mass, the conservation of classical kinetic energy leads to the equation $P^2 = {p_1}^2 + {p_2}^2$, which means that the triangle whose sides are the vectors **P**, $\mathbf{p}_1$, and $\mathbf{p}_2$ is a right-angle triangle.

If θ_1, θ_2 are the angles between the $\mathbf{p}_1$, $\mathbf{p}_2$ directions and $\mathbf{P}$ direction (see Figure 6.4), then, according to Eq. (6.48), it is

$$\mathrm{tg}\theta_1 = \left|\frac{p_{1y}}{p_{1x}}\right| = \frac{\sin\chi}{\gamma(U_c)(\cos\chi + 1)} \tag{6.53}$$

since $\mathbf{p}_2 = \mathbf{P} - \mathbf{p}_1$, it is

$$\mathrm{tg}\theta_2 = \left|\frac{p_{2\,y}}{p_{2\,x}}\right| = \left|\frac{p_{1\,y}}{P - p_{1\,x}}\right| \tag{6.54}$$

If both particles are equal, their energies will be equal in the center-of-momentum frame (because they have equal and opposite velocities). According to Eq. (6.47) it is $\mathcal{E}_c = 2E_{1c} = 2m_1\gamma(U_c)c^2$. Then $\mathbf{P} = \gamma(U_c)c^{-2}\mathcal{E}_c\mathbf{U}_c = 2\gamma(U_c)^2 m_1 \mathbf{U}_c$. Substituting in Eq. (6.54) it results

$$\mathrm{tg}\theta_2 = \left|\frac{p_{1\,y}}{P - p_{1\,x}}\right| = \frac{\sin\chi}{\gamma(U_c)(-\cos\chi + 1)}$$

Note that $\mathrm{tg}\theta_1\ \mathrm{tg}\theta_2 = \gamma(U_c)^{-2}$. So, in the reference system where one of the (equal) particles is initially at rest, the (acute) angle $\theta_1 + \theta_2$ between $\mathbf{p}_1$ and $\mathbf{p}_2$ is

$$\mathrm{tg}(\theta_1 + \theta_2) = \frac{\mathrm{tg}\theta_1 + \mathrm{tg}\theta_2}{1 - \mathrm{tg}\theta_1 \mathrm{tg}\theta_2} = \frac{2c^2}{U_c^2\,\gamma(U_c)\sin\chi}$$

The minimum value of $\theta_1 + \theta_2$ occurs when $\chi = 90°$. In this case it is $\theta_1 = \theta_2$ (see in Figure 6.5 the example on the right at the top), and the value is

$$\mathrm{tg}\theta_1 = \gamma(U_c)^{-1} = \mathrm{tg}\theta_2$$

6.9. INTERACTION BETWEEN ELECTROMAGNETIC RADIATION AND MATTER

At the beginning of the twentieth century it became apparent that electromagnetic radiation transfers energy to matter in a discontinuous way. In its 1905 article on the *photoelectric effect*, Einstein (1905a) explained the characteristics of the phenomenon by proposing *light quanta* (packets of energy $h\nu$, ν being the light wave frequency[19]) as the basic units for the energy exchange between the electromagnetic radiation and matter. In photoelectric effect, discovered by H. Hertz in 1887, the light extracts electrons from a metal surface. The wave model of light suggests that the larger the luminous intensity, the larger the kinetic energy of *photoelectrons*; but in practice the increase of intensity has

photoelectric effect

[19] $h = 6.626086 \times 10^{-34}$ J s $= 4.135667 \times 10^{-15}$ eV s is Planck constant.

the only effect of rising the number of photoelectrons without modifying their kinetic energies. On the other hand, the maximum kinetic energy of photoelectrons linearly increases with the light frequency, but the phenomenon stops if the light frequency becomes lower than a certain threshold frequency, which is different for each metal. These features of photoelectric effect cannot be understood within the context of the wave model of light. However, the phenomenon can be explained by assuming that light transfers energy to matter by means of quanta whose energy is proportional to the frequency of the light wave. In this case, the phenomenon will be described as the interaction of a quantum with an electron. If the quantum energy is high enough (i.e., if light frequency is larger than a threshold frequency that depends on the metal properties), then the electron is released. The photoelectron kinetic energy is equal to the energy $h\nu$ carried by the absorbed quantum, minus the energy employed in releasing the photoelectron (metal *work function*[20]). In this way the photoelectron kinetic energy linearly increases with frequency. A larger luminous intensity means a larger number of light quanta, then a larger quantity of extracted photoelectrons.

photon

At first, Einstein proposed the energy quantum without introducing a particle provided with momentum. Einstein seemed to resist this idea, but in 1917 he convinced himself—starting from a mechanical-statistical reasoning—about the necessity that the quantum of energy $E = h\ \nu$ be provided with the momentum $\mathbf{p} = h\ \nu\ c^{-1}\hat{\mathbf{n}} = h\ \lambda^{-1}\hat{\mathbf{n}}$ (Einstein, 1917a). The rest of the scientific community remained skeptic about the reality of the quantum of electromagnetic radiation until A.H. Compton experimented in 1923 with *X-rays* scattered by quasi-free electrons in graphite. He concluded that "The experimental support of the theory indicated very convincingly that the radiation quantum carries with it directed momentum as well as energy."[21] In 1924, Einstein stated that "Compton's experiment proves that radiation behaves—not only with respect to energy transfer, but also with respect to collision interactions—as if it consisted of projectiles of energy," and, alluding to the coexistence of the wave model with this renovated corpuscular model: "There are therefore now two theories of light, both indispensable, and... without any logical connection." In 1926, G.N. Lewis called the light quantum *photon*.

On the other hand, in 1924, L. de Broglie proposed that the necessity of both the corpuscular model and the wave model to describe the totality of luminous phenomena—the *wave-particle duality*—should also be a characteristic of the behavior of material particles. According to de Broglie, a particle of mass m is associated with a frequency ν and a wavelength λ such that $E = h\ \nu$ and $\mathbf{p} = h\ \lambda^{-1}\hat{\mathbf{e}}$. The relation (6.25) between E and p so becomes

---

[20] The alkali metals (Li, Na, K, Rb, Cs) have little bound valence electrons; nearly 2 eV are required to release them (the typical energy of visible light quanta). In other metals, ultraviolet light has to be employed in order to produce the photoelectric effect.

[21] J. Stark had already used the concept in 1909.

a dispersion relation: $\nu(\lambda) = (c^2\,\lambda^{-2} + h^{-2}\,m^2\,c^4)^{1/2}$. The group velocity $u \equiv \mathrm{d}\nu/\mathrm{d}\lambda^{-1} = \mathrm{d}E/\mathrm{d}p$ corresponds to the particle velocity (see Eq. (6.26)) and results to be equal to $c^2/(\lambda\nu)$ or c^2/w. As it was remarked in §3.17, this dispersion relation guarantees that the wave propagation direction transforms in the same way as the ray direction.[22] Afterward quantum theory conciliated the corpuscular and wave facets. Although Einstein agreed that quantum theory results are clearly satisfactory, he refused to consider the theory as a definitive description of physical reality.

The photon energy $E = h\nu$, together with its momentum $\mathbf{p} = h\,\nu\,c^{-1}\hat{\mathbf{n}}$, lead to the relation

$$\mathbf{p}_{\text{photon}} = E_{\text{photon}}\;c^{-1}\hat{\mathbf{n}} \tag{6.55}$$

which is characteristic of a particle having the speed of light (see Eq. (6.15)). In this case the energy–momentum invariant is null:

$$E^2_{\text{photon}} - p^2_{\text{photon}}c^2 = 0 \tag{6.56}$$

which means that the particle has zero mass. The dispersion relation (6.55) leads to the following energy transformation:

$$E' = \gamma(V)(E - V\,p_x) = \gamma(V)(E - \mathbf{V}\cdot\mathbf{p}) = \gamma(V)(1 - c^{-1}\,\mathbf{V}\cdot\hat{\mathbf{n}})E$$

which is equal to the transformation (3.58a) for the frequency of a light plane wave. This means that the proportionality between photon energy and frequency is consistent with Lorentz transformations.

Compton effect

The above-mentioned *Compton effect* is produced in the elastic scattering of a photon by a free electron. The photon energy (frequency) after scattering is smaller than the initial one. In the reference system where the electron is initially at rest, the initial energy and momentum are

$$\mathcal{E}_{\text{initial}} = h\nu_{\text{i}} + m_e c^2 \qquad \mathbf{P}_{\text{initial}} = h\nu_{\text{i}}\,c^{-1}\,\hat{\mathbf{n}}_{\text{i}}$$

After the photon is scattered, the respective final values are

$$\mathcal{E}_{\text{final}} = h\nu_{\text{f}} + E_{\text{e}} \qquad \mathbf{P}_{\text{final}} = h\nu_{\text{f}}\,c^{-1}\,\hat{\mathbf{n}}_{\text{f}} + \mathbf{p}_{\text{e}}$$

where E_e and $\mathbf{p}_e$ correspond to the final electron energy and momentum. We shall obtain information about the final photon energy (frequency) using the energy and momentum conservation laws. With the aim of eliminating the unknown

[22] In the proper frame of a free particle, the de Broglie's "pilot-wave" is nothing but an oscillation in phase in all the space. In fact, the pilot-wave phase is $2\pi\nu(t - w^{-1}\,x)$; by transforming it to the frame S' that moves with the particle velocity $u = c^2\,w^{-1}$ it results $2\pi\nu(t - c^{-2}u\,x) = 2\pi\nu\gamma(u)^{-1}t'$. Thus the oscillation frequency is $\nu_o = \nu\gamma(u)^{-1}$ (which coincides with the relation between E_o and E).

E_e and $\mathbf{p}_e$, we solve them from the conservation laws and then we form the electron energy-momentum invariant:

$$E_e = h\nu_i + m_e c^2 - h\nu_f \qquad \mathbf{p}_e = h\nu_i\, c^{-1}\,\hat{\mathbf{n}}_i - h\nu_f\, c^{-1}\,\hat{\mathbf{n}}_f$$

$$\begin{aligned} \Rightarrow \quad m_e{}^2 c^4 &= E_e{}^2 - p_e{}^2 c^2 \\ &= (h\nu_i + m_e c^2 - h\nu_f)^2 - h^2(\nu_i\,\hat{\mathbf{n}}_i - \nu_f\,\hat{\mathbf{n}}_f)\cdot(\nu_i\,\hat{\mathbf{n}}_i - \nu_f\,\hat{\mathbf{n}}_f) \\ &= m_e{}^2 c^4 + 2\,h(\nu_i m_e c^2 - h\nu_i\nu_f - m_e c^2\ \nu_f + h\nu_i\,\nu_f\hat{\mathbf{n}}_i\cdot\hat{\mathbf{n}}_f) \end{aligned}$$

Calling ϕ the angle between the initial and final propagation directions, and dividing the former equation by $\nu_i\,\nu_f$, it results

$$\frac{1}{h\nu_f} - \frac{1}{h\nu_i} = \frac{1}{m_e c^2}(1-\cos\phi) \quad \text{or} \quad \lambda_f - \lambda_i = \frac{h}{m_e c}(1-\cos\phi) \qquad (6.57)$$

Conservation laws just establish a relation between energy and propagation direction of the scattered photon. For a given final propagation direction ϕ, Eq. (6.57) indicates which is the corresponding frequency; then $\mathbf{p}_e$ will result from the conservation of **P**. In order to know the probability distribution for the different final propagation directions as a function of the initial photon energy, and the importance of Compton effect relative to other phenomena associated with the interaction between photons and electrons, it is necessary to turn to *quantum electrodynamics* which is the theory describing the interaction.

Equation (6.57) shows that $\Delta\nu/\nu$ is meaningful whenever the photon energy is at least comparable with the electron rest energy $m_e c^2$, which is equal to 0.511 MeV. In other words, the radiation wavelength should be comparable or smaller than the electron *Compton wavelength*, $\lambda_C \equiv h/(m_e c) = 2.43 \times 10^{-12}$ m. In the case of visible light, the photon energy is too small; therefore its frequency does not undergo an apparent change.[23]

As remarked in §6.8, the invariant magnitude $E_1\,E_2 - \mathbf{p}_1\cdot\mathbf{p}_2 c^2$ is conserved in an elastic collision between two particles (see Eq. (6.41)). In the reference system where the electron is initially at rest, the invariant is equal to $E_{\text{photon}}\ m_e c^2$. The conservation of the invariant in Compton effect means that the electron "sees" the photon with the same energy (frequency) before and after the scattering. This property can be verified by transforming the frequency ν_f, given by Eq. (6.57), to the electron proper frame S' and obtaining $\nu_f{}' = \nu_i$. Notice that $\nu_f{}' > \nu_f$ even if $\mathbf{u}_f\cdot\hat{\mathbf{n}}_f > 0$, which corresponds to a typical relativistic effect of the frequency transformation (3.56a).

[23] The low energy limit $h\nu \ll m_e c^2$, in which the photon frequency does not appreciably change after the scattering, is called *Thomson scattering*. J.J. Thomson considered, within the context of the wave model, that the electron oscillates when the wave passes, so radiating energy with the same frequency.

According to quantum theory, the electrons in an atom are organized in discontinuous orbitals, so providing the atom internal energy with a discrete set of feasible values. Because of this reason, the photons absorbed or emitted by an atom have characteristic energies that are the atom fingerprint and form the absorption and emission atomic spectral lines. When a photon interacts with an isolated atom, different processes can occur. If the photon energy is low, and not sufficient to produce an electronic transition between two atomic energy levels, then the electron will be elastically scattered by the atom without appreciable change of frequency since the energy taken by the atom can be neglected (*Rayleigh scattering*).[24] If the photon energy is sufficient to produce an electronic transition between two atomic energy levels, then the electron can be inelastically scattered; the atom undergoes a transition to an "excited" quantum state, and the photon energy (its frequency) will diminish to compensate the change of the atomic internal energy (*Raman scattering*). If the photon energy is similar to the difference between two levels of the atomic internal energy, then the photon can be *resonantly* absorbed by the atom; in this way the atom will go to an excited state from which it will decay later by emitting photons. If the photon energy exceeds the atom ionization energy then a *photoionization* may happen: the photon is absorbed, and an electron is pulled out of the atom structure and released. The necessary energy to ionize an atom varies from 4 or 5 eV for alkali elements up to 25 eV for helium, so it belongs to the range of *ultraviolet radiation* (which goes up to about 200 eV).

Photoionization (or photoelectric effect in a wide sense) is the main way of interaction of electromagnetic radiation and matter for ultraviolet light and X-rays of energy no larger than 100 keV, but its importance decreases for larger frequencies and it eventually disappears. In fact, the electron is seen as essentially free if $h\nu$ is much larger than its bound energy: a free electron cannot absorb a light quantum because the energy conservation would be violated (the electron does not have internal degrees of freedom to absorb the light quantum energy by modifying its rest energy; so in the center-of-momentum frame the final state would result to be incompatible with the energy conservation). At energies $h\nu$ of the order of 1 MeV, Compton effect becomes the dominant way of attenuation of electromagnetic radiation in matter. For even greater energies, the phenomenon of *pair creation* appears: a photon of energy greater than 1.022 MeV has enough energy to create an electron–positron pair, because the rest energy of the electron and its anti-particle is 0.511 MeV.[25] In spite of this fact, the pair creation by a unique photon is not possible because it would violate the momentum conservation; in fact the electron–positron pair has a center-of-momentum frame, while the momentum of a unique photon is non-null in any reference system. Therefore, at least two photons having different propagation directions are required to create

pair creation

[24] Since the photon wavelength is larger than the atom size - and, therefore, much larger than the electron Compton wavelength—it can be said that the atom electrons oscillate in phase and radiate with the same frequency than the incident wave.

[25] Particle–antiparticle pair creation conserves electric charge, lepton number, etc.

a pair; both should add the necessary threshold energy. Nevertheless, in the presence of matter, a unique photon could produce a particle–antiparticle pair in the vicinity of an atom nucleus that takes part in the interaction; in this case, the nucleus makes possible the existence of a center-of-momentum frame. The created positron quickly annihilates with another electron giving rise to a pair of photons propagating in different directions. In the center-of-momentum frame of the annihilated pair, the produced photons propagate along opposite directions and each one has a minimum energy of 0.511 MeV. The annihilated positron and electron masses are completely converted into electromagnetic radiation energy.

pair annihilation

Let us now search for the constraints imposed by the energy and momentum conservation on the processes of absorption and emission of a photon by an atom. When a photon is absorbed by an atom, the atom is left in an excited state of its internal structure. The internal excitation energy is not exactly equal to the energy $h\nu_a$ of the absorbed photon: in the frame where the atom is initially at rest, a part of the photon energy is used to conserve the momentum. Concretely, if E_f and $\mathbf{p}_\mathrm{f}$ are the final energy and momentum of the excited atom, respectively, then

absorption of photons

$$h\nu_a c^{-1}\,\hat{\mathbf{n}} = \mathbf{p}_\mathrm{f} \qquad h\nu_a + mc^2 = E_\mathrm{f} \tag{6.58}$$

where m is the initial atom mass. The mass M of the excited atom is

$$M^2c^4 = E_\mathrm{f}{}^2 - p_\mathrm{f}{}^2c^2 = (h\nu_a + mc^2)^2 - (h\nu_a)^2 = m^2c^4\left(1 + \frac{2\,h\nu_a}{mc^2}\right)$$

i.e.,

$$M\,c^2 = mc^2\sqrt{1 + \frac{2\,h\nu_a}{mc^2}} < mc^2 + h\nu_a \tag{6.59}$$

The internal structure of an atom is such that the energy $h\nu_a$ that can be absorbed is much smaller than its rest energy mc^2. Then the former expression can be approximated as

$$Mc^2 \cong mc^2 + h\nu_a - \frac{1}{2}\frac{(h\nu_a)^2}{mc^2} \tag{6.60}$$

The velocity acquired by the excited atom is

$$\mathbf{u}_\mathrm{f} = \frac{\mathbf{p}_\mathrm{f}\;c^2}{E_\mathrm{f}} = \frac{h\nu_a c}{h\nu_a + mc^2}\hat{\mathbf{n}} = \frac{c\hat{\mathbf{n}}}{\dfrac{mc^2}{h\nu_a} + 1} \cong \frac{h\nu_a}{mc}\hat{\mathbf{n}} \tag{6.61}$$

Inversely, when an atom decays from an excited quantum state of mass M, to a quantum state of mass m, the internal energy variation $(M - m)c^2$ is not completely transferred to the emitted photon energy $h\nu_\mathrm{e}$, since the atom

emission of photons

recoils to conserve the momentum. In the frame where the atom is initially at rest (center-of-momentum frame), the conservation laws say that

$$0 = h\nu_e c^{-1}\hat{\mathbf{n}} + \mathbf{p}_f \qquad Mc^2 = h\nu_e + E_f$$

These equations coincide with Eq. (6.58), apart from the change of h for $-h$, and m for M. This remark allows us to avoid solving again the equations system. The atom mass m and velocity $\mathbf{u}_f$ after the emission are then

$$mc^2 = Mc^2\sqrt{1 - \frac{2h\nu_e}{Mc^2}} \cong Mc^2 - h\nu_e - \frac{1}{2}\frac{(h\nu_e)^2}{Mc^2} \tag{6.62}$$

$$\mathbf{u}_f = -\frac{c\,\hat{\mathbf{n}}}{\dfrac{Mc^2}{h\nu_e} - 1} \cong -\frac{h\nu_e}{Mc}\hat{\mathbf{n}} \tag{6.63}$$

The range of energies $h\nu$ involved in processes of atomic absorption and emission starts from infrared radiation, visible light (1.75–3 eV), and ultraviolet radiation (up to 200 eV) for light atoms, and reaches the X-rays of around 100 keV for heavy atoms. As it can be seen in Eqs. (6.60) and (6.62), the recoil energy is approximately $(h\nu)^2/2E_o$, E_o being the atom rest energy. Then, $|\Delta\nu|/\nu \sim h\nu/E_o$. For light atoms with rest energies varying between 10^9 eV and 10^{10} eV, whose emission and absorption energies are of some electron-volts, the recoil energy is very small and results $|\Delta\nu|/\nu \sim 10^{-8}$–$10^{-9}$. This effect is negligible, and comparable to the *natural width* of spectral lines. If atoms take part of a gas at the temperature T, then the atom random motion associated with thermal agitation will introduce a Doppler shift of the emission and absorption frequencies obtained in the previous treatment; this effect adds a shift $|\Delta\nu|/\nu \sim u/c$, where u is the velocity of gas atoms. Since the mean kinetic energy of gas particles is of the order of $k_B T$, [26] it results that $|\Delta\nu|/\nu \sim (k_B T/E_o)^{1/2}$. At the temperature of 300 K it is $k_B T \sim 0.025$ eV; for atoms with rest energy $E_o \sim 10^{10}$ eV, the contribution of thermal agitation is $|\Delta\nu|/\nu \sim 10^{-6}$, being much more important than the recoil effect above computed. The recoil energy becomes important for heavy atoms, whose internal structures allow emissions in the range of X-rays.[27] Besides, a greater rest energy diminishes the incidence of thermal agitation.

Also nuclei have a discrete structure of energy levels, and can undergo transitions emitting or absorbing photons; these *γ-rays* have energies that vary between 0.001 and 8 MeV,[28] and give rise to significant recoil energies. Nevertheless, in some crystals the recoil energy can be negligible because not only the

[26] $k_B = 1.3806 \times 10^{-23}\,\mathrm{J\,K^{-1}} = 8.6173 \times 10^{-5}\,\mathrm{eV\,K^{-1}}$ is Boltzmann constant.

[27] Note that the energy levels have values approximately proportional to the squared atomic number Z, while masses are approximately proportional to Z.

[28] The name γ-rays is used for photons coming from nuclear transitions, but also for any photon whose energy is greater than about 100 keV.

emitter atom but the entire crystal recoils to conserve the momentum. In such a case, it is the crystal mass which is involved in Eq. (6.62), and the emitted photon energy $h\nu_e$ results to be practically equal to the internal energy change due to the quantum transition. In these conditions, the emitted photon can be resonantly absorbed by another similar crystal. The emission and absorption lines of these crystalline solids are extraordinarily well defined, and their widths are very close to the natural width associated with the half-life time of the excited state. This effect, discovered by R.L. Mössbauer in 1958, is useful for high-resolution spectroscopic determinations in solid state physics. It is also relevant for detecting perturbations on the emitter–absorber system, since the extreme definition of the emission line implies that any perturbation will destroy its reabsorption (the width of the 14.4 keV line of a ^{57}Fe nucleus in a crystal is such that $|\Delta\nu|/\nu \approx 3 \times 10^{-13}$).

Mössbauer effect and Pound-Rebka experiment

Mössbauer effect was exploited by Pound and Rebka (1960) to verify that the frequency of a photon going up a distance H in a gravitational field g, undergoes a change $\Delta\nu/\nu \approx -gH/c^2$. This change was measured in the terrestrial gravitational field along a distance $H = 22.6\,\text{m}$ $(\Delta\nu/\nu \approx -2 \times 10^{-15})$. Although this frequency shift is even smaller than the natural width of the ^{57}Fe line employed in the experiment, Pound and Rebka did success in showing the effect of the gravitational field on the reabsorption of the γ emission. They introduced a slight oscillating motion between the emitter and receiver crystals which produced a reabsorption varying in time. Except for the faint gravitational effect, the reabsorption should be maximum whenever the relative velocity emitter–receiver is zero. The weak frequency shift owing to the gravitational field produces a phase shift between the reabsorption curve and the emitter–receiver relative motion; this phase shift was determined by Pound and Rebka. The relativistic behavior of electromagnetic radiation and particles in the presence of a gravitational field is described within the framework of General Relativity, and will be tackled in Chapters 8 and 9.

7

Covariant Formulation

7.1. FOUR-TENSORS

The fundamental laws of Physics cannot change their form under spatial rotations, since space is supposed to be isotropic. This symmetry of physical laws finds its natural realization in vector language. For instance, Maxwell's laws (1B.1) and (1B.2) are written in terms of vectors **E** and **B**, and vector operators such as rotor and divergence. The form of these equations is not affected by a rotation of Cartesian axes, because the operations they involve are independent of the orientation of axes. This property is also valid for relations including scalar or vector products between vectors, or other tensor operations among tensor magnitudes like the inertia tensor and the stress tensor. In all these cases, only the *components* of vectors and tensors are modified when Cartesian axes rotate; but the equations or laws expressing relations among them keep their form. In addition, Maxwell's laws also keep their form under Lorentz transformations (this is the essence of Special Relativity), and the same happens with the laws of relativistic Dynamics like the relativistic energy and momentum conservation laws. However, this *covariance* of relativistic fundamental physics laws under Lorentz transformations is not evident in the language hitherto used. Minkowski developed the *four-tensor* language, which is the natural language for writing the laws of relativistic physics. Using this language, the covariance of the laws of Physics under Lorentz boosts (3.37–3.39), spatial rotations, or any combination of them (i.e., under any Lorentz group transformation) will become evident.

The Lorentz group transformations are linear. Therefore, the action of any group element on event coordinates t, x, y, z can be written as

$$x^{j'} = \sum_{j=0}^{3} \Lambda^{j'}{}_{j}\, x^j \tag{7.1}$$

where x^j are Cartesian coordinates of an event in the frame S,

$$x^0 \equiv ct \qquad x^1 \equiv x \qquad x^2 \equiv y \qquad x^3 \equiv z \tag{7.2}$$

and $x^{j'}$ are the respective coordinates of the same event in the frame S'. The quantities $\Lambda^{j'}{}_{j}$ are the (independent of x^j) coefficients characterizing the Lorentz group transformation under consideration. For instance, the coefficients $\Lambda^{j'}{}_{j}$ belonging to a boost along the x axis (3.37–3.39) are the components of the matrix (see Eq. (4.16))

$$\Lambda^{j'}{}_{j} = \begin{pmatrix} \gamma(\beta) & -\gamma(\beta)\,\beta & 0 & 0 \\ -\gamma(\beta)\,\beta & \gamma(\beta) & 0 & 0 \\ 0 & 0 & 1 & 0 \\ 0 & 0 & 0 & 1 \end{pmatrix} \tag{7.3}$$

where $\beta = V/c$. Instead, when considering a rotation in the x–y plane then the components of the matrix in Eq. (4.19) have to be used.

Einstein convention

From now on we adopt *Einstein convention*, which omits the summation symbol in any equation where the sum is performed on an index that appears repeated in the way of a pair subindex–superindex, as it happens in Eq. (7.1). In the case of Eq. (7.1), the index j appears both as subindex and supraindex, and the sum is performed on the four possible values of j. Index j is called a *dummy* index because, unlike j', j does not survive as an index after the sum is done. By using Einstein convention, Eq. (7.1) is written in the way

$$x^{j'} = \Lambda^{j'}{}_{j}\, x^j \tag{7.4}$$

For each transformation $\Lambda^{j'}{}_{j}$, changing coordinates in the frame S to the respective coordinates in frame S', there exists another Lorentz group transformation performing the inverse process, coming back from S' to S. This inverse transformation will be denoted with the symbol $\Lambda^{j}_{j'}$ (we remark that the same symbol Λ is used for both transformations, since both of them belong to the Lorentz group; the positions of primed and non-primed indexes indicate whether the transformation acts from S to S' or in the opposite direction):

$$x^j = \Lambda^{j}{}_{j'}\, x^{j'} \tag{7.5}$$

where the sum on the dummy index j' is implied. The notation adopted for the inverse transformation allows passing from transformation (7.4) to its inverse (7.5) by the mere exchange of primed and non-primed indexes. Since $\Lambda^{j'}{}_{j}$ and $\Lambda^{j}{}_{j'}$ are each of them inverse of the other, they fulfill[1]

$$\Lambda^{j'}{}_{j}\, \Lambda^{j}_{k'} = \delta^{j'}_{k'} \qquad\qquad \Lambda^{j}{}_{j'}\, \Lambda^{j'}{}_{k} = \delta^{j}_{k} \tag{7.6}$$

The inverse transformation of a boost is obtained by replacing the boost velocity with the opposite vector. In the case of transformation (7.3) it results

$$\Lambda^{j}{}_{j'} = \begin{pmatrix} \gamma(\beta) & \gamma(\beta)\,\beta & 0 & 0 \\ \gamma(\beta)\,\beta & \gamma(\beta) & 0 & 0 \\ 0 & 0 & 1 & 0 \\ 0 & 0 & 0 & 1 \end{pmatrix} \tag{7.7}$$

[1] *Kronecker symbol* δ^j_k (the components of identity) is equal to 1 if $j = k$, and zero in other case.

A *contravariant four-vector* is any object with Cartesian components A^j, which transforms under Lorentz group transformations according to

contravariant four-vector

$$A^{j'} = \Lambda^{j'}{}_{j} \, A^j \tag{7.8}$$

Since the Lorentz group includes rotations, the three spatial Cartesian components of a contravariant four-vector—A^1, A^2, A^3—compound an ordinary vector $\mathbf{A} = (A_x, A_y, A_z)$:

$$A^j = (A^0, A^1, A^2, A^3) = (A^0, \mathbf{A}) \tag{7.9}$$

In the previous chapters we could identify several examples of contravariant vectors; they are included in Table 7.1, together with the reference to the equations showing their transformations under Lorentz boosts:

Table 7.1. Examples of contravariant four-vectors

event coordinates	x^j	$(ct, \mathbf{r})$	Eqs. (3.37–3.39)
wave four-vector[2]	k^j	$2\pi\nu(c^{-1}, w^{-1}\,\hat{\mathbf{n}})$	Eqs. (3.58, 3.62, 3.63)
charge–current density	j^j	$(\rho\, c, \mathbf{j})$	Eq. (5.11)
electromagnetic potential	A^j	$(c^{-1}\phi, \mathbf{A})$	Eq. (5.19)
energy–momentum	p^j	$(c^{-1}E, \mathbf{p})$	Eqs. (6.20–6.21)

Thus, the intertwined energy and momentum conservation laws correspond to the conservation of a unique geometric object: the energy–momentum four-vector.

A particle world line (see Figure 4.6) can be parametrized with the proper time evaluated along itself (Eq. (4.8)). In this way a particle world line results to be characterized by four parametric equations

$$x^j = x^j(\tau) \qquad j = 0, 1, 2, 3 \tag{7.10}$$

Because the coefficients $\Lambda^{j'}{}_{j}$ of a Lorentz group transformation do not depend on the event coordinates, then the differentiation of Eq. (7.4) results in $\mathrm{d}x^{j'} = \Lambda^{j'}{}_{j}\, \mathrm{d}x^j$. Since the proper time is invariant, then the quantities

$$U^j \equiv \frac{\mathrm{d}x^j}{\mathrm{d}\tau}(\tau) = \left(c\,\frac{\mathrm{d}t}{\mathrm{d}\tau}, \frac{\mathrm{d}\mathbf{r}}{\mathrm{d}\tau}\right)(\tau) \tag{7.11}$$

four-velocity

transform like the components of a contravariant four-vector, which is called *four-velocity* and is tangent to the particle world line. Taking into account the relation (4.7), $\mathrm{d}\tau = \gamma(u)^{-1}\,\mathrm{d}t$, then

$$U^j = \gamma(u)\,(c, \mathbf{u}) \tag{7.12}$$

[2] Multiply Eqs. (3.62, 3.63) by Eq. (3.58), and note that $\cos\vartheta$ and $\sin\vartheta$ are the $\hat{\mathbf{n}}$ components parallel and transversal to the boost direction.

The four-velocity components satisfy that

$$U^{0^2} - |\mathbf{U}|^2 = \gamma(u)^2 \, (c^2 - u^2) = c^2 \tag{7.13}$$

Equation (7.13) indicates that the four-velocity components are not independent but they are related through the equation of a hyperboloid. In fact, a particle has three degrees of freedom, and its state of motion is described by the three components of vector $\mathbf{u} = \gamma^{-1}\,(u)\,\mathbf{U}$; when the vector $\mathbf{U}$ is given, the temporal component U^0 is fixed by Eq. (7.13).

According to Eqs. (6.22) and (6.23), the product of mass m (which is invariant) and the four-velocity gives the energy–momentum four-vector:

energy–momentum four-vector

$$p^j = m\, U^j \tag{7.14}$$

Any four-vector defined along a particle world line can be differentiated with respect to the invariant proper time parametrizing the world line. The result is another four-vector. The derivative of the energy–momentum four-vector is

$$\frac{\mathrm{d}p^j}{\mathrm{d}\tau} = \gamma(u)\left(c^{-1}\frac{\mathrm{d}E}{\mathrm{d}t},\ \frac{\mathrm{d}\mathbf{p}}{\mathrm{d}t}\right)$$

and, according to Eqs. (6.26) and (6.27), it must be equaled to the *four-force*

four-force

$$K^j \equiv \gamma(u)\,(\mathbf{F}\cdot\frac{\mathbf{u}}{c},\ \mathbf{F}) \tag{7.15}$$

whose temporal component contains the power $\mathbf{F}\cdot\mathbf{u}$.

A *covariant four-vector* is any object with Cartesian components C_j, which transforms under Lorentz group transformations according to

covariant four-vector

$$C_{j'} = C_j\ \Lambda^j{}_{j'} \tag{7.16}$$

Notice that the covariant four-vectors components change from S to S' following the transformation which is inverse of the one changing the contravariant four-vectors components. As a consequence of this behavior, the combination $C_j\, A^j$ (sum on j) between covariant and contravariant four-vectors results to be invariant (independent of the reference system):

$$C_{j'}\, A^{j'} = C_{j'}\, \Lambda^{j'}{}_j\, A^j = C_j\, A^j \tag{7.17}$$

The operation described in Eq. (7.17) is called *contraction*; the contraction always involves a pair of covariant–contravariant indexes.

Cartesian partial derivatives can be treated as components of a covariant four-vector ∂_j called *gradient*:

$$\partial_j \equiv \frac{\partial}{\partial x^j} \tag{7.18}$$

In fact, Eq. (7.5) says that

$$\partial_{j'} = \frac{\partial}{\partial x^{j'}} = \frac{\partial x^j}{\partial x^{j'}}\frac{\partial}{\partial x^j} = \Lambda^j_{\ j'}\ \partial_j$$

The contraction of the gradient and a contravariant four-vector field $A^j(x)$ is a particular case of Eq. (7.17), and leads to the invariant[3] called *divergence* of field A^j:

divergence

$$\partial_j A^j = \frac{1}{c}\frac{\partial A^0}{\partial t} + \frac{\partial A^1}{\partial x} + \frac{\partial A^2}{\partial y} + \frac{\partial A^3}{\partial z} = \frac{1}{c}\frac{\partial A^0}{\partial t} + \nabla\cdot\mathbf{A} \tag{7.19}$$

The divergence appears in continuity equations, which express the local conservation law of some physical magnitude. For instance, continuity equation (1B.3), which tells about the electric charge local conservation, can be written in four-vector language as

$$\partial_i\, j^i = 0 \tag{7.20}$$

Since the divergence of a contravariant four-vector is invariant under Lorentz group transformations, then $\partial_i\, j^i$ is zero in any inertial frame; i.e., the electric charge local conservation is verified in any inertial reference system. Thus the law (7.20) satisfies the Principle of relativity under Lorentz transformations. Four-vector language makes possible to verify this property by means of the mere inspection of the physical law.

Contravariant and covariant four-vectors, together with invariants, are particular cases of objects called *four-tensors*, whose Cartesian components are identified through r contravariant indexes and s covariant indexes, and change under Lorentz group transformations according to

four-tensors

$$T^{i'\,j'\ldots}{}_{k'\,l'\,m'\ldots} = \Lambda^{i'}{}_i\,\Lambda^{j'}{}_j\ldots\Lambda^k_{k'}\,\Lambda^l_{l'}\,\Lambda^m_{m'}\ldots T^{ij\ldots}{}_{klm\ldots} \tag{7.21}$$

It is clear that any linear combination, with invariant coefficients, of tensors of the same type (having the same quantity of covariant and contravariant indexes) results in another tensor of the same type.

The *tensor product* is an operation where several tensors form another tensor with a larger number of indexes, by multiplying their components:

tensor product

$$T^{ijkl}{}_{mnp} = O\,B^{ij}\,C_m\,D^k_{np}\,E^l \tag{7.22}$$

(O is an invariant). The components $T^{ijkl}{}_{mnp}$ of the object defined in Eq. (7.22) transform according to the rule (7.21). In particular, the Cartesian derivatives $\partial_m D^j{}_{np}$ of Cartesian components $D^j{}_{np}$ of a tensor generate the Cartesian components of another tensor (see Note 3).

The operation of index contraction (7.17) can be defined for a pair of indexes of any tensor (the two indexes to be contracted must always be of *different* type). For instance

$$T^{ijkl}{}_{knp} = M^{ijl}{}_{np} \tag{7.23}$$

contraction

[3] The operator ∂_j can be regarded as the component of a covariant four-vector just under linear coordinate transformations, like those of Lorentz group. In a more general case, the coefficients $\Lambda^{j'}{}_j$ in (7.8) are not constant, so their derivatives affect the transformation of $\partial_j A^j$. See §8.4.

To prove that quantities $M^{ijl}{}_{np}$ in Eq. (7.23) effectively transform as the components of a tensor, the relations (7.6) have to be used (index contraction in transformation (7.21) generates Kronecker symbols δ^j_k).

symmetric and antisymmetric

A tensor can be *symmetric* or *antisymmetric* in a given pair of indexes of the *same type*. For example, a tensor $M^{ijl}{}_{np}$ is symmetric in its first and third contravariant indexes if

$$M^{ijl}{}_{np} = M^{lji}{}_{np} \tag{7.24}$$

and it is antisymmetric in its two only covariant indexes if

$$M^{ijl}{}_{np} = -M^{ijl}{}_{pn} \tag{7.25}$$

The properties of symmetry and antisymmetry of pair of indexes are not affected by linear transformations:

$$M_{j'k'} = \Lambda^j_{j'}\,\Lambda^k_{k'}\,M_{jk} = \pm\,\Lambda^j_{j'}\,\Lambda^k_{k'}\,M_{kj}$$
$$= \pm\,\Lambda^k_{j'}\,\Lambda^j_{k'}\,M_{jk} = \pm\,M_{k'j'}$$

(in the last line, we take advantage that j, k are dummy indexes; so nothing changes if j is called k and k is called j).

tensor language and Principle of relativity

The example of the continuity equation (7.20) shows that any physical law that can be written as a four-tensor equation will automatically keep its form under Lorentz group transformations, and satisfy the Principle of relativity. In particular, such laws can be written as a tensor equaled to zero (by carrying all its terms to the same side); according to Eq. (7.22), whenever all the components of a tensor are zero in some reference system, then they will be zero in any reference system. Therefore, the physical law will say, in this case, that the components of a given tensor are zero in any inertial reference system, so manifestly putting all the inertial frames on an equal footing.

7.2. METRIC

A *scalar product* between two contravariant vectors A^i, B^j is an operation $A\cdot B$ assigning an invariant to the pair of vectors. The operation must have the following properties:

(i) *commutativity:* $A\cdot B = B\cdot A$,
(ii) $(a\,A + c\,C)\cdot B = a\,(A\cdot B) + c\,(C\cdot B) \quad \forall a,\ c$ *reals*

A scalar product between contravariant vectors can be defined by means of the contraction of the pair of vectors and a 2-index symmetric tensor:

$$A\cdot B \equiv g_{ij}\,A^i\,B^j \tag{7.26}$$

The commutativity of the scalar product (7.26) is guaranteed by the symmetry of tensor g_{ij}:

$$A \cdot B = g_{ij}\, A^i\, B^j = g_{ji}\, A^i\, B^j = g_{ji}\, B^j\, A^i = B \cdot A$$

The symmetric tensor g_{ij} is called *metric tensor*; it will be chosen in such a way that the product $A \cdot A$ be equal to the invariant quadratic form that any four-vector possesses:

$$A \cdot A \equiv A^{0^2} - A^{1^2} - A^{2^2} - A^{3^2} = A^{0^2} - |\mathbf{A}|^2 \qquad (7.27)$$

where $|\mathbf{A}|^2 = \mathbf{A} \cdot \mathbf{A}$ entails the ordinary scalar product in three-dimensional Euclidean space. The invariance of quadratic form (7.27) under Lorentz boosts is an already proven property: it corresponds to the invariance of the interval when $A^i = \Delta x^i$, and it is also valid for any four-vector because all of them transform in the same way. On the other hand, it is clear that spatial rotations do not affect the product $A \cdot A$, since each term in the right side of Eq. (7.27) is invariant under spatial rotations. The value of the invariant $A \cdot A$ is characteristic of the four-vector under consideration; for instance, according to Eq. (7.13) the four-velocity invariant is $U \cdot U = c^2$. Table 7.2 shows the results for some known four-vectors.

$A \cdot A$ in Eq. (7.27) is not positive definite: as it was shown in Chapter 4, the separation between two events can be timelike ($\Delta s^2 > 0$), spacelike ($\Delta s^2 < 0$), or null ($\Delta s^2 = 0$). This classification is applicable to any four-vector:

timelike four-vector	$A \cdot A > 0$	$(\lvert A^0 \rvert > \lvert \mathbf{A} \rvert)$
spacelike four-vector	$A \cdot A < 0$	$(\lvert A^0 \rvert < \lvert \mathbf{A} \rvert)$
null four-vector	$A \cdot A = 0$	$(\lvert A^0 \rvert = \lvert \mathbf{A} \rvert)$

This classification separates four-vectors being, respectively, internal, external, and tangent to the light cone (see §4.3, §4.4 and §4.6).

Some four-vectors have always the same character. For instance, the particle four-velocity and energy–momentum four-vector are always timelike, while the four-force is always spacelike, and the wave four-vector of a light ray is always null. Other four-vectors, like the separation between a pair of events, can have diverse characters, depending on the examined problem. For example, in the charge and current configuration of Fig. (5.5) the charge–current density j^i is spacelike in the upper line but is timelike in the lower line.[6]

Table 7.2. Invariant quadratic form for some four-vectors

event separation $\Delta x^i = x_2{}^i - x_1{}^i$	$\Delta x \cdot \Delta x \equiv \Delta s^2$
wave four-vector of a light ray ($w = c$)	$k \cdot k = 0$
four-velocity[4]	$U \cdot U = c^2$
energy–momentum[5]	$p \cdot p = m^2\, c^2$

[4] $U \cdot U$ can be also obtained in the following way: $U \cdot U = \frac{\mathrm{d}x}{\mathrm{d}\tau} \cdot \frac{\mathrm{d}x}{\mathrm{d}\tau} = \left(\frac{\mathrm{d}s}{\mathrm{d}\tau}\right)^2 = c^2$.
[5] $c^2(p \cdot p) = c^2(E^2 c^{-2} - |\mathbf{p}|^2)$ is the energy–momentum invariant of §6.3.
[6] In spite of this, the charge–current density is timelike whenever it is verified that $\mathbf{j} = \rho\, \mathbf{u}$.

The metric tensor involved in the definition of internal product (7.27) is

$$g_{ij} = \begin{pmatrix} 1 & 0 & 0 & 0 \\ 0 & -1 & 0 & 0 \\ 0 & 0 & -1 & 0 \\ 0 & 0 & 0 & -1 \end{pmatrix} \tag{7.28}$$

metric tensor

In an Euclidean geometry the metric tensor is diagonal as well, but the diagonal components are all equal. It is said, then, that Minkowski space-time is provided with a *pseudo-Euclidean* geometry. Equation (7.26) then becomes

scalar product

$$\begin{aligned} A \cdot B &= g_{ij}\, A^i\, B^j \\ &= A^0\, B^0 - A^1\, B^1 - A^2\, B^2 - A^3\, B^3 = A^0\, B^0 - \mathbf{A} \cdot \mathbf{B} \end{aligned} \tag{7.29}$$

The scalar product between two timelike four-vectors, or between a timelike four-vector and a null one, results to be strictly positive if $sign(A^0) = sign(B^0)$, and strictly negative in other cases. If two four-vectors have zero scalar product and one of them is timelike, then the other is spacelike. As an example, let us differentiate $U \cdot U = c^2$ along the world line of a particle whose four-velocity is U^j:

$$0 = \frac{\mathrm{d}}{\mathrm{d}\tau}(U \cdot U) = \frac{\mathrm{d}}{\mathrm{d}\tau}(U^0\ U^0 - \mathbf{U} \cdot \mathbf{U}) = 2\,(U^0 \frac{\mathrm{d}U^0}{\mathrm{d}\tau} - \mathbf{U} \cdot \frac{\mathrm{d}\mathbf{U}}{\mathrm{d}\tau}) = 2\ U \cdot \frac{\mathrm{d}U}{\mathrm{d}\tau}$$

τ being the particle proper time. Therefore the scalar product of the four-velocity and the *four-acceleration*

four-acceleration

$$a^j \equiv \frac{\mathrm{d}U^j}{\mathrm{d}\tau} \qquad U \cdot a = 0 \tag{7.30}$$

is always zero. The four-acceleration is then spacelike. The four-force acting on a particle is equal to its mass multiplied by its four-acceleration,

$$K^j = \frac{\mathrm{d}p^j}{\mathrm{d}\tau} = \frac{\mathrm{d}}{\mathrm{d}\tau}(mU^j) = m\frac{\mathrm{d}U^j}{\mathrm{d}\tau} = m\, a^j \tag{7.31}$$

and fulfills that $K \cdot U = 0$. The spacelike character of the four-force K^j is clearly exhibited in its definition (7.15): by computing the invariant $K \cdot K$ in the frame $S_{o(t)}$ where the particle is instantaneously at rest, it is obtained $-\,|\mathbf{F}_o|^2$.

The relation (7.31) between four-force and four-acceleration can be combined with Eq. (6.29) and definition (7.15) to write a^j as a function of $\boldsymbol{a}$ and $\mathbf{u}$:

$$a^j = \gamma(u)^4\,(\frac{\mathbf{u} \cdot \boldsymbol{a}}{c},\ \boldsymbol{a}_{\text{longitudinal}} + \gamma(u)^{-2}\,\boldsymbol{a}_{\text{transversal}}) \tag{7.32}$$

where *longitudinal* and *transversal* allude to the $\mathbf{u}$ direction. Then

$$a \cdot a = -\gamma(u)^4\,(\gamma(u)^2\, a_{\text{longitudinal}}{}^2 + a_{\text{transversal}}{}^2) \tag{7.33}$$

The existence of a metric tensor makes possible to establish a biunivocal relation between covariant and contravariant four-vectors. This relation also requires of the symmetric tensor g^{ij} that is inverse of the metric tensor:

$$g^{ij}\, g_{jk} = \delta^i_k \tag{7.34}$$

Due to the elemental structure of metric tensor (7.28), it results that components of g^{ij} are coincident with the ones of g_{ij}:

$$g^{ij} = \begin{pmatrix} 1 & 0 & 0 & 0 \\ 0 & -1 & 0 & 0 \\ 0 & 0 & -1 & 0 \\ 0 & 0 & 0 & -1 \end{pmatrix} \tag{7.35}$$

index raising and lowering

The "raising and lowering" index operation is performed by the metric tensor and its inverse tensor in the following way: for a given contravariant four-vector A^j, its covariant "partner" A_i is such that

$$A_i = g_{ij}\, A^j \qquad \Leftrightarrow \qquad A^j = g^{jk}\, A_k \tag{7.36}$$

i.e.,

$$A_0 = A^0 \qquad A_\mu = -A^\mu \qquad \mu = 1,\, 2,\, 3 \tag{7.37}$$

Through this association between covariant and contravariant four-vectors, the internal product can be written in several different manners:

$$A \cdot B = g_{ij}\, A^i\, B^j = A_j\, B^j = g^{jk}\, A_j\, B_k = A^k\, B_k \tag{7.38}$$

Indexes belonging to any kind of tensor can be raised or lowered. For instance:

$$T^j{}_{klm} = g_{il}\, T^j{}_k{}^i{}_m \qquad T^j{}_k{}^i{}_m = g^{il}\, T^j{}_{klm} \tag{7.39}$$

In particular, Eq. (7.34) says that $g^i{}_j = \delta^i{}_j$.

The form diag $(1, -1, -1, -1)$ of Cartesian components of the metric tensor and its inverse tensor does not change under Lorentz group transformations. In fact, the spatial block is isotropic, so it does not change under spatial rotations. Neither the components change under arbitrary Lorentz boosts, what can be straightforwardly verified by using rule (7.21) and transformation (7.3) (and noticing that a generic boost is nothing but the composition of boost (7.3) and a rotation). This property of metric tensor is evident in the equality $c^2 \Delta t^2 - |\Delta \mathbf{r}|^2 = c^2 \Delta t'^2 - |\Delta \mathbf{r}'|^2$, which was proved in §4.1 and corresponds to the equality $g_{ij}\, \Delta x^i\, \Delta x^j = g_{i'j'}\, \Delta x^{i'} \Delta x^{j'}$ with $g_{ij} = g_{i'j'}$.

7.3. FOUR-VECTOR "NORM" AND ARGUMENT

Vectors in the plane have a polar representation in terms of its modulus and the angle between the vector and a given direction. Analogously, any non-null contravariant four-vector (A^0, $\mathbf{A}$) possesses a similar representation in the plane defined by the x^0 axis and the $\mathbf{A}$ direction. The magnitude

$$|A| \equiv |A \cdot A|^{1/2} \tag{7.40}$$

is not strictly a *norm*, since it does not satisfy the triangle inequality because the scalar product (7.29) is not positive definite. This does not prevent us from defining an *argument* Θ for the four-vector A^j. By choosing the x axis parallel to $\mathbf{A}$ direction, Θ is defined according to the timelike or spacelike A^j character as

$$\textit{timelike four-vector} \quad \cosh\Theta = \frac{\varepsilon A^0}{|A|}, \quad \sinh\Theta = \frac{\varepsilon A^x}{|A|} \tag{7.41a}$$

$$\textit{spacelike four-vector} \quad \cosh\Theta = \frac{\varepsilon A^x}{|A|}, \quad \sinh\Theta = \frac{\varepsilon A^0}{|A|} \tag{7.41b}$$

where $\varepsilon = \pm 1$, as appropriate, in order to guarantee the positive value for cosh Θ: $\varepsilon A^0 > 0$ for timelike four-vectors,[7] and $\varepsilon A^x > 0$ for spacelike four-vectors. Equations (7.41) are consistent with relation $\cosh^2\Theta - \sinh^2\Theta = 1$. Figure 7.1 shows a timelike and a spacelike four-vector; the hyperbolas indicate the places of the end of all those four-vectors having equal "norm," and the radial lines measure the argument Θ.

In Figure 7.1 the scalar product of four-vectors T^i and E^j is

$$T \cdot E = |T||E|\,(\cosh\Theta_T \sinh\Theta_E - \sinh\Theta_T \cosh\Theta_E) = |T||E| \sinh(\Theta_E - \Theta_T)$$

so it vanishes when $\Theta_T = \Theta_E$, i.e., when the bisectrix of the angle between T^i and E^j is a light ray.

The transformation of Cartesian four-vector components under a Lorentz boost involves the transformation of the argument Θ (on the contrary, the "norm" is invariant). Let us consider a boost along the x-axis direction – which is the $\mathbf{A}$ direction in Eqs. (7.41a,b)—written in terms of its velocity parameter Θ_V like in Eq. (4.14). Let us substitute the Cartesian components A^0 and A^x as functions of the argument Θ. Component $T^{0'}$ corresponding to a timelike four-vector is

$$\begin{aligned} |T|\cosh\Theta'_T &= |T|\cosh\Theta_T \cosh\Theta_V - |T|\sinh\Theta_T \sinh\Theta_V \\ &= |T|\cosh(\Theta_T - \Theta_V) \end{aligned}$$

[7] In the case of the four-velocity U^j, Θ_U is the rapidity (4.15).

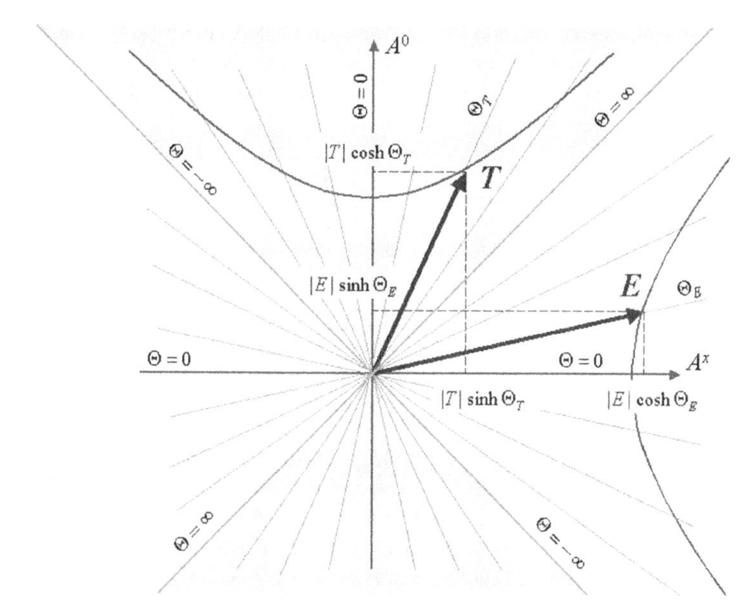

Figure 7.1. Graphic representations of a timelike four-vector T and a spacelike four-vector E, and their arguments Θ_T and Θ_E.

and similarly for $T^{x'}$. For a spacelike four-vector, $E^{0'}$ results to be

$$|E| \sinh\Theta'_E = |E| \sinh\Theta_E \cosh \Theta_V - |E| \cosh\Theta_E \sinh \Theta_V = |E| \sinh (\Theta_E - \Theta_V)$$

In both cases we obtain that

$$\Theta' = \Theta - \Theta_V \tag{7.42}$$

This transformation of arguments, which was already seen in the case of the velocity parameter, is equally valid for timelike or spacelike four-vectors. The aspect of the axes of frame S' in Figure 3.17 is just the required for subtracting Θ_V from the four-vector arguments, in agreement with Eq. (7.42).

7.4. ANGULAR MOMENTUM

The components of particle angular momentum $\boldsymbol{l} = \mathbf{r} \times \mathbf{p}$ do not transform like a part of a four-vector, but they belong to the antisymmetric four-tensor

angular momentum four-tensor

$$l^{ij} \equiv x^i\, p^j - x^j\, p^i \qquad l^{ji} = -l^{ij} \tag{7.43}$$

In fact, the Cartesian components of angular momentum with respect to the coordinate origin are

angular momentum pseudovector

$$\boldsymbol{l} = (l^{23},\, l^{31},\, l^{12}) \tag{7.44}$$

The angular momentum of an isolated physical system must be conserved. If the system is composed by N particles that locally interact (by means of collisions), then the conserved angular momentum will be the addition of all the particle angular momenta[8]

$$\mathbf{L} \equiv \sum_{N\ \text{particles}} \boldsymbol{l} \tag{7.45}$$

The components of **L** take part in the four-tensor

$$L^{ij} \equiv \sum_{N\ \text{particles}} l^{ij} \tag{7.46}$$

In order that **L** conservation be valid in any inertial reference system, not only those components of L^{ij} associated with **L** have to be conserved. Since the L^{ij} components intermingle under Lorentz transformations, it will be necessary that the complete four-tensor must be conserved. This condition resembles the one leading to the relativistic energy conservation: in order that the relativistic momentum be conserved in all inertial frames, then the energy has to be conserved as well (and vice versa); in other words, the complete energy–momentum four-vector has to be conserved. An antisymmetric four-tensor has six independent components (the four diagonal components are zero, and the components at each side of the diagonal just differ in the sign). Only three of the six L^{ij} independent components come from the angular momentum. The remaining three components,

$$L^{\mu 0} = \sum_{\text{particles}} (x^{\mu} p^{0} - x^{0} p^{\mu}) = \sum_{\text{particles}} (x^{\mu} c^{-1} E - ct\, p^{\mu}) \qquad \mu = 1, 2, 3$$

form an ordinary vector $\pounds = (L^{10}, L^{20}, L^{30})$ where the total relativistic momentum and the center of inertia position (6.39) can be recognized:

$$\pounds = \sum_{\text{particles}} (\mathbf{r}\, c^{-1} E - ct\, \mathbf{p}) = c^{-1} \mathcal{E} \mathbf{R}_S(t) - ct\, \mathbf{P} \tag{7.47}$$

$\mathcal{E}$ and **P** of an isolated system are conserved. Then

$$\frac{\mathrm{d}\pounds}{\mathrm{d}t} = c^{-1} \mathcal{E} \frac{\mathrm{d}\mathbf{R}_S}{\mathrm{d}t} - c\,\mathbf{P} = c^{-1} \mathcal{E} \mathbf{U}_C - c\,\mathbf{P} = 0.$$

Therefore $\pounds$ of an isolated system is also conserved.[9] According to Eq. (7.47), $\pounds = c^{-1} \mathcal{E} \mathbf{R}_S(t = 0)$ is a "momentum" of the total energy at $t = 0$, and can be cancelled by means of a proper choice of the coordinate origin.

The behavior of components of the angular momentum four-tensor under a Lorentz group transformation leads to relations among angular momentum and centers of inertia referred to different frames. The application of a boost along

[8] If, besides, a field mediates interactions "at a distance," then the field angular momentum will take part in the conserved total angular momentum (see §6.1).

[9] The comment about **P** in Note 2 of Chapter 6 is also applicable to **L** and $\pounds$.

the x axis gives rise to the following transformation of the components of an antisymmetric four-tensor:

$$L^{i'j'} = \begin{pmatrix} 0 & L^{01} & \gamma(L^{02}-\beta L^{12}) & \gamma(L^{03}-\beta L^{13}) \\ \cdots & 0 & \gamma(L^{12}-\beta L^{02}) & \gamma(L^{13}-\beta L^{03}) \\ \cdots & \cdots & 0 & L^{23} \\ \cdots & \cdots & \cdots & 0 \end{pmatrix} \tag{7.48}$$

The transformation (7.48) can be written like transformations of vector $\pounds = (L^{10}, L^{20}, L^{30})$ and pseudovector $\mathbf{L} = (L^{23}, L^{31}, L^{12})$:

$$\begin{array}{ll} \pounds'_{||} = \pounds_{||} & \pounds'_{\perp} = \gamma(V)\,(\pounds_{\perp} - \boldsymbol{\beta} \times \mathbf{L}) \\ \mathbf{L}'_{||} = \mathbf{L}_{||} & \mathbf{L}'_{\perp} = \gamma(V)\,(\mathbf{L}_{\perp} + \boldsymbol{\beta} \times \pounds) \end{array} \tag{7.49}$$

where $||$ and $\perp$ allude to directions being parallel and orthogonal to $\boldsymbol{\beta} = \mathbf{V}/c$.[10] A special attention deserves the transformation from the center-of-momentum frame S_c to an arbitrary frame S; in such a case it is $\mathbf{V} = -\mathbf{U}_c$. Substituting $\pounds_c = c^{-1}\varepsilon_c \mathbf{R}_c$ in (7.49), and considering that $\varepsilon = \gamma(U_c)\ \varepsilon_c$, it is obtained

$$\mathbf{R}_S(t=0)_{||} = \gamma(U_c)^{-1}\,\mathbf{R}_{c||} \tag{7.50}$$

$$\mathbf{R}_{S\perp} = \mathbf{R}_{c\perp} + \varepsilon_c^{-1}\,\mathbf{U}_c \times \mathbf{L}_c \tag{7.51}$$

$$\mathbf{L}_{||} = \mathbf{L}_{c||} \qquad \mathbf{L}_{\perp} = \gamma(U_c)\,\mathbf{L}_{c\perp} + \mathbf{R}_c \times \mathbf{P} \tag{7.52}$$

Equation (7.50) is nothing but the result of the relativistic length contraction ($\mathbf{R}_S$ is the center of inertia referred to S and *measured* in S). In Eqs. (7.51) and (7.52), $\mathbf{L}_c$ is the angular momentum in the center-of-momentum frame. Since the total momentum is zero in S_c, the value of $\mathbf{L}_c$ does not depend on the choice of the S_c coordinate origin. $\mathbf{L}_c$ is the *intrinsic angular momentum.*

Equations (7.50–7.51) become simpler with a proper choice of coordinate origin. Remember that, as a matter of convention, Lorentz transformations are written in such a way that the frames share the spatial coordinate origin at $t = 0 = t_c$. We can choose the coordinate origin coinciding with the (fixed) position of the center of inertia in S_c : $\mathbf{R}_c = 0$. In this case it is

$$\mathbf{R}_S(t=0) = \frac{\mathbf{U}_c \times \mathbf{L}_c}{\varepsilon_c} \tag{7.53}$$

In frame S, the center of inertia moves with velocity $\mathbf{U}_c$. Unless $\mathbf{U}_c$ is parallel to the intrinsic angular momentum, the world line belonging to the center of inertia referred to S does not go through the coordinate origin. Therefore it differs from the world line covered by the center of inertia referred to S_c. As it was

[10] Note the similarity with the transformations of electric and magnetic fields in §5.2, which indicates that fields $\mathbf{E}$ and $\mathbf{B}$ form a two indexes antisymmetric four-tensor, as we shall see in §7.7. This analogy also implies that $\pounds \cdot \mathbf{L}$ and $\pounds^2 - L^2$ are invariant (compare with Eqs. (5.29) and (5.30)).

explained in §6.7, the notion of center of inertia depends on the reference system. Equation (7.53) expresses how much the center of inertia referred to S goes away from the center of inertia referred to S_c. Since the separation is perpendicular to $\mathbf{U}_c$, both frames agree about the magnitude of the gap. An example of this separation was shown in §6.7. The maximum gap is $\mathcal{E}_c^{-1}\, c\, L_c$. If at a given time the physical system of interest were entirely contained in a certain sphere in S_c – i.e., if there were not any kind of energy associated with the system outside this sphere – then the sphere radius could not be smaller than $\mathcal{E}_c^{-1}\, c\, L_c$, because the center of inertia in any frame must be inside the physical system. This means that the size of a physical system provided with intrinsic angular momentum is not arbitrary but it is bounded from below.

size of a physical system provided with intrinsic angular momentum

7.5. VOLUME AND HYPERSURFACES

In an Euclidean space of n dimensions, n linearly independent vectors, **A**, **B**, **C**, ... delimit an $n-$volume which can be calculated with the determinant of the $n \times n$ matrix whose rows are constituted by the Cartesian components of each one of the n vectors:

$$n - volume = \det \begin{pmatrix} A^1 & A^2 & A^3 & \cdots \\ B^1 & B^2 & B^3 & \cdots \\ C^1 & C^2 & C^3 & \cdots \\ \cdots & \cdots & \cdots & \\ \cdots & \cdots & & \end{pmatrix} \tag{7.54}$$

In fact:

- for dimension $n = 2$, the result is

$$2 - volume = A^1\, B^2 - A^2\, B^1$$

  In three dimensions this result would be the vector product $\mathbf{A} \times \mathbf{B}$. If α is the angle between **A** and **B**, then the modulus of $\mathbf{A} \times \mathbf{B}$ is $|\mathbf{A}|\ |\mathbf{B}|\ \sin\, \alpha$, and it is equal to the area of the parallelogram defined by both vectors.

- for dimension $n = 3$, the result is

$$\begin{aligned} 3 - volume = {} & A^1\, B^2\, C^3\ - A^1\, B^3\, C^2 + A^2\, B^3\, C^1 \\ & - A^2\, B^1\, C^3\ + A^3\, B^1\, C^2\ - A^3\, B^2\, C^1 \end{aligned}$$

  which is equal to the product $\mathbf{C} \cdot (\mathbf{A} \times \mathbf{B})$, and it corresponds to the volume of the parallelepiped defined by the three vectors.

The determinant (7.54) completely antisymmetrizes the tensor product of the n vectors and generates, in this way, the expected properties of a volume: if any of the vectors is linearly dependent of the others then the volume will be zero; if one or several vectors are multiplied by given factors then the volume will result multiplied by the same factors. The order of the n vectors determines

orientation

the volume sign; the permutation of any two vectors (i.e., two rows in (7.54)) changes the volume sign. It is said that the order of the n vectors determines the volume *orientation*.

The example corresponding to $n = 3$ shows that the notion of volume is related with an operation among the vectors whose result is independent of the choice of Cartesian axes. Cartesian components A^i with respect to given axes are transformed into Cartesian components $A^{i'}$ with respect to other axes by means of rotations. Rotations are orthogonal matrices, and their determinants are equal to 1. In general, any linear transformation $A^i \rightarrow A^{i'} = \Lambda^{i'}{}_i\, A^i$ with determinant equal to 1 leaves the determinant (7.54) invariant. In fact, the matrix in Eq. (7.54) changes to

$$\begin{pmatrix} \Lambda^{1'}{}_i A^i & \Lambda^{2'}{}_i A^i & \Lambda^{3'}{}_i A^i & \cdots \\ \Lambda^{1'}{}_i B^i & \Lambda^{2'}{}_i B^i & \Lambda^{3'}{}_i B^i & \cdots \\ \Lambda^{1'}{}_i C^i & \Lambda^{2'}{}_i C^i & \cdots & \\ \cdots & \cdots & \cdots & \\ \cdots & \cdots & & \end{pmatrix}$$

$$= \begin{pmatrix} A^1 & A^2 & A^3 & \cdots \\ B^1 & B^2 & B^3 & \cdots \\ C^1 & C^2 & \cdots & \\ \cdots & \cdots & \cdots & \\ \cdots & \cdots & & \end{pmatrix} \begin{pmatrix} \Lambda^{1'}{}_1 & \Lambda^{2'}{}_1 & \Lambda^{3'}{}_1 & \cdots \\ \Lambda^{1'}{}_2 & \Lambda^{2'}{}_2 & \Lambda^{3'}{}_2 & \cdots \\ \Lambda^{1'}{}_3 & \Lambda^{2'}{}_3 & \cdots & \\ \cdots & \cdots & \cdots & \\ \cdots & \cdots & & \end{pmatrix}$$

(the matrix in the last factor is the transpose matrix of the transformation coefficients). The determinant of a product of matrices is equal to the product of determinants of the respective matrices; besides, the determinant does not change under transposition. Thus it results

$$\det \begin{pmatrix} A^{1'} & A^{2'} & A^{3'} & \cdots \\ B^{1'} & B^{2'} & B^{3'} & \cdots \\ C^{1'} & C^{2'} & \cdots & \\ \cdots & \cdots & & \end{pmatrix} = \det\left(\Lambda^{i'}{}_i\right) \cdot \det \begin{pmatrix} A^1 & A^2 & A^3 & \cdots \\ B^1 & B^2 & B^3 & \cdots \\ C^1 & C^2 & \cdots & \cdots \\ \cdots & \cdots & & \end{pmatrix} \tag{7.55}$$

Therefore, whenever the linear transformations of interest have determinant equal to 1, the n-volume definition (7.54) will be appropriate. In fact, it will be invariant; so it will have a geometrical meaning, independent of the basis used to evaluate the vector components.[11]

Special Relativity is formulated in a four-dimensional pseudo-Euclidean space-time. In Minkowski space-time the linear transformations of our interest are those of Lorentz group, which leave the scalar product (7.29) invariant: spatial rotations, boosts, and their combinations. All of these transformations have determinant equal to 1 (excluding coordinate reflections). Thus the notion of n-volume (7.54) is also applicable to define the *four-volume* in Minkowski space-time, since its invariance under Lorentz group transformations is guaranteed.

---

[11] The orientation of n-volume changes under the reflection of a coordinate, because $\det\left(\Lambda^{i'}_i\right) = -1$.

In particular the four-volume element delimited by four infinitesimal vectors directed along the Cartesian axes

$$A^i=(\Delta x^0, 0, 0, 0), \quad B^i=(0, \Delta x^1, 0, 0), \quad C^i=(0, 0, \Delta x^2, 0), \quad D^i=(0, 0, 0, \Delta x^3)$$

is

four-volume element

$$\Delta^4 x = \Delta x^0 \, \Delta x^1 \, \Delta x^2 \, \Delta x^3 = c \; \Delta t \; \Delta x \; \Delta y \; \Delta z \tag{7.56}$$

which coincides with the four-volume of an Euclidean geometry. The orientation of four-volume (7.56), which depends on the order chosen for vectors A^i, B^i, C^i, D^i is the one induced by the coordinate system.

As the $n=2$ and $n=3$ examples show, determinant (7.54) can be written with the help of Levi-Civita symbol

$$\underbrace{\varepsilon_{ijkl\ldots}}_{n\text{ indexes}} = \begin{cases} 1 & \text{if} \quad ijkl\ldots\ldots \text{ is an even permutation of } 0123\ldots\ldots \\ -1 & \text{if} \quad ijkl\ldots\ldots \text{ is an odd permutation of } 0123\ldots\ldots \\ 0 & \text{in other cases} \end{cases} \tag{7.57}$$

In fact, determinant (7.54) is equal to

$$det \begin{pmatrix} A^1 & A^2 & A^3 & \cdots \\ B^1 & B^2 & B^3 & \cdots \\ C^1 & C^2 & \cdots & \\ \cdots & \cdots & & \end{pmatrix} = \underbrace{\varepsilon_{ijk\ldots}}_{n\text{ indexes}} \underbrace{A^i \, B^j \, C^k \ldots}_{n\text{ vectors}} \tag{7.58}$$

In particular $\varepsilon_{ijk\ldots} \, \Lambda^i_{\;i'} \, \Lambda^j_{\;j'} \, \Lambda^k_{\;k'} \ldots$ is equal to $\pm$ det $(\Lambda^l_{\;l'})$, according to $i' \, j' \, k' \ldots$ is an even or odd permutation of the natural succession, and zero in other cases. This result can be written in the following way:

$$\varepsilon_{i'\,j'\,k'\ldots} = \det{(\Lambda^l_{\;l'})}^{-1} \; \varepsilon_{ijk\ldots} \; \Lambda^i_{\;i'} \; \Lambda^j_{\;j'} \; \Lambda^k_{\;k'} \ldots \tag{7.59}$$

where $\varepsilon_{i'j'k'} \ldots$ is defined in the same way of (7.57). For transformations belonging to the *proper* Lorentz group (those that do not reflect an odd number of coordinates) it is det $(\Lambda^l_{\;l'}) = 1$; under such transformations the Levi-Civita symbol then behaves like a completely antisymmetric n-index tensor (antisymmetric in all pair of indexes). Instead, whenever a coordinate is reflected it is det $(\Lambda^l_{\;l'}) = -1$, which reveals the pseudotensor character of Levi-Civita symbol under Lorentz transformations.[12]

Let us now consider an $n-1$ dimensional surface, which will be called *hypersurface*. A hypersurface element (i.e., an elemental $(n-1)$-volume) can be characterized by $n-1$ linearly independent infinitesimal vectors tangent to the hypersurface Δx^i_A, Δx^i_B, Δx^i_C, ... The covariant vector

$$\Delta\Sigma_i \equiv \varepsilon_{ijkl\ldots} \underbrace{\Delta x^j_A \, \Delta x^k_B \, \Delta x^l_C \ldots}_{n-1\text{ vectors}} \tag{7.60}$$

[12] A pseudotensor is an object whose components change under arbitrary linear transformations in the following way: $P_{i'\ldots}{}^{j'k'\ldots} = \text{sign}[\det \Lambda^{l'}{}_l] \, \Lambda^i{}_{i'} \ldots \Lambda^{j'}{}_j \, \Lambda^{k'}{}_k \ldots P_{i\ldots}{}^{jk\ldots}$.

is *normal* to the hypersurface because the contraction with any vector tangent to the hypersurface results to be zero, since all tangent vectors are necessarily linear combinations of $\Delta x^i{}_A$, $\Delta x^i{}_B$, $\Delta x^i{}_C$, ...[13]

- *Example*: for $n = 3$, let us consider the hypersurface element defined by the sides $\Delta x^i{}_A = (\Delta x, 0, 0)$ and $\Delta x^i{}_B = (0, \Delta y, 0)$; then

$$\Delta\Sigma_x = 0, \qquad \Delta\Sigma_y = 0, \qquad \Delta\Sigma_z = \Delta x \, \Delta y$$

Let $a, b, c, \ldots$ be coordinates on a hypersurface, and let us examine an hypersurface element whose sides $\Delta x^i{}_A$, $\Delta x^i{}_B$, $\Delta x^i{}_C$, ...are tangent to the coordinate lines $a, b, c, \ldots$ respectively, so delimiting a hypersurface element with infinitesimal sides Δa, Δb, Δc, ... Thus

$$\Delta\Sigma_i = \varepsilon_{i\ j\ k\ l\ldots} \frac{\Delta x^j_A}{\Delta a} \frac{\Delta x^k_B}{\Delta b} \frac{\Delta x^l_C}{\Delta c} \cdots \Delta a \, \Delta b \, \Delta c \ldots$$

If differences go to zero, then the quotients in the right side will become partial derivatives of coordinates x^j with respect to the coordinates defined on the hypersurface.[14] The contraction with Levi-Civita symbol combines them to form the determinant of a partial derivatives $(n-1) \times (n-1)$ matrix, which is the Jacobian $\partial(x^j, x^k, x^l, \ldots)/\partial(a, b, c, \ldots)$ of the coordinate change between the $n-1$ coordinates $a, b, c, \ldots$ and an equal number of other coordinates that result from subtracting the ith coordinate from the set $\{x^j\}$. The result is

$$d\Sigma_i = \varepsilon_{i|j\,k\,l\ldots|} \frac{\partial(x^j, x^k, x^l, \ldots)}{\partial(a, b, c, \ldots)} \, da \, db \, dc \ldots \tag{7.61}$$

hypersurface element

where $\varepsilon_{i|j\ k\ l\ldots|}$ realizes the absence of the ith coordinate in the Jacobian, and bars in $|j\ k\ l\ \ldots\ |$ indicate that indexes $j\ k\ l\ \ldots$ labeling the Jacobian columns are increasingly ordered.[15] The orientation of the hypersurface element (7.61), which comes from the order of its sides $\Delta x^i{}_A$, $\Delta x^i{}_B$, $\Delta x^i{}_C$, ..., is determined by the coordinate system $a, b, c, \ldots$ defined on the hypersurface.

Example: For $n = 3$, let us consider a spherical surface of radius R, and angular coordinates θ, φ defined on it (this coordinate ordering corresponds to the *exterior* orientation[16]):

$$x = R \sin\theta \cos\varphi, \qquad y = R \sin\theta \sin\varphi, \qquad z = R \cos\theta$$

---

[13] If n^i is the unitary vector normal to the hypersurface, then $\Delta\sigma = n^i \, \Delta\Sigma_i = \varepsilon_{i\ j\ k\ l\ldots} n^i \Delta x^j{}_A \, \Delta x^k{}_B \, \Delta x^l{}_C$ ...has the value of the *induced* $(n-1)$-volume on the hypersurface, since it is the n-volume of a right parallelepiped whose base is the hypersurface element and its height is unitary.

[14] The components of $\Delta x^j{}_A$ are the differences of coordinates x^j when coordinate a undergoes an increase Δa while $b, c, \ldots$ remain constant.

[15] If there were not bars, the Jacobian would be repeated $(n-1)!$ times. Equivalently: $d\Sigma_i = \frac{1}{(n-1)!} \varepsilon_{i\ j\ k\ l\ldots} \partial(x^j\, x^k\, x^l \ldots)/\partial(a\, b\, c \ldots) \, da \, db \, dc \ldots$

[16] $d\Sigma_i$ has exterior orientation if $V^i \, d\Sigma_i > 0$ whenever V^i is directed toward the outside.

$$d\Sigma_x = \frac{\partial(y,z)}{\partial(\theta,\varphi)} d\theta\, d\varphi = \det\begin{pmatrix} \frac{\partial y}{\partial\theta} & \frac{\partial z}{\partial\theta} \\ \frac{\partial y}{\partial\varphi} & \frac{\partial z}{\partial\varphi} \end{pmatrix} d\theta d\varphi = R^2 \sin^2\theta \cos\varphi d\theta d\varphi$$

$$d\Sigma_y = \frac{\partial(z,x)}{\partial(\theta,\varphi)} d\theta d\varphi = -\det\begin{pmatrix} \frac{\partial x}{\partial\theta} & \frac{\partial z}{\partial\theta} \\ \frac{\partial x}{\partial\varphi} & \frac{\partial z}{\partial\varphi} \end{pmatrix} d\theta d\varphi = R^2 \sin^2\theta \sin\varphi\, d\theta d\varphi$$

$$d\Sigma_z = \frac{\partial(x,y)}{\partial(\theta,\varphi)} d\theta\, d\varphi = \det\begin{pmatrix} \frac{\partial x}{\partial\theta} & \frac{\partial y}{\partial\theta} \\ \frac{\partial x}{\partial\varphi} & \frac{\partial y}{\partial\varphi} \end{pmatrix} d\theta d\varphi = R^2 \sin\theta \cos\theta\, d\theta d\varphi$$

A hypersurface in a pseudo-Euclidean geometry is said to be *spacelike* if all the vectors tangent to the hypersurface have spacelike character. The vector $d\Sigma_i$ of a spacelike hypersurface is timelike, because the contraction of $d\Sigma_i$ and any (spacelike) vector tangent to the hypersurface is null.

The contraction $V^i\, d\Sigma_i$ of a contravariant vector and a hypersurface element is an invariant called *flux*. This denomination is extended to contractions of the hypersurface element with any tensor.

flux

7.6. ENERGY–MOMENTUM TENSOR OF CONTINUOUS MEDIA

In an extensive body we can distinguish two types of forces between parts. Forces associated with pressures, deformations, and viscosities are due to interactions between close molecules; these are short-range forces because they act in the range of intermolecular distances. In a macroscopic approximation, where the molecular structure is neglected and the body is regarded as a *continuous medium*, the short-range forces are treated as "contact" interactions that happen on the surfaces separating contiguous *volume elements* in the three-dimensional space.[17] On the other hand, there exist interactions "at a distance" as the ones coming, for instance, from macroscopic charge distributions in the medium. In Relativity, however, the Newtonian interactions at a distance are substituted by local interactions where two distant matter portions influence each other through an intermediate field that carries energy–momentum. Matter locally interacts with the intermediate field exchanging energy–momentum; the intermediate field propagates and partially transfers energy–momentum to another matter portion through an equally local interaction. As discussed in §6.1, the locality of interactions is an essential requirement in Relativity in order to make possible the formulation of conservation

[17] The size of a volume element is "infinitesimal" compared with the scale of distances characteristic of the macroscopic approximation, although it is big enough to contain a large number of molecules.

laws fulfilled in any reference system. Conservation of energy and momentum of an isolated system cannot be the result of compensations among simultaneous variations at distant places since simultaneity has not an absolute meaning, but must be locally accomplished (at each event in space-time). But this implies that the intermediate field is a part of the isolated system under consideration, and its energy–momentum must enter in the balance of conservation laws.

The model for the relativistic conservation of a continuously distributed magnitude is the continuity equation (7.20), which expresses the electric *charge* local conservation. The structure of this equation is repeated in any other local conservation law. In particular, the energy–momentum conservation in a continuous medium is expressed in a similar way. So it will be very useful to stop here to analyze this structure:

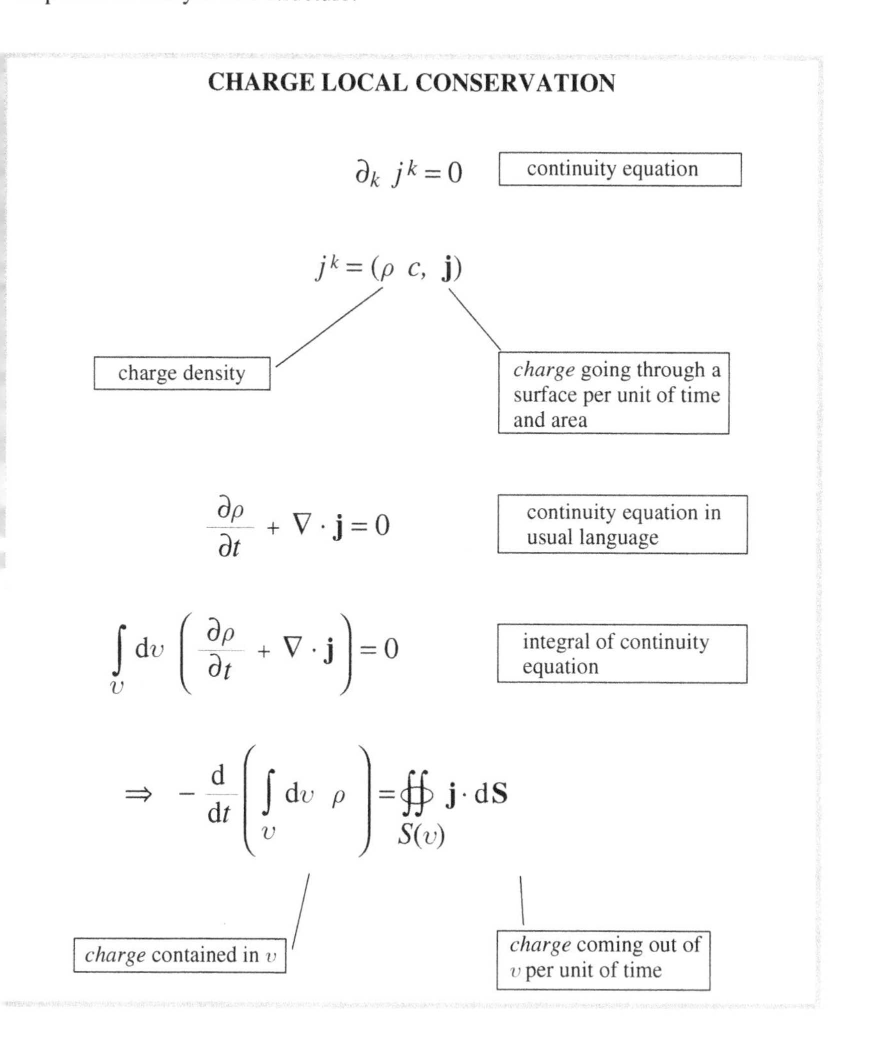

The last equation, which has been obtained by means of the divergence theorem in three dimensions, states that the variation in time of the *charge* contained in v is exclusively due to the incoming or outgoing *charge* through the v surface. *Charge* is not created or annihilated in v; *charge* is locally conserved. This integral equation, which is equivalent to the starting differential equation, can also be written in a manifestly covariant form by taking into account that the surface element $\Delta\mathbf{S}$ together with the elapsed time Δt define a hypersurface element. In fact, we know that the area $\mathbf{S}$ of the parallelogram delimited by two vectors $\mathbf{A}$ and $\mathbf{B}$ is equal to the vector product $\mathbf{A} \times \mathbf{B}$, whose Cartesian components are $\varepsilon_{\lambda\mu\nu} A^\mu B^\nu = \varepsilon_{0\lambda\mu\nu} A^\mu B^\nu$ ($\lambda = 1, 2, 3$). If $\Delta x^i{}_A = (0, \Delta x^\mu{}_A)$ and $\Delta x^i{}_B = (0, \Delta x^\nu{}_B)$ are sides of $\Delta\mathbf{S}$, we shall build a hypersurface element associated with $\Delta\mathbf{S}$ with the help of a third side $\Delta x^i{}_T = (c\,\Delta t, 0, 0, 0)$:

$$\Delta\Sigma_k = \varepsilon_{kijl}\, \Delta x^i_A\, \Delta x^j_T\, \Delta x^l_B = c\,\Delta t\, \varepsilon_{0k\mu\nu}\, \Delta x^\mu_A\, \Delta x^\nu_B = (0,\ c\,\Delta t\, \Delta\mathbf{S}) \qquad (7.62)$$

By integrating in time the last equation in the former box, it is obtained

$$\Delta\, charge = -\int c^{-1} j^k\, \mathrm{d}\Sigma_k$$

variation of *charge* contained in v = *charge* going through S (v)

$$\mathrm{d}\Sigma_k = (0,\ c\, \mathrm{d}t\, \mathrm{d}\mathbf{S})$$

hypersurface element

Also the (invariant) *charge* contained in v can be expressed in a manifestly covariant way. If $\Delta x^i{}_A$, $\Delta x^i{}_B$, and $\Delta x^i{}_C$ are three linearly independent four-vectors such that $\Delta x^0{}_A = \Delta x^0{}_B = \Delta x^0{}_C = 0$, then they define a Δv and a hypersurface element

$$\Delta\Sigma_k = \varepsilon_{k\lambda\mu\nu}\, \Delta x^\lambda_A\, \Delta x^\mu_B\, \Delta x^\nu_C = (\Delta v,\ 0,\ 0,\ 0) \qquad (7.63)$$

The *charge* contained in the 3-volume v is written as

$$charge = \int_v \mathrm{d}v\ \rho = \int c^{-1}\, j^k\, \mathrm{d}\Sigma_k$$

charge contained in v

$$\mathrm{d}\Sigma_k = (\mathrm{d}v, 0, 0, 0)$$

hypersurface element

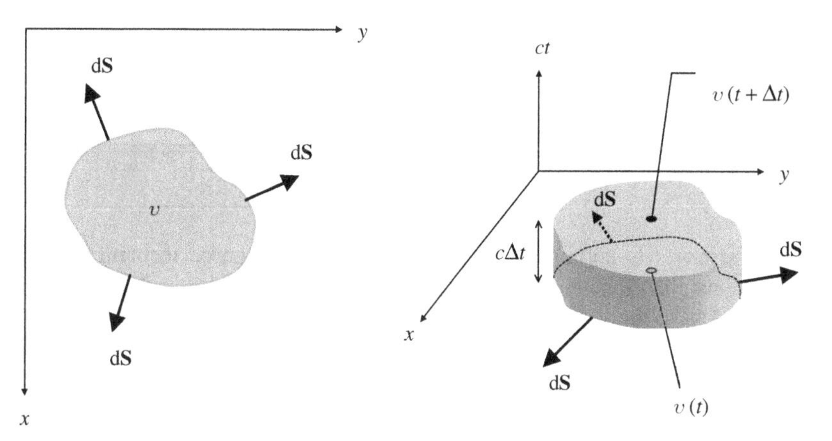

Figure 7.2. A fixed volume and its surface as they are seen at a given time t (left), and in the space-time (right). Coordinate z has been suppressed.

Thus, the variation of the *charge* contained in the 3-volume v, which was indicated with Δ *charge* in a former box, can also be written as a difference between two integrals of this last type evaluated at times t and $t+\Delta t$. Figure 7.2 shows a 3-volume v and its surface $S(v)$ (coordinate z was suppressed), together with its development in Minkowski space-time. Taking into account the exterior orientations for the hypersurfaces indicated in Figure 7.3, the charge balance in covariant language results to be

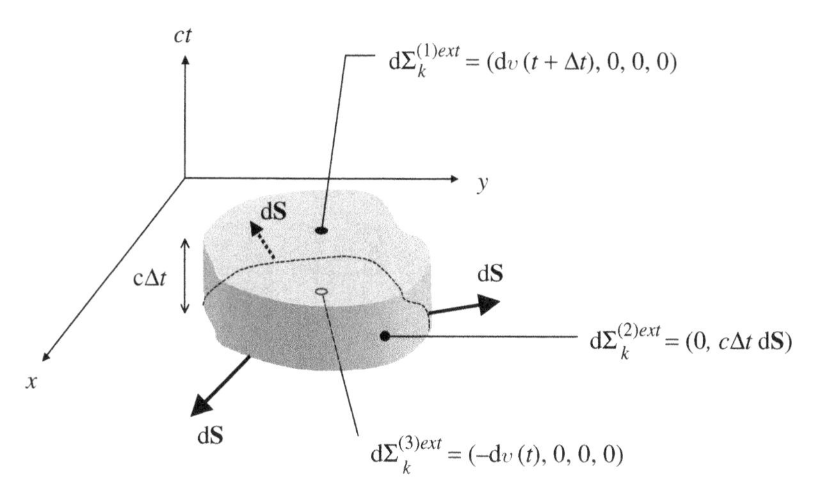

Figure 7.3. Components of externally oriented hypersurface elements.

$$\int_{\Sigma^{(1)}} c^{-1} j^k \mathrm{d}\Sigma_k^{(1)ext} + \int_{\Sigma^{(3)}} c^{-1} j^k \mathrm{d}\Sigma_k^{(3)ext} = - \int_{\Sigma^{(2)}} c^{-1} j^k \mathrm{d}\Sigma_k^{(2)ext}$$

$$\Rightarrow \oint_{\Sigma} j^k \mathrm{d}\Sigma_k = 0$$

the flux of j^k through a closed hypersurface is zero

This last result is nothing but a corollary of the divergence theorem in four dimensions: the vanishing of (four) divergence of j^k, which is our starting point, is equivalent to the vanishing of the j^k flux through any closed hypersurface.

Since we now know the structure of a local conservation law, we only have to replace the word *charge* with the conserved four-tensor magnitude of interest, and the four-vector j^k with the corresponding density.

energy–momentum local conservation

In this section we shall describe the local conservation of the energy–momentum p^i for a continuous medium. In the absence of external influences, the energy–momentum p^i contained in a 3-volume v can only vary as the result of local interactions happening at the surface $S(v)$[18]. These local interactions cause energy–momentum exchanges between contiguous volumes. The exchanged energy–momentum per unit of time and area plays the role of vector **j**, while the energy–momentum per unit of volume plays the role of ρ; however, because the conserved magnitude is not a scalar Q but a four-vector p^i, it will be necessary a four-tensor T^{ik} $(t, \mathbf{r})$ in the place of the four-vector j^k $(t, \mathbf{r})$ in order to describe the densities and flux of each p^i component:

$$charge \longrightarrow p^i$$
$$j^k \longrightarrow T^{ik}$$

energy–momentum tensor

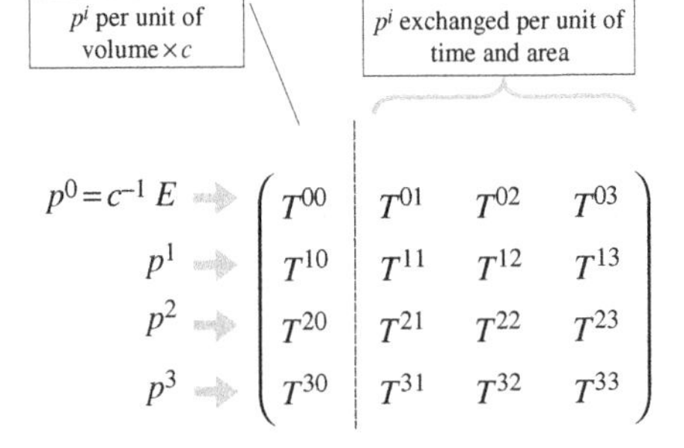

Four–tensor T^{ik} is called *energy–momentum tensor*. In a given reference system, T^{ik} components have the following meanings:

[18] p^i includes the energy–momentum of matter and intermediate fields.

- $c^{-1}\, T^{i0}$ is the density of the ith component of four-vector p^i. In particular T^{00} is the energy density. The three quantities $c^{-1}\, T^{\mu 0}$ ($\mu = 1, 2, 3$) behave like the components of an ordinary vector under spatial rotations; they describe the density of vector $\mathbf{p}$.
- $T^{i\nu}$ ($\nu = 1, 2, 3$) is the exchanged energy–momentum p^i per unit of time and area through a surface having normal $\hat{\mathbf{x}}^\nu$. In particular $c\; T^{0\nu}$ are the three components of the ordinary vector describing the energy exchange per unit of time and area, while the quantities $T^{\mu\nu}$ take into account the exchanges of momentum components p^μ. Following the analogy with *charge* conservation, let us write the variation of the momentum contained in a fixed 3-volume v; then we shall divide by Δt to obtain the force on the 3-volume v due to the interaction with the contiguous volumes:

$$\Delta p^\mu = \int c^{-1} T^{\mu k}\, \mathrm{d}\Sigma_k$$

$$\mathrm{d}\Sigma_k = (0, -c\mathrm{d}t\, \mathrm{d}\mathbf{S}) = (0, -c\mathrm{d}t\, \hat{\mathbf{n}}\, \mathrm{d}S)$$

$$\Rightarrow \quad F^\mu = -\oiint_{s(v)} T^{\mu\nu}\, n_\nu \mathrm{d}S \qquad \boxed{\text{force on 3-volume } v}$$

stress tensor

Therefore $-T^{\mu\nu} n_\nu$ is the force component μ exerted from the exterior of a surface $\Delta\mathbf{S} = \hat{\mathbf{n}}\Delta S$ (exterior means the region toward the normal $n_\nu = \hat{\mathbf{n}}$ points to). – $T^{\mu\nu}$ behaves like a tensor under rotations in the three-dimensional space, and is called *stress tensor.*

Completing the analogy, we shall write the equation accomplished by T^{ik} which expresses the local conservation of the energy–momentum of an isolated system:

$$\partial_k T^{ik} = 0 \Leftrightarrow \oint_\Sigma T^{ik}\, \mathrm{d}\Sigma_k = 0 \tag{7.64}$$

where Σ is any fixed closed hypersurface. Equation (7.64) must be a consequence of the dynamical laws of any isolated system.

If the system is not isolated, then the *charge* balance in the last equation of the first box of this section must be modified to include external contributions. In particular, if $f_{\text{ext}}{}^i$ is the external force per unit of volume acting on the system, the energy–momentum balance results in

$$\frac{\mathrm{d}}{\mathrm{d}t}\int_v \mathrm{d}v\, c^{-1}\, T^{i0} = -\oiint_{s(v)} T^{i\nu}\, n_\nu\, \mathrm{d}S + \int_v \mathrm{d}v\, f^i_{\text{ext}}$$

$\boxed{p^i \text{ contained in } v}$ $\qquad$ $\boxed{p^i \text{ transferred per unit of time through } S(v)}$

which expresses that the variation of the energy–momentum contained in a fixed 3-volume υ is due to the entrance of energy–momentum coming from contiguous volume elements belonging to the same physical system, plus the contribution of agents foreign to the system. Since the 3-volume υ does not change in time, the derivative d/dt can enter the integral as ∂_0. On the other hand, the three-dimensional divergence theorem transforms the surface integral into a volume integral so yielding

$$\int_{\upsilon} \mathrm{d}\upsilon \left(\partial_0 T^{i0} + \partial_\nu T^{i\nu}\right) = \int_{\upsilon} \mathrm{d}\upsilon\, f^{\,i}_{\text{ext}}$$

i.e.,[19]

$$\partial_k T^{ik} = f^{\,i}_{\text{ext}} \tag{7.65}$$

Finally, the energy–momentum tensor has a property that is required for the angular momentum conservation: T^{ik} is symmetric. In fact, the angular momentum of a continuous medium without spin, contained in 3-volume υ, is written by means of the energy–momentum density T^{i0}:

$$L^{ij} = c^{-1} \int_{\upsilon} (x^i T^{j0} - x^j T^{i0})\, \mathrm{d}\upsilon \tag{7.66}$$

The condition for the local conservation of the angular momentum of an isolated system straightforwardly derives by noticing that the former expression can be written as

$$L^{ij} = \int_{\Sigma} c^{-1} M^{ijk}\, \mathrm{d}\Sigma_k \tag{7.67}$$

where Σ is the hypersurface t = *constant* (as shown, $\mathrm{d}\Sigma_k = (d\upsilon, 0, 0, 0)$), and M^{ijk} is a tensor containing the angular momentum density and the exchanged angular momentum per unit of time and area through surfaces oriented along the Cartesian axes:

$$M^{ijk} \equiv x^i T^{jk} - x^j T^{ik} \tag{7.68}$$

Local conservation of angular momentum is satisfied if this tensor has divergence equal to zero. Taking into account Eq. (7.64), the result is

$$0 = \partial_k M^{ijk} = \partial_k (x^i T^{jk} - x^j T^{ik}) = \delta^i_k T^{jk} - \delta^j_k T^{ik} = T^{ji} - T^{ij}$$

so the energy–momentum tensor must be symmetric

$$T^{ji} = T^{ij} \tag{7.69}$$

[19] Note the four-vector character of the force per unit of volume $f^i = (\mathbf{f} \cdot \mathbf{u}\, c^{-1}, \mathbf{f})$.

In particular, the equality $T^{0\nu} = T^{\nu 0}$ means that the energy exchanged in a time Δt through a surface $\Delta \mathbf{S} = \hat{\mathbf{n}} \Delta S$,

$$\Delta E = T^{0k}\, \Delta\Sigma_k = -T^{0\nu}\, n_\nu\, c\, \Delta t\, \Delta S$$

is always equal to c multiplied by the momentum contained in a 3-volume of base ΔS and height $c\,\Delta t$ projected on the normal $\hat{\mathbf{n}}$. This general result means that any energy transport, whatever its form is – particle flux, heat conduction, electromagnetic radiation, etc. – entails momentum.

flux of free particles

As an example, let us consider a flux of free particles having the same velocity $\mathbf{u}$ and energy E.[20] During the time Δt certain number of these particles will go through the surface $\Delta \mathbf{S}$ carrying some energy. The particles going through $\Delta \mathbf{S}$ are the ones contained in the parallelepiped formed by the base ΔS and the side $\mathbf{u}\,\Delta t$, whose volume is $\mathbf{u} \cdot \Delta \mathbf{S}\ \Delta t$. If there are n particles per unit of volume, then the energy transported through $\Delta \mathbf{S}$ is equal to $n\, E\, \mathbf{u} \cdot \Delta \mathbf{S}\, \Delta t$. Due to the relation $E\,\mathbf{u} = \mathbf{p}\, c^2$ (see Eq. (6.15)), the result is also equal to $n\, c\, \mathbf{p} \cdot \Delta \mathbf{S}\, c\, \Delta t$, which is the projection on the $\Delta \mathbf{S}$ direction of the momentum contained in the volume $\Delta S\, c\, \Delta t$ (times c), so satisfying the equality $T^{0\nu} = T^{\nu 0}$. The components of the energy–momentum tensor of this physical system are simpler in the frame moving with the particles. In the *comoving* frame the particles are at rest: the momentum is zero everywhere, and there is no energy transfer. Therefore the only non-null component is T^{00}, which is the proper energy density $\rho_o = n_o\, E_o$

$$T^{ik} = \begin{pmatrix} \rho_o & 0 & 0 & 0 \\ 0 & 0 & 0 & 0 \\ 0 & 0 & 0 & 0 \\ 0 & 0 & 0 & 0 \end{pmatrix}$$

Because the particles are at rest, their four-velocity is $U^i = (c, 0, 0, 0)$; so it is valid that

$$T^{ik} = \rho_o\, c^{-2}\, U^i\, U^k \tag{7.70}$$

Since the tensor product $U^i\, U^k$ transforms in the same way as T^{ik}, the Eq. (7.70) will remain valid in any other reference system.[21] Therefore, in another frame where the particle four-velocities are $U^i = \gamma(u)\,(c,\ \mathbf{u})$, the energy density will be

$$\rho = T^{00} = \rho_o\, \gamma(u)^2 \tag{7.71}$$

The two factors $\gamma(u)$ come from the transformations of the energy and the volume: $E = \gamma(u)\, E_o$, $v = \gamma(u)^{-1}\, v_o$.

---

[20] An aggregate of non-interacting particles is called *dust*.

[21] Note the analogy with the convective current $j^k = \rho_o\, U^k$ in Eq. (5.12). If U^k is a null four-vector, then Eq. (7.70) describes a pure radiation field.

perfect fluid

As a second example, we shall obtain the energy–momentum tensor for a perfect fluid. By definition, a perfect fluid has no viscosity: in the frame where the fluid is at rest (comoving frame) the forces are perpendicular to the surfaces separating contiguous volumes, whatever the direction of the surface element $\mathbf{\Delta S}$ is. This property implies that the stress tensor $T^{\mu}{}_{\nu}$ is isotropic in the comoving frame: $T^{\mu}{}_{\nu} = -p_o\, \delta^{\mu}{}_{\nu}$, where p_o is the *pressure*. On the other hand, there is no heat conduction in a perfect fluid; the only allowed energy flux is due to the macroscopic movement of matter. Thus, in the comoving frame it is $T^{0\nu} = 0 = T^{\nu 0}$, so it results

$$T^{ik} = \begin{pmatrix} \rho_o & 0 & 0 & 0 \\ 0 & p_o & 0 & 0 \\ 0 & 0 & p_o & 0 \\ 0 & 0 & 0 & p_o \end{pmatrix} = \begin{pmatrix} \rho_o + p_o & 0 & 0 & 0 \\ 0 & 0 & 0 & 0 \\ 0 & 0 & 0 & 0 \\ 0 & 0 & 0 & 0 \end{pmatrix} + \begin{pmatrix} -p_o & 0 & 0 & 0 \\ 0 & p_o & 0 & 0 \\ 0 & 0 & p_o & 0 \\ 0 & 0 & 0 & p_o \end{pmatrix}$$

which can be written in the following way:

$$T^{ik} = (\rho_o + p_o)\, c^{-2}\, U^i\, U^k - p_o\, g^{ik} \tag{7.72}$$

The dust studied in the previous example can be regarded as a perfect fluid without pressure. Due to its tensor character, Eq. (7.72) is valid not only in the comoving frame but in any reference system. Therefore, in a frame where the fluid moves with velocity $\mathbf{u}$, it results

$$\rho = T^{00} = (\rho_o + p_o)\, \gamma(u)^2 - p_o = \gamma(u)^2\, (\rho_o + u^2\, c^{-2}\, p_o) \tag{7.73}$$

Remarkably, the approximation of this expression for small velocities

$$\rho \cong \rho_o + (\rho_o + p_o)\, \frac{u^2}{c^2} \tag{7.74}$$

exhibits a contribution of the pressure to the inertial mass (the expected factor 1/2 for the kinetic energy contained in a volume v results from the relation between v and the proper volume v_o: $v \cong (1 - u^2\, c^{-2}/2)\, v_o$). So, the Newtonian approximation requires not only velocities much smaller than the speed of light, but also $p_o << \rho_o$.

The contribution of p_o to the inertial mass can be noticed in the dynamical equation resulting from the spatial components of Eq. (7.64):

$$\partial_k \left[(\rho_o + p_o)\, c^{-2}\, U^\mu\, U^k - p_o\, g^{\mu k}\right] = 0$$

$$\Rightarrow \quad U^\mu\, \partial_k \left[(\rho_o + p_o)\, c^{-2}\, U^k\right] + (\rho_o + p_o)\, c^{-2}\, U^k\, \partial_k U^\mu - g^{\mu\nu}\, \partial_\nu\, p_o = 0$$

The first term is zero in the comoving frame. In the second term, the expression $U^k\, \partial_k\, U^\mu = (\mathrm{d}x^k/\mathrm{d}\,\tau)\, \partial_k\, U^\mu = \mathrm{d}U^\mu/\mathrm{d}\tau$ gives the spatial components of the four-acceleration a^j, which is equal to $\boldsymbol{a}$ in the comoving frame (see Eq. (7.32)). Then the dynamical equation results in

$$(\rho_o + p_o)\, c^{-2}\boldsymbol{a} + \nabla\, p_o = 0 \tag{7.75}$$

which only acquires the Newtonian form if $p_o << \rho_o$.

The temporal component of Eq. (7.64) can be extracted by contracting it with U_i. Remembering that $U_i\,U^i = c^2$, the result is

$$\partial_k\left[(\rho_o + p_o)\,U^k\right] + (\rho_o + p_o)\,c^{-2}\,U_i\,a^i - U^i\,\partial_i\,p_o = 0$$

The second term is null: $U_i\,a^i \equiv 0$ (see Eq. (7.30)). On the other hand, if the number of particles is conserved then the four-vector $n_o\,U^k$, describing the particle density and current (compare with four-vector j^k in Eq. (5.12)), must fulfill a continuity equation:

$$\partial_k(n_o\,U^k) = 0 \qquad (7.76)$$

Using this result in the previous equation, and taking into account that $U^k\,\partial_k = (\mathrm{d}x^k/\mathrm{d}\,\tau)\,\partial_k = \mathrm{d}/\mathrm{d}\,\tau$ (the derivative along the world lines of fluid elements), it is obtained

$$n_o\,\frac{\mathrm{d}}{\mathrm{d}\tau}\left[\frac{\rho_o + p_o}{n_o}\right] - \frac{\mathrm{d}p_o}{\mathrm{d}\tau} = 0 \qquad \Rightarrow \frac{\mathrm{d}}{\mathrm{d}\tau}\left(\frac{\rho_o}{n_o}\right) = -p_o\,\frac{\mathrm{d}}{\mathrm{d}\tau}\left(\frac{1}{n_o}\right) \qquad (7.77)$$

which means that the variation of the energy per particle is equal to the pressure–volume work per particle, hence describing the energy conservation in the absence of heat transfer.

7.7. ELECTROMAGNETISM

The four-tensor language makes possible to write the laws of electromagnetism in a manifestly covariant way.[22] Taking into account the four-vector behavior of the electromagnetic potential $A^i = (c^{-1}\phi, \mathbf{A})$, then the antisymmetric magnitude

$$F_{ij} \equiv \partial_i\,A_j - \partial_j\,A_i \qquad (7.78)$$

field tensor

has a guaranteed tensor character. F_{ij} is called *field tensor*. According to Eq. (5.18), the components of the field tensor are

$$F_{ij} = \begin{pmatrix} 0 & c^{-1}E_x & c^{-1}E_y & c^{-1}E_z \\ -c^{-1}E_x & 0 & -B_z & B_y \\ -c^{-1}E_y & B_z & 0 & -B_x \\ -c^{-1}E_z & -B_y & B_x & 0 \end{pmatrix} \qquad (7.79)$$

[22] Formulae in this Section can be converted into the Gaussian System by means of the substitutions $\mu_0 \to 4\,\pi\,c^{-2}$, $\mathbf{B} \to \mathbf{B}\,c^{-1}$, $\mathbf{A} \to \mathbf{A}\,c^{-1}$, $A_i \to A_i\,c^{-1}$, $F_{ij} \to F_{ij}\,c^{-1}$.

The *dual* tensor is [23]

$$^*F_{ij} \equiv \frac{1}{2}\,\varepsilon_{ijkl}\,F^{kl} = \begin{pmatrix} 0 & -B_x & -B_y & -B_z \\ B_x & 0 & -c^{-1}E_z & c^{-1}E_y \\ B_y & c^{-1}E_z & 0 & -c^{-1}E_x \\ B_z & -c^{-1}E_y & c^{-1}E_x & 0 \end{pmatrix} \tag{7.80}$$

The following invariants can be built with tensors F_{ij} and $^*F_{ij}$ [24]

$$^*F_{ij}\,F^{ij} = 4\,c^{-1}\,\mathbf{E}\cdot\mathbf{B} \tag{7.81}$$

$$-F_{ij}\,F^{ij} = {}^*F_{ij}\,{}^*F^{ij} = 2\,(c^{-2}\,E^2 - B^2) \tag{7.82}$$

By using these tensors, the laws of electromagnetism can be written in a manifestly covariant way. The equations without sources (1B.1a–b) are contained in

$$\partial_i\,F_{jk} + \partial_j\,F_{ki} + \partial_k\,F_{ij} = 0 \qquad \text{or} \qquad \partial_j\,{}^*F^{ij} = 0 \tag{7.83}$$

In any of its two manners, Eq. (7.83) involves four independent equations, which correspond to the scalar equation (1B.1a) and the three components of vector equation (1B.1b). In particular, the equation at left consists in the vanishing of the components of a completely antisymmetric 3-index tensor; in the four-dimensional space-time, such a tensor has only four independent components (components with two repeated indexes are identically null; there are four ways to choose three different indexes among four possible values). Both forms of Eq. (7.83) are mutually dual. As it happens with Eqs. (1B.1a–b), Eq. (7.83) will reduce to an identity if the fields are substituted as functions of the potentials (in this case, by using the definition (7.78)).
The four equations with sources (1B.2a–b) are expressed as

$$\partial_i{}^*F_{jk} + \partial_j{}^*F_{ki} + \partial_k{}^*F_{ij} = \mu_o{}^*j_{ijk} \qquad \text{or} \qquad \partial_j\,F^{ij} = -\mu_o\,j^i \tag{7.84}$$

The continuity equation (1B.3) or (7.20) immediately results from the former equation by noticing that $\partial_i\,\partial_j\,F^{ij} \equiv 0$ (because it is the contraction between a symmetric tensor operator and an antisymmetric tensor).

The law for potential A^i is obtained by using definition (7.78) in Eq. (7.84):

$$-\mu_o\,j_i = g^{kj}\,\partial_j\,F_{ik} = g^{kj}\,\partial_j\,(\partial_i\,A_k - \partial_k\,A_i) = \partial_i\,\partial_j\,A^j - \Box A_i \tag{7.85}$$

where $\Box$ is the d'Alembertian operator

d'Alembertian

$$\Box \equiv g^{ij}\,\partial_i\,\partial_j = \frac{1}{c^2}\frac{\partial^2}{\partial t^2} - \frac{\partial^2}{\partial x^2} - \frac{\partial^2}{\partial y^2} - \frac{\partial^2}{\partial z^2} \tag{7.86}$$

---

[23] The dual of a completely antisymmetric r-index tensor is defined as the completely antisymmetric $(n-r)$-index (pseudo) tensor that results from contracting the tensor with Levi-Civita symbol and dividing by $r!$.

[24] $^*F_{ij}\,F^{ij}$ is a pseudo-scalar.

Since partial derivatives commute, then the field tensor does not change under a *gauge transformation*

$$A_i \to A_i + \partial_i \xi \qquad F_{ij} \to F_{ij} \tag{7.87}$$

We can take advantage of this gauge freedom by choosing ξ in such a way that the A_i four-divergence results to be zero. This choice is called *Lorentz gauge*:

$$\partial_i A^i = 0 \tag{7.88}$$

Lorentz gauge

In the Lorentz gauge the Eq. (7.85) reduces to

$$\Box A^i = \mu_o j^i \tag{7.89}$$

Let us try a monochromatic plane wave to solve the homogeneous ($j^i = 0$) equation: $A^i = A_o{}^i \cos(k_j x^j)$, where $A_o{}^i$ does not depend on the coordinates and $k^j = 2\pi\nu(c^{-1}, w^{-1}\hat{\mathbf{n}})$ is the wave four-vector (the invariant $k_j x^j$ is the wave *phase*). Using this wave in Eq. (7.89), it results that the wave four-vector is null: $k_j k^j = 0$, which means $w = c$. The fields are $\mathbf{E} = -\partial\mathbf{A}/\partial t - \nabla\phi = 2\pi\nu(\mathbf{A}_o - c^{-1}\hat{\mathbf{n}}\phi_o)\sin(k_j x^j)$, and $\mathbf{B} = \nabla\times\mathbf{A} = 2\pi\nu c^{-1}\mathbf{A}_{o\perp}\sin(k_j x^j)$, where $\mathbf{A}_{o\perp} = \hat{\mathbf{n}}\times\mathbf{A}_o = \mathbf{A}_o - \hat{\mathbf{n}}(\hat{\mathbf{n}}\cdot\mathbf{A}_o)$. On the other hand, Lorentz gauge (7.88) requires that $k_j A_o{}^j = 0$, i.e., $\phi_o = c\hat{\mathbf{n}}\cdot\mathbf{A}_o$. Thus, $\mathbf{E} = 2\pi\nu\mathbf{A}_{o\perp}\sin(k_j x^j)$. Therefore, the fields **E** and **B** only depend on the transversal potential. This means that only two genuine degrees of freedom are contained in the four-potential, which correspond to the two independent transversal polarization directions. Equation (7.88) does not completely fix the potential gauge freedom, since it only determinates a relation between temporal and longitudinal A^i components; but it is sufficient to express the fields in terms of authentic degrees of freedom.[25]

The four-force (7.15) associated with the force exerted by the electromagnetic field on a charged particle, the Lorentz force $\mathbf{F} = q(\mathbf{E} + \mathbf{u}\times\mathbf{B})$, is

$$K^i = q F^i{}_j U^j \tag{7.90}$$

If charged particles are distributed in a volume element with charge density ρ and current density **j**, then the force per unit of volume exerted by the field on the element is

$$\mathbf{f} = \rho\mathbf{E} + \mathbf{j}\times\mathbf{B} \tag{7.91}$$

This force per unit of volume is the spatial part of the four-vector

$$f^i = F^i{}_k j^k = (c^{-1}\mathbf{E}\cdot\mathbf{j}, \mathbf{f}) \tag{7.92}$$

four-force per unit of volume

Equation (7.84) shows that the four-vector f_i is equal to

$$f_i = F_{ij} j^j = -\mu_o{}^{-1} F_{ij}\partial_l F^{jl} = \mu_o{}^{-1}[\partial_l(F_{ij}F^{lj}) + F^{jl}\partial_l F_{ij}]$$

The last term can be rewritten as a divergence. With this aim, we shall decompose it into two equal terms, using the antisymmetric character of field tensor:

$$F^{jl}\partial_l F_{ij} = \frac{1}{2}F^{jl}(\partial_l F_{ij} + \partial_j F_{li})$$

[25] A gauge transformation (7.87) such that $\Box\xi = 0$ does not alter the Lorentz gauge. Thus, the temporal and longitudinal A^i components can be canceled by using $\xi = -(2\pi\nu)^{-1}\phi_o\sin(k_j x^j)$.

So, due to Eq. (7.83) it results

$$F^{jl}\,\partial_l\,F_{ij} = -\frac{1}{2}\,F^{jl}\,\partial_i\,F_{jl} = -\frac{1}{4}\,\partial_i\,(F^{jl}\,F_{jl})$$

Then

$$f_i = \frac{1}{\mu_o}\,\partial_k\left[F_{ij}\,F^{kj} - \frac{1}{4}\,\delta_i^k\,F^{jl}\,F_{jl}\right] \qquad (7.93)$$

According to Eq. (7.65), the dynamics of charged particles is governed by the equation

$$f^i = \partial_k\,T^{ik}_{\text{particles}} \qquad (7.94)$$

Substituting Eq. (7.94) in Eq. (7.93), a local conservation law results:

$$\partial_k\left[T^{ik}_{\text{particles}} + \frac{1}{\mu_o}\left(-F^i{}_j\,F^{kj} + \frac{1}{4}\,g^{ik}\,F^{jl}\,F_{jl}\right)\right] = 0 \qquad (7.95)$$

The system composed by charges and electromagnetic field is an isolated system. Then, its energy–momentum has to be conserved. Therefore, the symmetric tensor in the second term of Eq. (7.95) is the energy–momentum tensor of the electromagnetic field:

electromagnetic energy–momentum tensor

$$T^{ik}_{\text{field}} = \frac{1}{\mu_o}\left(-F^i{}_j\,F^{kj} + \frac{1}{4}\,g^{ik}\,F^{jl}\,F_{jl}\right) \qquad (7.96)$$

Since the *trace* $g^i{}_i$ of metric tensor is equal to 4 ($g^i{}_i = \delta^i{}_i = 4$), it results that the trace $T^i{}_i$ of the electromagnetic energy–momentum tensor is zero. The components of $T^{ik}{}_{\text{field}}$, when written as functions of fields, are

$$T^{ik} = \left(\begin{array}{c|ccc} \rho_{em} & & c^{-1}\mathbf{S} & \\ \hline & \dfrac{-c^{-2}E_x^{\,2} - B_x^{\,2}}{\mu_0} + \rho_{em} & \dfrac{-c^{-2}E_xE_y - B_xB_y}{\mu_0} & \dfrac{-c^{-2}E_xE_z - B_xB_z}{\mu_0} \\ c^{-1}\mathbf{S} & \dfrac{-c^{-2}E_yE_x - B_yB_x}{\mu_0} & \dfrac{-c^{-2}E_y^{\,2} - B_y^{\,2}}{\mu_0} + \rho_{em} & \dfrac{-c^{-2}E_yE_z - B_yB_z}{\mu_0} \\ & \dfrac{-c^{-2}E_zE_x - B_zB_x}{\mu_0} & \dfrac{-c^{-2}E_zE_y - B_zB_y}{\mu_0} & \dfrac{-c^{-2}E_z^{\,2} - B_z^{\,2}}{\mu_0} + \rho_{em} \end{array}\right)$$

where ρ_{em} is the electromagnetic energy density

$$\rho_{em} = \frac{1}{2\,\mu_o}\,(c^{-2}\,\mathbf{E}\cdot\mathbf{E}+\mathbf{B}\cdot\mathbf{B}) \tag{7.97}$$

and **S** is the *Poynting vector*, which measures the energy transfer per unit of time and area:

$$\mathbf{S} = \frac{1}{\mu_o}\mathbf{E}\times\mathbf{B} \tag{7.98}$$

The momentum density vector **g** is formed with components $c^{-1}\,T^{\nu 0}$:

$$\mathbf{g} = c^{-2}\,\mathbf{S} \tag{7.99}$$

The stress (tri) tensor $-T^{\mu\nu}$ is $\mu_o^{-1}\,(c^{-2}\,\mathbf{E}\otimes\mathbf{E}+\mathbf{B}\otimes\mathbf{B})-\mathbf{I}\,\rho_{em}$, where **I** is the identity tensor and symbol $\otimes$ denotes the tensor product between ordinary vectors in the three-dimensional Euclidean space.

As an example, the energy–momentum tensor for a plane wave propagating along the x axis with linear polarization along the z axis ($\mathbf{E} = E_z\,(\mathbf{r},\,t)\,\hat{\mathbf{z}}$, $\mathbf{B} = c^{-1}\,\hat{\mathbf{x}}\times\mathbf{E}$) is

$$T^{ik} = \frac{1}{\mu_o}\begin{pmatrix} c^{-2}\,E^2 & c^{-2}\,E^2 & 0 & 0 \\ c^{-2}\,E^2 & c^{-2}\,E^2 & 0 & 0 \\ 0 & 0 & 0 & 0 \\ 0 & 0 & 0 & 0 \end{pmatrix}$$

Note that the plane wave energy and momentum densities fulfill the relation $\rho_{em} = |\mathbf{g}|\,c$, which is consistent with the energy–momentum relation for a zero mass particle. Besides, the stress tensor says that this field configuration only has pressure along the propagation direction, whose value is coincident with that of ρ_{em}. If a plane wave is normally incident on a material which completely absorbs it, then the wave pressure will translate into a mechanical pressure on the material surface (*radiation pressure*). For a monochromatic plane wave, E^2 can be averaged in a period (the mean value of sine function is 1/2), to obtain that the mean radiation pressure is equal to $(\mu_o^{-1}\,c^{-2}/2)$ times the squared electric field amplitude.[26]

photon gas

Another simple example is an electromagnetic field with random fluctuations. In this case, the field mean values are zero, $\langle\mathbf{E}\rangle = 0 = \langle\mathbf{B}\rangle$, but the mean values of the squared field are non-null: $\langle\mathrm{E}^2\rangle \neq 0 \neq \langle\mathrm{B}^2\rangle$. If fluctuations are isotropic, then $\langle E_x{}^2\rangle = \langle E_y{}^2\rangle = \langle E_z{}^2\rangle = (1/3)\,\langle E^2\rangle$, and the same is valid for the **B** components. If, moreover, the fluctuations of different components are not

[26] Field invariants (7.81) and (7.82) are null for the plane wave. In these cases the energy–momentum tensor cannot be diagonalized by applying a boost. Instead, if at least one of the invariants is non-null, then at each event there will exist frames where $T^i{}_k = \mathrm{diag}\,(\rho_{em},\,-\rho_{em},\,-\rho_{em},\,\rho_{em})$. In these frames the Poynting vector is zero at the event, either because **E** and **B** result to be parallel ($\mathbf{E}\cdot\mathbf{B}\neq 0$ case) or because one of them is zero ($\mathbf{E}\cdot\mathbf{B} = 0$ case); see §5.9. In the precedent diagonal form of $T^i{}_k$, the x^3 axis was chosen to coincide with the field direction in such frames.

correlated, it will result $\langle E_x\ E_y\rangle = 0$, etc. Thus, the energy–momentum tensor for a *photon gas* is obtained:

$$< T^{ik} >= \begin{pmatrix} \rho_{\text{em}} & 0 & 0 & 0 \\ 0 & \frac{1}{3}\rho_{\text{em}} & 0 & 0 \\ 0 & 0 & \frac{1}{3}\rho_{\text{em}} & 0 \\ 0 & 0 & 0 & \frac{1}{3}\rho_{\text{em}} \end{pmatrix}$$

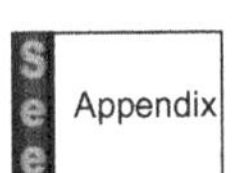

7.8. FERMI–WALKER TRANSPORT

In this section we shall find again the Thomas precession underwent by the series of proper frames $S_{o(t)}$ belonging to an accelerated particle, which was already studied in §4.8. By definition, the particle proper frame S_o is such that its temporal axis has the four-velocity direction. S_o is completed with three mutually orthogonal spacelike directions on the hypersurface normal to the four-velocity. These four directions can be characterized by means of a *tetrad* composed by four orthonormal four-vectors $e^i{}_a$ ($a = 0, 1, 2, 3$ is a label that identifies each one of the four-vectors):

$$e^i{}_0 \equiv c^{-1}\, U^i \tag{7.100}$$

$$e_a \cdot e_b = \eta_{ab} \tag{7.101}$$

where $\eta_{ab} = \text{diag}\,(1, -1, -1, -1)$. If the particle moves with constant velocity, then its proper frame S_o will remain unchanged; thus, the chosen tetrad does not change in time. On the contrary, the particle acceleration gives rise to a series of proper frames, and it becomes necessary to indicate how the frame $S_{o\,(t+dt)}$ is built from $S_{o\,(t)}$. The rule must be such that Eqs. (7.100) and (7.101) remain valid at any time. On the other hand, the transformation going from $S_{o(t)}$ to $S_{o\,(t+dt)}$ has to be merely an infinitesimal Lorentz boost with velocity $\boldsymbol{\beta}\, c = \boldsymbol{a}_o(\tau)\, d\tau$, i.e., the relative velocity between both frames ($\boldsymbol{a}_o$ is the proper acceleration). The spatial rotations of $S_{o\,(t+dt)}$ Cartesian axes with respect to the $S_{o\,(t)}$ ones have to be excluded. We shall show that these requirements are accomplished if the series of tetrads $\{e^i{}_a(\tau)\}$ along the particle world line is governed by the equation

$$\frac{d}{d\tau}\, e^i{}_a(\tau) = c^{-2}\,(a^i\, U_j - U^i\, a_j) e^j{}_a(\tau) \tag{7.102}$$

Equation (7.102) is consistent with (7.100), because the substitution of (7.100) in (7.102) leads to the identity $dU^i/d\tau \equiv a^i$ (use (7.30)). On the other hand, any pair of four-vectors governed by Eq. (7.102) keeps constant its scalar product along the particle world line:

$$\frac{\mathrm{d}}{\mathrm{d}\tau}(g_{ij}A^iB^j) = g_{ij}\frac{\mathrm{d}}{\mathrm{d}\tau}(A^iB^j) = B_i\frac{\mathrm{d}A^i}{\mathrm{d}\tau} + A_i\frac{\mathrm{d}B^i}{\mathrm{d}\tau}$$
$$= c^{-2}(a_i U_j - U_i a_j)(B^i A^j + A^i B^j) \equiv 0$$

(because the contraction of a symmetric tensor and an antisymmetric tensor is null). Therefore the evolution (7.102) conserves the orthonormality (7.101) along the particle world line.

Since the spacelike elements of the tetrad are orthogonal to the four-velocity, then Eq. (7.102) for these elements becomes

$$\frac{\mathrm{d}}{\mathrm{d}\tau}e^i{}_\alpha(\tau) = -c^{-2}U^i a_j e^j{}_\alpha(\tau) \qquad \alpha = 1, 2, 3 \tag{7.103}$$

In the proper frame $S_{o(t)}$ it is

$$a^i = (0, \boldsymbol{a}_o) \qquad U^i = c\,e^i{}_0 = (c, 0, 0, 0) \quad e^i{}_\alpha = (0, \hat{\mathbf{e}}_\alpha)$$

and Eq. (7.103) becomes

$$\Delta e^0{}_\alpha = c^{-1}\boldsymbol{a}_o \cdot \hat{\mathbf{e}}_\alpha \Delta\tau = \boldsymbol{\beta}\cdot\hat{\mathbf{e}}_\alpha \qquad \Delta e^\mu{}_\alpha = O(\Delta\tau^2)$$

which is nothing but an infinitesimal Lorentz boost with velocity $\boldsymbol{\beta}c = \boldsymbol{a}_o\Delta\tau$ (at the first order in β, the spatial Cartesian axes transversal to $\boldsymbol{\beta}$ do not undergo changes, while the longitudinal versor acquires a temporal component equal to β, which is necessary to keep the orthogonality with $e^i{}_0 = c^{-1}U^i$, as it can be seen in Figure 3.17). In this way, a tetrad that evolves according to Eq. (7.102) will satisfy all the requirements imposed to the series of proper frames.

Equation (7.102) defines the *Fermi–Walker transport* of a four-vector along a particle world line. This is the way of carrying a four-vector without causing it a spatial rotation in the proper frame. In particular, if the world line were straight (inertial movement), then Fermi–Walker transport would be nothing but the parallel transport in a pseudo-Euclidean geometry, and the Cartesian components of the transported four-vector will remain unchanged. Fermi–Walker transport is the way of carrying the intrinsic angular momentum (*spin*) when it is free of torque and, therefore, its spatial direction is conserved in the proper frame.[27]

evolution of a free of torque spin

[27] The intrinsic angular momentum $\mathbf{L}_c$ of Eqs. (7.51) and (7.52) is contained in Pauli–Lubanski four-vector $W_i = -1/2\,\varepsilon_{ijkl}L^{jk}P^l$. For a physical system having center-of-momentum frame S_c, it is $P^l = (c^{-1}\mathcal{E}_c, 0, 0, 0)$ in S_c, and $W_i = 1/2\,c^{-1}\mathcal{E}_c\,\varepsilon_{0ijk}L^{jk} = c^{-1}\mathcal{E}_c\,(0, \mathbf{L}_c)$ (alternatively, for such physical systems it is defined the spin four-vector $S_i = -1/2\,\varepsilon_{ijkl}L^{jk}U_c{}^l c^{-1}$). Pauli–Lubanski and energy–momentum four-vectors are always orthogonal: $W_i P^i \equiv 0$.

If Fermi–Walker transport is examined from an arbitrary inertial frame S, it results that the transported four-vector does undergo a spatial rotation. Taking into account the form of four-vectors U^i and a^i in an arbitrary inertial frame S (see Eqs. (7.12) and (7.32)), it is obtained that

$$c^{-2}\,(a^i\,U_j - U^i\,a_j) = \left(\begin{array}{c|ccc} 0 & & c^{-1}\gamma(u)^3\boldsymbol{a} & \\ \hline & 0 & -\omega_z & \omega_y \\ c^{-1}\gamma(u)^3\boldsymbol{a} & \omega_z & 0 & -\omega_x \\ & -\omega_y & \omega_x & 0 \end{array}\right)$$

where

$$\boldsymbol{\omega} = \gamma(u)^3\,\frac{\boldsymbol{a}\times\mathbf{u}}{c^2} \tag{7.104}$$

Equation (7.102) for the Fermi–Walker transport of a four-vector $V^i = (V^0, \mathbf{V})$ then decompose as

$$\frac{\mathrm{d}\,V^0}{\mathrm{d}t} = c^{-1}\,\gamma(u)^2\boldsymbol{a}\cdot\mathbf{V} \tag{7.105a}$$

$$\frac{\mathrm{d}\mathbf{V}}{\mathrm{d}t} = c^{-1}\gamma(u)^2\,V^0\,\boldsymbol{a} + \gamma(u)^{-1}\,\boldsymbol{\omega}\times\mathbf{V} \tag{7.105b}$$

($\mathrm{d}/\mathrm{d}\,\tau = \gamma(u)\,\mathrm{d}/\mathrm{d}t$). The orientation change $\mathrm{d}\,\varphi$ underwent by the ordinary vector $\mathbf{V}$ in frame S results from the component of $\mathrm{d}\mathbf{V}$ which is transversal to $\mathbf{V}$,

$$\mathrm{d}\,\boldsymbol{\varphi} = \frac{\mathbf{V}\times\mathrm{d}\mathbf{V}}{|\mathbf{V}|^2}$$

then, using Eq. (7.105b):

$$\frac{\mathrm{d}\,\boldsymbol{\varphi}}{\mathrm{d}t} = \gamma(u)^{-1}\,\frac{\mathbf{V}\times(\boldsymbol{\omega}\times\mathbf{V})}{|\mathbf{V}|^2} + c^{-1}\,\gamma(u)^2\,V^0\,\frac{\mathbf{V}\times\boldsymbol{a}}{|\mathbf{V}|^2} \tag{7.105c}$$

We shall analyze Eqs. (7.105a–c) for the world line belonging to a uniform circular motion with angular velocity $\boldsymbol{\Omega}$. In this way we should find again the Thomas precession resulting from Wigner rotation, which was studied in §4.8 and applied to the behavior of an atomic electron spin in Complement 4A. In a uniform circular motion, vectors $\mathbf{u}$ and $\boldsymbol{a}$ have constant modulus and always remain on the orbit plane, so rotating with angular velocity $\boldsymbol{\Omega}$. Vector $\boldsymbol{\omega}$ in Eq. (7.104) is perpendicular to the orbit plane (it has $\hat{\mathbf{z}}$ direction); therefore, it results in (7.105b) that V_z is constant. For simplicity, let us choose $V_z = 0$, and consider the behavior of a vector $\mathbf{V}$ contained in the orbit plane. In such a case Eq. (7.105c) results to be

uniform circular motion

$$\frac{\mathrm{d}\boldsymbol{\varphi}}{\mathrm{d}t} = \gamma(u)^{-1}\,\boldsymbol{\omega} + c^{-1}\,\gamma(u)^2\,V^0\,\frac{\mathbf{V}\times\boldsymbol{a}}{|\mathbf{V}|^2} \tag{7.106}$$

We shall assume that $u << c$. In a uniform circular motion it is $a = \Omega\, u$, then $a << \Omega\, c$. According to Eq. (7.104), vectors $\boldsymbol{\Omega}$ and $\boldsymbol{\omega}$ are such that

$$\boldsymbol{\omega} = -\gamma(u)^3\left(\frac{a}{\Omega\, c}\right)^2 \boldsymbol{\Omega} \quad \Rightarrow \quad \omega << \Omega \tag{7.107}$$

In addition, we shall choose V^0 to be initially zero. Given the periodic character of the movement, it can be expected that V^0 oscillates around zero with a small amplitude if $u << c$. We shall then assume that $V^0 << |\mathbf{V}|$ (the consistency of this assumption will be checked at the end of the calculation). In particular, since Fermi–Walker transport preserves the four-vector norm, it can be stated that $|\mathbf{V}|^2 = (V^0)^2 - V\cdot V$ is approximately constant along the transport. In summary, vector $\mathbf{V}$ just precesses, with an angular velocity much smaller than Ω given by Eq. (7.106). Since vector $\mathbf{V}$ precesses very little during each orbit, then the temporal evolution of $\boldsymbol{a}\cdot\mathbf{V}$ and $\boldsymbol{a}\times\mathbf{V}$ in Eqs. (7.105a) and (7.106) is basically caused by the rotation of the acceleration $\boldsymbol{a}$; so we can approximate

$$\boldsymbol{a}\cdot\mathbf{V} \approx a\,|\mathbf{V}|\,\cos\Omega\, t \qquad \mathbf{V}\times\boldsymbol{a} \approx a\,|\mathbf{V}|\,\sin\Omega\, t\,\hat{\mathbf{z}}$$

Thus, the solution of Eq. (7.105a) is

$$V^0 \approx \frac{a}{\Omega\, c}|\mathbf{V}|\,\sin\Omega t$$

which is a small amplitude oscillation that is consistent with the performed approximations. Equation (7.106) becomes

$$\frac{\mathrm{d}\boldsymbol{\varphi}}{\mathrm{d}t} \approx -\left(\frac{a}{\Omega\, c}\right)^2 \left(1-\sin^2\Omega\, t\right)\boldsymbol{\Omega} \approx \left(1-\sin^2\Omega\, t\right)\boldsymbol{\omega} \tag{7.108}$$

Equation (7.108) expresses the velocity to which the $\mathbf{V}$ direction changes; this velocity oscillates between 0 and ω around the mean value

$$<\frac{\mathrm{d}\boldsymbol{\varphi}}{\mathrm{d}t}> \approx \frac{1}{2}\,\boldsymbol{\omega} \approx \frac{1}{2}\,\frac{\boldsymbol{a}\times\mathbf{u}}{c^2} = \boldsymbol{\omega}_{\mathrm{T}} \tag{7.109}$$

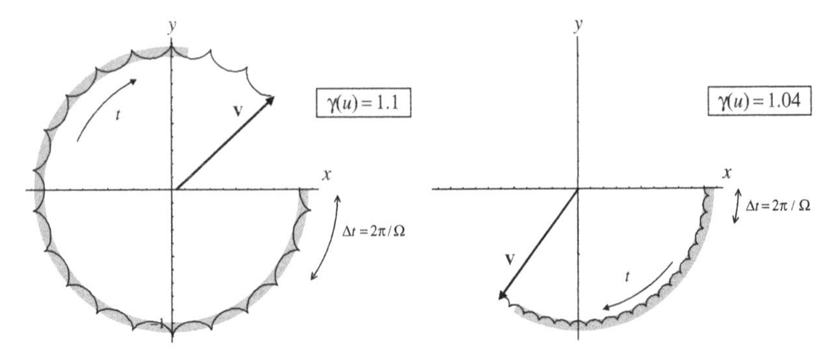

Figure 7.4. Evolution of a vector undergoing a Fermi–Walker transport along the world line of a particle performing a uniform circular motion with angular velocity Ω. It can be observed the **V** precession, and the $|\mathbf{V}|$ oscillations with a period similar to the one of the orbit. The gray fringe indicates the Thomas precession approximated value (7.109). The graphics were made with different velocities u, but equal Ω and total elapsed time.

which corresponds to Thomas precession. As shown in §4.8, Thomas precession appears from the composition of boosts linking the successive proper frames of a particle with a non-rectilinear movement: the composition of two boosts along different directions results equal to a unique boost, whose velocity is the relativistic addition of the velocities of the composed boosts, together with a spatial rotation called Wigner rotation. This rotation implies that the composition of boosts linking successive proper frames gives rise to a precession of them, when they are regarded from an arbitrary inertial frame S. The spatial orientation of vector **V**, which remains without changes in the proper frame by definition of Fermi–Walker transport, will undergo the same precession if it is seen from frame S. However, this is not the only alteration of the **V** orientation in S, because the transformation to the frame S implies a boost associated with the particle velocity **u** relative to S. As it was seen in the examples given in Complement 3A (Problem 1) and §3.14, a boost also distorts orientations. Thus, the changes of the **V** orientation with respect to S receives a contribution from the periodic particle movement, which oscillates with the movement frequency and comes from the boost between $S_{o(t)}$ and S.

Figure 7.4 compares the precession (7.109) with the evolution of vector **V** that would result from solving Eqs. (7.105a–b) in an exact way. Large values of u/c were used in order that the periodic contribution to Eq. (7.108) becomes apparent. The figure also shows that the approximated Thomas precession value (7.109) is smaller than the exact value (4.23).

8

Inertia and Gravity

8.1. THE CRITICISM OF ABSOLUTE MOTION

In §1.1 we mentioned the position sustained by Leibniz and Huygens against the notion of absolute motion defended by Newton. The *relationists* demanded a science describing only relative motion among the bodies, since they thought that this is the only physically meaningful type of motion. On his part, Newton was aware that the equivalence among inertial frames weakened the figure of absolute space. According to the Principle of inertia, two free-of-force particles move with constant velocity with respect to absolute space; but the observation just reveals the relative motion with constant velocity, not being possible to determine whether one of the particles is at absolute rest or not. This indistinguishability of inertial movements is the content of Galilean Principle of relativity. Anyway Newton mitigated this weakness of his theory by emphasizing that absolute motion could be evidenced by dynamical methods, even if it does not by kinematic methods: if two bodies move with relative acceleration, it would be possible to recognize the one having an inertial motion (absence of *relativity* for movements having relative acceleration). Newton resorted to the rotation movement to exemplify this statement. In the *Principia* Newton describes the behavior of a rotating vessel of water hanging from a rope. The vessel communicates to the water its movement, and the water surface acquires the characteristic paraboloid shape. According to Newton, the centrifugal effect leading to the paraboloid shape reveals the water absolute motion, whatever the relative motion water-vessel is.[1]

Mach's criticism

Newton's argument was criticized by Ernst Mach (1838–1916), who considered that the only conclusion that could be inferred from Newton's thought experiment is that the paraboloid shape is a consequence of the relative rotation

[1] The relative motion water-vessel goes through several stages. First, some time elapses from the starting of vessel rotation until this motion is completely communicated to the water and the paraboloid is formed. If the vessel is suddenly stopped, the paraboloid shape will not immediately disappear, since water remains rotating for a time. There is, then, relative motion water-vessel with and without paraboloid, and relative rest with and without paraboloid.

water-rest of the universe, while the relative rotation water-vessel does not produce any appreciable effect on the water surface:

> *Newton's experiment with the rotating vessel of water simply informs us, that the relative rotation of the water with respect to the sides of the vessel produces no noticeable centrifugal forces, but that such forces are produced by its relative rotation with respect to the mass of the earth and the other celestial bodies. No one is competent to say how the experiment would turn out if the sides of the vessel increased in thickness and mass till they were ultimately several leagues thick.*
>
> E. Mach, 1883

Supporter of the *positivism*, Mach advocated a science describing *facts* in the more economical and simple way. He detested any theory or hypothesis containing elements not directly coming from experience. For Mach, the absolute motion was a metaphysical excess, just "fictions of our imagination." According to Mach, Mechanics had to be reformulated in such a way that the situation where Earth "rotates" and the rest of celestial bodies remain "fixed," resulted to be indistinguishable from the situation where Earth is "fixed" and the rest of the universe "rotates" around it. For Mach, absolute space does not exist, thus both mentioned situations are identical because the relative motion is the same in both cases. In 1872, Mach remarked that while the Earth (absolute) rotation has tangible effects in Newtonian mechanics—polar flatness, Foucault pendulum, etc.—instead there are not predicted observable effects on the earth for the case where Earth were fixed and the celestial bodies (absolutely) rotated around it. Mach then demanded a revision of the Principle of inertia, in a way that such kind of effects results to be associated with relative rotations instead of absolute ones. For Mach, the Principle of inertia should be just a statement about the movement of "free" particles with respect to the rest of the universe (or the frame of "fixed stars with the time of Earth's rotation").

It is evident that Mach's criticism of Newtonian mechanics can be extended to Special Relativity, since accelerations are absolute in both theories. Both theories base the Dynamics on the same Principle of inertia, which can only be applied in a family of privileged reference systems—the inertial frames. Neither of them can ascribe the inertial frames privilege to a physical reason, but to their states of motion with respect to absolute space and time or absolute space-time.

Although Mach was clear and convincing to criticize the Newtonian mechanics, he was vague and obscure to propose an alternative. Anyway, this criticism was the starting point to see that the inertial behavior of a body could be the result of a sort of interaction with the rest of the universe. Einstein advanced on this idea in 1912, declaring himself influenced by Mach's criticism and a W. Hofmann's work (1904) where a kind of relational mechanics was proposed.[2] In 1918, Einstein ascribed to Mach the demand for an inertia derived from an

[2] In Hofmann's relational mechanics, the kinetic energy belonging to a two "free" particles system is a bilinear function of the their velocities, so having the form of an interaction energy. H. Reissner

interaction among bodies, calling *Mach's Principle* the statements of his *General Relativity Theory* that he considered as the expression of the idea.

Mach's Principle

In the introduction of his work of 1916, which displays General Relativity in its definitive form, Einstein paraphrased Newton thought experiment and Mach's criticism by considering two fluid bodies of equal extent and constitution suspended in the space, very distanced each other—so the gravitational interaction between them can be ignored—whose only relative motion is a uniform rigid rotation around the line joining both bodies. Einstein formulated the following question: if a measure determined that one of the bodies is spherical and the other one is an ellipsoid of revolution, which would be the reason for such a difference? Einstein demanded that the cause of the difference be an *observable fact of experience*, and added,

> *In classical mechanics, and no less in the special theory of relativity, there is an inherent epistemological defect which was, perhaps for the first time, clearly pointed out by Ernst Mach...*
>
> *...Newtonian mechanics does not give a satisfactory answer to this question. It pronounces as follows:—The laws of mechanics apply to the space* [reference system] R_1, *in respect to which the body* S_1 *is at rest, but not to the space* R_2, *in respect to which the body* S_2 *is at rest. But the privileged space* R_1 *of Galileo, thus introduced, is a merely* factitious *cause, and not a thing that can be observed...*
>
> *...The only satisfactory answer must be that the physical system consisting of* S_1 *and* S_2 *reveals within itself no imaginable cause to which the differing behaviour of* S_1 *and* S_2 *can be referred. The cause must therefore lie outside the system. We have to take it that the general laws of motion, which in particular determine the shapes of* S_1 *and* S_2, *must be such that the mechanical behaviour of* S_1 *and* S_2, *is partly conditioned, in quite essential respects, by distant masses which we have not included in the system under consideration. These distant masses and their motions relative to* S_1 *and* S_2 *must then be regarded as the seat of the causes (which must be susceptible to observation) of the different behaviour of our two bodies* S_1 *and* S_2. *They take over the rôle of the factitious cause* R_1. *Of all imaginable spaces* R_1, R_2, *etc., in any kind of motion relatively to one another, there is none which we may look upon as privileged a priori without reviving the above-mentioned epistemological objection.* The laws of physics must be of such a nature that they apply to systems of reference in any kind of motion. *Along this road we arrive at an extension of the postulate of relativity.*
>
> A. Einstein, 1916

The extension of the Principle of relativity that Einstein is thus proclaiming does not pretend the indistinguishability of bodies S_1 and S_2, as it happens with bodies animated with inertial movements in Newtonian mechanics and Special

(1914) and E. Schrödinger (1925) deepened in the research on similar models. This kind of theories lead to anisotropies for the inertial mass of a body, when the body interacts with an anisotropic matter distribution (for instance, Milky Way's distribution). This possible effect has been experimentally discarded by Hughes, Robinson, and Beltrán-López (1960).

Relativity. On the contrary, it aims at recognizing the *differences* between S_1 and S_2(equatorial oblateness of one of them); but, instead of explaining them by resorting to absolute space-time, these differences will be attributed to the states of motion of the bodies with respect to the rest of the universe.[3,4] What kind of equivalence will be then established between the reference systems associated with two relatively accelerated bodies? How will the inertial frames family shed its artificial privilege for reappearing in the new theory? In what way will the laws of Physics give expression to the influence of the rest of the universe on the inertial behavior of a body?

See § 9.7

8.2. THE PRINCIPLE OF EQUIVALENCE

The construction of 1905's Special Relativity achieved the objective of including Maxwell's electromagnetism in the set of fundamental laws of Physics which are valid in any inertial frame, by means of the reform of the notions of space and time. However, the other fundamental interaction known at that time, the gravitational interaction, was left outside this framework. A formulation of a theory of gravitation that harmonized with the new physics was then necessary. In 1907, Einstein proposed the *equivalence* between gravitational forces and inertial forces (centrifugal force, Coriolis force, etc.) based on the known equality between gravitational mass and inertial mass. This meant that two apparently different problems—the extension of the Principle of relativity to non-inertial frames (where inertial forces appear) and the reformulation of the theory of gravity—were unified in a sole problem: the one of establishing relativistic laws for a unique field: the inertial–gravitational field.

The *inertial mass* m_i of a body is nothing but the mass m taking part in the Second Law of Dynamics (1.8), and gives the measure of the body resistance to be accelerated by the application of a force. The *gravitational mass* m_g is the body property measuring the magnitude of the gravitational force acting on the body in the presence of a given gravitational field **g**:

$$\mathbf{F} = m_g\, \mathbf{g} \tag{8.1}$$

As a consequence of the equality between m_i and m_g, the movement of a particle in a gravitational field does not depend on the properties of the particle. In fact, when the gravitational interaction force (8.1) is substituted in the Second Law of Dynamics (1.8), the masses cancel out in the equation of movement, hence resulting

$$\mathbf{a} = \mathbf{g} \tag{8.2}$$

[3] This strategy is called *Machianization* in M. Friedman (1986) (see Bibliography), where the differences among *relativity, Machianization, and general covariance* are deeply analyzed.

[4] It can be concluded from the cited Einstein's paragraph that the difference between bodies S_1 and S_2 in relative rotation should disappear if S_1 and S_2 were the only bodies in the universe. The remarkable disagreement between this conclusion and General Relativity results will be commented in §9.7.

The fact that body movements in a gravitational field do not depend on the body properties is well known from Galileo's time ("all bodies fall with the same acceleration **g**"), the issue being elucidated in the Tower of Pisa experiment. It should be emphasized that the situation is quite different in other types of interactions. For instance, the electrostatic force is not proportional to mass but to the electric charge q of the particle that is being acted by the electrostatic field $\mathbf{E}\,(\mathbf{F} = q\,\mathbf{E})$. Therefore, the movement of a charged particle in an electric field *does not* result to be independent of its properties, but its acceleration is proportional to the quotient q/m.

In spite of the singular character of gravitational interaction compared with other interactions, the equality between inertial mass and gravitational mass was considered as a fortuitous fact until the beginning of the twentieth century, without any additional scientific assessment.[5] Contrarily, Einstein perceived the equality between inertial and gravitational masses as a fundamental property involving a deep meaning in which the extension of the Principle of relativity could be based on.

As already known, in Newtonian mechanics it is possible to simulate the use of the Second Law of Dynamics (1.8) in non-inertial frames by introducing *inertial forces*, which are not genuine forces, in the sense that they do not come from interactions with other bodies (the reaction force is located in no place). If we know the state of absolute acceleration of a non-inertial frame, we can decompose the body absolute acceleration **a** in an acceleration **a**′ relative to the non-inertial frame plus additional terms. In such case, Eq. (1.8) is written as

$$\mathbf{F} = m_i\ (\mathbf{a}' + \text{additional terms}) \tag{8.3}$$

The additional terms can pass to the left side simulating forces that are called "inertial forces." These inertial forces just contain the information about that part of (absolute) particle dynamics that is not reflected in acceleration **a**′ relative to the non-inertial frame. As can be seen in Eq. (8.3), the inertial forces are proportional to the particle inertial mass m_i.[6]

---

[5] The equality between inertial and gravitational masses has been experimentally verified by means of torsion balances which compare the quotient between inertial and gravitational forces for different materials (Eötvös, 1889; Eötvös, Pékar, and Fekete, 1922; Roll, Krotkov, and Dicke, 1964). The tests included heavy chemical elements, which have significant nuclear binding energies, so proving that the nuclear binding energy contributes in equal magnitude to inertial and gravitational masses. The difference between quotients $m_i\,/\,m_g$ for different materials is smaller than a part in 10^{11}.

[6] In order to calculate the additional terms in Eq. (8.3), the particle position must be decomposed as the sum of position **R** of the non-inertial frame coordinate origin plus position **r**′ relative to the non-inertial frame. Thus $\mathbf{a} = \ddot{\mathbf{r}} = \ddot{\mathbf{R}} + \ddot{\mathbf{r}}'$ (each point indicates a derivative with respect to time). Let be $\mathbf{r}' = x^{\alpha'}\,\mathbf{e}_{\alpha'}$ where $x^{\alpha'}$ and $\mathbf{e}_{\alpha'}$ are coordinates and Cartesian versors belonging to the non-inertial frame. If the non-inertial frame rotates with velocity $\boldsymbol{\Omega}$, then it is $\dot{\mathbf{e}}_{\alpha'} = \Omega \times \mathbf{e}_{\alpha'}$, and $\ddot{\mathbf{r}}' = \ddot{x}^{\alpha'}\,\mathbf{e}_{\alpha'} + 2\,\dot{x}^{\alpha'}\,\Omega \times \mathbf{e}_{\alpha'} + x^{\alpha'}[\Omega \times (\Omega \times \mathbf{e}_{\alpha'}) + \dot{\Omega} \times \mathbf{e}_{\alpha'}] = \mathbf{a}' + 2\,\Omega \times \mathbf{u}' + \Omega \times (\Omega \times \mathbf{r}') + \dot{\Omega} \times \mathbf{r}'$ The two terms after **a**′ give rise to Coriolis force and centrifugal force when they pass to the left side in Eq. (8.3); besides, a contribution due to $\dot{\Omega}$ is added to them. These inertial forces produced by the rotation have to be added with the term $\mathbf{F}_{\text{inertia}} = -m_i\ddot{\mathbf{R}}$ coming from the acceleration of the non-inertial frame origin.

On the other hand, the gravitational force $\mathbf{F} = m_g\,\mathbf{g}$ is proportional to the gravitational mass. For Einstein, the identity between both types of masses meant the identity between both types of forces. Although inertial forces in Newtonian mechanics are not legitimate forces, Mach's criticism allows a revision of the nature of these "forces" to consider them now to be the result of a sort of interaction between a body and the rest of the universe. This requalification of inertial forces then closed the circle of their identification with gravitational forces: both types of forces could be regarded as coming from interactions among masses. The extension of the Principle of relativity will then be based on postulating the indistinguishability between an ordinary gravitational field and the field of inertial forces characterizing a "non-inertial" frame. In this way, the aim of achieving a relativistic description of a gravitational field in an "inertial" frame does not differ from the aim of extending the Principle of relativity to "non-inertial" frames. Indeed, it should be more proper to refer to an *inertial–gravitational* field.[7] In this approach, the dichotomy between reference systems that are "bad" or "good" to formulate the laws of Physics disappears. In each reference system there is an inertial–gravitational field, and the point is to formulate the laws describing the interaction between the inertial–gravitational field and the rest of the physical systems. These laws should be applicable in any reference system, hence generalizing the Principle of relativity.

In this context, the only reference systems that can be said privileged are those where the inertial–gravitational field vanishes. To exemplify this situation let us consider, from a Newtonian point of view, the simplest case of a non-rotating laboratory (reference system) whose coordinate origin has uniform acceleration $\ddot{\mathbf{R}}$, in the presence of a uniform gravitational field $\mathbf{g}$. Since the laboratory does not rotate, the only inertial force that is present is $\mathbf{F}_{\text{inertia}} = -m_i\ddot{\mathbf{R}}$. Therefore, the net force on a particle that is only acted by the inertial–gravitational force is $\mathbf{F} = m_g\,\mathbf{g} - m_i\,\ddot{\mathbf{R}} = m(\mathbf{g} - \ddot{\mathbf{R}})$ in the laboratory frame. The inertial–gravitational field vanishes if $\ddot{\mathbf{R}} = \mathbf{g}$, i.e., if the laboratory is in free fall. In this case the particle is "free" of forces in the laboratory, and its motion will be rectilinear and uniform with respect to the freely falling laboratory. This means that the identity between inertial and gravitational masses prevents an experimenter who works inside the laboratory from recognizing his accelerated state of motion. On the contrary, the experimenter will be inclined to think that he is in an inertial reference system, because he will consider that the particle is genuinely free since there is no indication at all of forces acting on it, and its motion observes the Principle of inertia.

> *This assumption of exact physical equivalence* [between a uniform gravitational field and a uniform acceleration] *makes it impossible for us to speak of the absolute acceleration of the system of reference, just as the usual theory of relativity forbids us to talk of the absolute velocity of a system.*
>
> A. Einstein, 1911

[7] Benedict and Immanuel Friedlaender (1896) anticipated this unified vision of inertia and gravity.

The fact that a non-rotating freely falling laboratory cannot be distinguished from what we should call an inertial frame is verified each time we watch television images from the interior of a spacecraft in terrestrial orbit. A spacecraft orbiting the Earth is nothing but a freely falling laboratory, i.e., a body just under the gravitational field action. Inside the spacecraft the objects "float" (they are at relative rest) or move with constant relative velocity. Instead, an observer who is external to the spacecraft would say that the objects inside the spacecraft are accelerated (in an absolute way or with respect to the fixed stars), and that the null relative acceleration obeys to the fact that all bodies fall with the same acceleration in a gravitational field.

The example of the orbiting laboratory has an ingredient that has not been taken into account yet: the terrestrial gravitational field is not uniform. If the gravitational field is not uniform then there actually exists a relative acceleration between different orbits (free fall trajectories). This effect is described in Figure 8.1, where the change of relative position between two particles tracing close orbits corresponding to a same initial velocity is shown. We can calculate the relative acceleration between two close free fall trajectories: according to Newtonian mechanics, if $\Phi(\mathbf{r})$ is the gravitational potential, then the free fall trajectories satisfy the equation

$$\frac{\mathrm{d}^2\mathbf{r}}{\mathrm{d}t^2} = -\nabla\Phi(\mathbf{r}) \tag{8.4}$$

Thus, by subtracting the equations for two close trajectories $\mathbf{r}_1(t)$ and $\mathbf{r}_2(t) = \mathbf{r}_1(t) + \delta\mathbf{r}(t)$, the equation for the relative acceleration is obtained.

$$\frac{\mathrm{d}^2}{\mathrm{d}t^2}\delta\mathbf{r} = -\nabla\Phi(\mathbf{r}_1 + \delta\mathbf{r}) + \nabla\Phi(\mathbf{r}_1) \cong -(\delta\mathbf{r}\cdot\nabla)\nabla\Phi(\mathbf{r}_1) \tag{8.5}$$

The right side in Eq. (8.5) can be regarded as a "force" (per unit of mass) that is responsible for the relative acceleration between close free fall trajectories. If the gravitational field is uniform then the gravitational potential is linear in

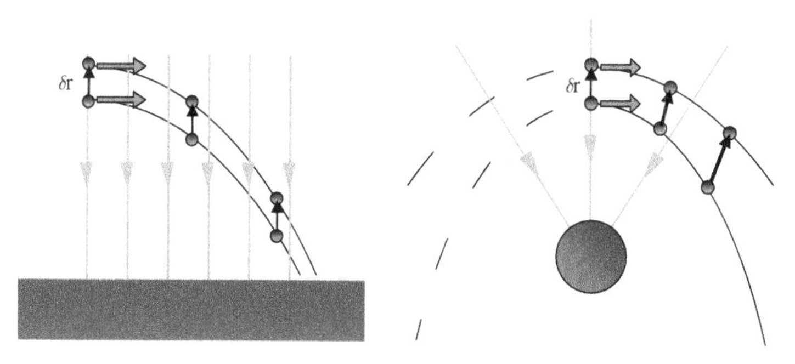

Figure 8.1. Close trajectories for two freely gravitating bodies. If the gravitational field is uniform (left) then the relative acceleration between the bodies will be null. On the contrary, in a non-uniform gravitational field the relative position $\delta\mathbf{r}$ undergoes an acceleration.

the Cartesian coordinates, and second derivatives of the potential in the right side of Eq. (8.5) vanish. In another case a relative acceleration appears that is proportional to the separation between trajectories, which is the effect producing the tides in our "spacecraft," the Earth.

This effect of relative acceleration is not exclusive of orbital movements; it also appears when particles move along the field lines (see Figure 8.2). Therefore, two "free" bodies having a same initial velocity with respect to the freely falling laboratory (spacecraft), for instance two bodies initially at rest relative to the laboratory, will undergo a relative acceleration; the larger the initial separation between the bodies, the larger their relative acceleration. This behavior affords an inner observer the possibility of detecting that he or she is in an accelerated laboratory in the presence of a gravitational field. In other words, if the gravitational field is not uniform then the inertial–gravitational field will not completely vanish in a freely falling laboratory. However, as the separation between the bodies decreases, the difference between a freely falling laboratory and a frame that could be called inertial becomes more negligible, since the right side in Eq. (8.5) decreases with the separation. Thus, it could be said that the cancellation of the inertial–gravitational field in a freely falling laboratory is a local property. A freely falling laboratory is *locally* indistinguishable from what is called inertial frame in Newtonian mechanics. In a freely falling laboratory the Principle of inertia is locally valid.

We shall raise these remarks about the local equivalence between gravitational forces and inertial forces to the rank of Principle. The *Principle of equivalence* will be the link between the laws of Physics in Special Relativity and what these laws should be in the presence of an inertial–gravitational field:

Principle of equivalence

At each event there exists a family of locally inertial frames where the laws of Physics adopt the known form they have in Special Relativity.

The Principle of equivalence allows the existence of privileged "locally inertial" reference systems. In order that this privilege be admissible, the theory

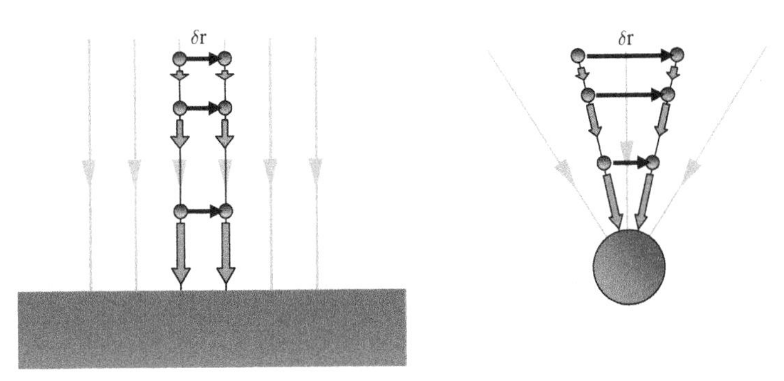

Figure 8.2. Freely gravitating bodies that follow field lines. The relative acceleration is zero in a uniform gravitational field, but different from zero in a non-uniform field. The figures show positions separated by equal elapsed times.

describing the inertial–gravitational field should meet the spirit of Mach's Principle, in the sense that the inertial–gravitational field should be exclusively determined by the content of matter in the universe. In such a case, the privilege of locally inertial reference systems, which is conferred by the local cancellation of the inertial–gravitational field, will come from the relation between this kind of frames and the rest of the universe. Thus, the privilege will be then natural because it is dictated by the matter distribution in the universe, therefore contrasting with the artificial privilege of inertial frames in Newtonian mechanics and Special Relativity that is based on the state of absolute motion of the frames.

8.3. THE INERTIAL-GRAVITATIONAL FIELD

The formulation of the Principle of equivalence goes beyond the mechanical considerations supporting it. In fact, it intends non-mechanical phenomena—like light propagation—to be governed in a locally inertial frame by the same laws governing them in Special Relativity. In other words, the equivalence between gravitational and inertial forces is extended to electromagnetic phenomena: if we want to know how an electromagnetic field behaves in presence of gravity, then we can get a good indication by looking at the behavior of an electromagnetic field in an accelerated frame. The study of a couple of cases of this last kind will lead us to conclude that what we call inertial–gravitational field is just the manifestation of space-time geometric properties.

The first case to be considered is the propagation of a light ray in an accelerated frame. In the absence of gravity, classical physics predicts that the ray trajectory is rectilinear in any inertial frame; Special Relativity agrees with this description. It does not matter which is the chosen inertial frame; the trajectory will be rectilinear in any of them, because the transformation from one of them to any other involves a composition of two finite constant velocities, so it will result a constant velocity. The change of inertial frame produces a change of the propagation direction (aberration) but the rectilinear character of the ray is not altered. On the contrary, the transformation to an accelerated frame implies the composition of the ray uniform movement with the non-uniform movement characterizing the accelerated frame. Thus, the result is that the ray trajectory is not rectilinear in a non-inertial frame (see Figure 8.3). According to the Principle of equivalence, a similar effect has to produce a gravitational field: the inertial–gravitational field deflects the light rays. The inclusion of electromagnetic phenomena in the scope of the application of the Principle of equivalence leads to conclude that the electromagnetic energy gravitates.

the inertial–gravitational field deflects light rays

We shall now study how the frequency is affected by the inertial–gravitational field. Figure 8.4 shows an accelerated laboratory where a source is emitting light pulses of frequency ν_{src}. The pulses are received by a sensor at another place of the laboratory. Owing to the finite speed of light and the laboratory acceleration, when the pulses reach the sensor they find it in a state

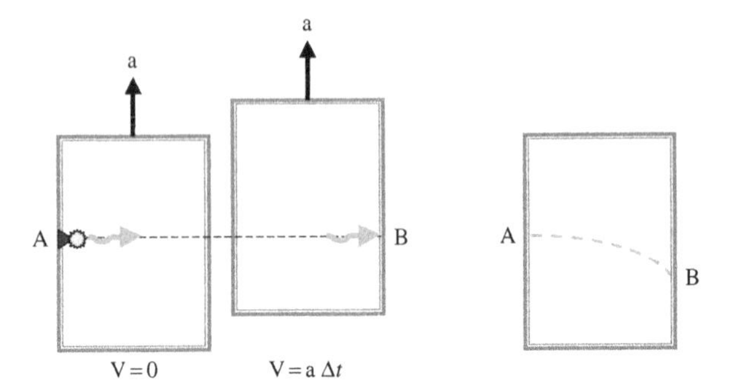

Figure 8.3. The trajectory of a light ray is rectilinear in an inertial frame (left) but it bends in an accelerated frame (right).

of motion that is different from one of the source at the time of emission. In this way, the frequency ν_{obs} measured by the sensor will be different from ν_{src} as a consequence of Doppler effect (see §3.15). It does not matter that the sensor had acquired such a velocity difference while the pulses were traveling or that the relative velocity sensor-source had not changed since source and sensor are fixed to the same laboratory. Only the source velocity at the time of emission and the sensor velocity at the time of reception are relevant, because neither the source motion after the emission nor the sensor motion before the reception play any role in Doppler effect. Since the time elapsed when the pulses cover the distance H between source and sensor is $\Delta t = H/c$, then the velocity difference the sensor acquires with respect to the initial source velocity is $\Delta V = a \Delta t = a H/c$. According to Eq. (3.54), frequency ν_{obs} is

$$\nu_{obs} \cong \left(1 - \frac{\Delta V}{c}\right) \nu_{src} = \left(1 - \frac{a H}{c^2}\right) \nu_{src}$$

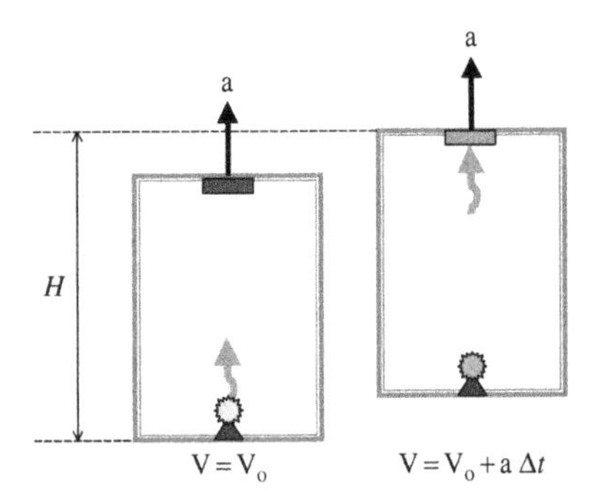

Figure 8.4. Light frequency undergoes a Doppler shift when is observed in an accelerated frame, because the frame velocity at the time of observation differs from the frame velocity at the time of emission.

so it is "redshifted" with respect to the source frequency. The Principle of equivalence forces us to accept that the same frequency shift will exist if the acceleration field is substituted by a gravitational field $\mathbf{g} = -\mathbf{a}$. In other words, the light frequency is redshifted when light "rises" a gravitational potential. Since the gravitational potential difference for a uniform field is $\Delta\Phi = g\,H$, then the gravitational frequency shift is

frequency redshifts when light rises a gravitational potential

$$\frac{\Delta\nu}{\nu} = \frac{\nu_{\text{obs}}}{\nu_{\text{src}}} - 1 \approx -\frac{\Delta\Phi}{c^2} \tag{8.6}$$

Frequency v measures the energy of photons taking part in the luminous phenomenon. When a particle–antiparticle pair is created, two or more photons convert their energy into masses and kinetic energies of the created particles. Vice versa, in a pair annihilation the masses and kinetic energies of annihilated particles are converted into energy of the resulting photons. Since any mass has a gravitational potential energy which depends on the position occupied in the gravitational field, the light frequency must consistently vary when light propagates in a gravitational field, in order to avoid a violation of energy conservation by means of successive pair creation and annihilation. When photons rise a gravitational field, their frequency diminishes, also diminishing their capability for creating mass. But the created mass will benefit from the acquired gravitational potential energy.

As explained in §6.9, the frequency shift caused by the inertial–gravitational terrestrial field was measured by Pound and Rebka in 1960.

The light deflection due to the solar gravitational field was first observed by two expeditions sent from London to Sobral (Brazil) and Principe Island (a Portuguese possession in the Gulf of Guinea at that time, nowadays Republic of Saint Tome and Principe), on occasion of the total solar eclipse in 1919. The first of them was supervised by A. Crommelin and C. Davidson, while the second one was organized by A.S. Eddington and E.T. Cottingham. The total eclipse makes possible the observation of those stars whose light reaches the Earth going past close to the solar disc, where the solar gravitational deflection is more important. The deflection alters the line of vision to the star, thus displacing its site in the sky. Therefore, the deflection is determined by comparing the normal star location (when the Sun occupies another region in the sky) with the "displaced" location of the same star. The experiment was proposed by Einstein in 1911. The agreement between the performed measurements and the result predicted by 1915's General Relativity (see §9.3) definitively established Einstein's theory.[8]

See § 9.3

In this section we aim at glimpsing the kind of mathematical structure that could describe the inertial–gravitational field, explaining the gravitational deflection of light rays and the gravitational frequency shift. Once this issue is elucidated, it will be necessary to find laws governing this inertial–gravitational field,

[8] The analysis of data obtained in Sobral and Principe was published in 1920 by Dyson, Eddington, and Davidson.

having as a guide Newton's law of universal gravitation. We are going to see an argument showing that the gravitational frequency shift demands a new revision of our assumptions about the nature of space-time. Concretely, the Euclidean geometry ascribed to space, or the pseudo-Euclidean geometry of Minkowski space-time, will not be suitable to explain the gravitational frequency shift required by the Principle of equivalence. This will force us to admit that space-time geometry is not a priori determined, but the inertial–gravitational field is precisely the manifestation of the geometric properties of space-time, which should be determined by the matter distribution in order to observe Mach's Principle.

According to the Principle of equivalence, the frequency of a light ray rising a gravitational potential results to be redshifted. Let us consider a system source–sensor in a *non-accelerated* laboratory in the presence of a static (independent of time) uniform gravitational field. Whatever the gravitational influence on the pulses covering the distance between source and sensor could be, it is clear that all successive pulses will be affected in the same way, since the field does not change in time and all the pulses perform the same trajectory. Therefore the traveling time of the first pulse will be identical to the traveling time of the second one. However, if the successive pulses cover the distance between source and sensor in the same time, then they will arrive to the sensor with a time difference identical to the one of emission. In other words, the observed period T_{obs} would be equal to the emission period T_{src} and there would not be gravitational frequency shift at all!

In the context of Special Relativity T_{src} and T_{obs} are proper times measured on the world lines of the source and the sensor respectively. In the frame fixed to the laboratory—the source and sensor proper frame—the spatial Cartesian coordinates of source and sensor are constant along their world lines. Then the relation between proper time $\Delta\tau$ and laboratory coordinate time Δt is

$$\Delta\tau = c^{-1}\,\Delta s = c^{-1}\sqrt{g_{ij}\,\Delta x^i\,\Delta x^j} = \sqrt{g_{00}}\,\Delta t$$

Obviously, in Special Relativity it is $g_{00} = 1$; therefore the proper time is coincident with the coordinate time measured in the proper frame. Since the coordinate times involved in the examined experiment are equal at the emission and the reception, then the gravitational frequency shift will force to consider a theory where the relation between proper time (the clock rate) and coordinate time result to be altered by the presence of a gravitational potential. In fact, the only difference between a clock placed next to the sensor and another clock placed next to the source is the different positions each one occupies in the gravitational potential (remarkably, the gravitational forces acting on both clocks are equal, because we assumed a uniform gravitational field). The former equation then suggests that g_{00} should contain information about the gravitational potential. In a stationary gravitational field, g_{00} should be a function of the position such that the expression

$$\frac{\nu_{obs}}{\nu_{src}} = \frac{T_{src}}{T_{obs}} = \frac{\sqrt{g_{00}(\mathbf{r}_{src})}\,\Delta t_{emission}}{\sqrt{g_{00}(\mathbf{r}_{obs})}\,\Delta t_{reception}} = \sqrt{\frac{g_{00}(\mathbf{r}_{src})}{g_{00}(\mathbf{r}_{obs})}} \tag{8.7}$$

reflects the result (8.6) coming from the Principle of equivalence. But result (8.6) is valid only if $|\Delta\Phi/c^2| << 1$. In the approximation of "weak" gravitational potential i.e., $|\Phi/c^2| << 1$—the result (8.6) and Eq. 8.7) are coincident at the lowest order if

$$g_{00} \approx 1 + \frac{2\,\Phi}{c^2} \qquad \frac{|\Phi|}{c^2} << 1 \tag{8.8}$$

Thus the gravitational potential would become a feature of metric tensor. But the metric tensor is a unique geometric object: its components intermingle under changes of reference system. Then, if we accept that the gravitational potential in a reference system is associated with g_{00}, we shall have to accept in general that the complete metric tensor—not only one of its components—contains the information about the inertial–gravitational field. In such a case the space-time geometry no longer will be an a priori given structure: it will become an attribute determined by the matter distribution in the universe, so accomplishing Mach's Principle. The space-time no longer will be pseudo-Euclidean, as it was so far assumed, because its geometry will become another dynamical variable. The properties commonly attributed to space, like isotropy and homogeneity, no longer will be taken as a truth, since they will be subjected to the dictates of the matter inside it. Before General Relativity the space-time was the stage where the physical phenomena took place. The stage influenced on the phenomena, because it was the only cause of the inertial properties of bodies; but it did not receive in return any influence, since its geometric properties were immutable. From General Relativity the space-time geometry no longer is the immutable stage but it is just another participant of physical phenomena described by Dynamics. The geometry influences on matter but, at the same time, it is determined by the matter distribution in the universe.

> *...it is contrary to the mode of thinking in science to conceive of a thing (the space-time continuum) which acts itself, but which cannot be acted upon. This is the reason why E. Mach was led to make the attempt to eliminate space as an active cause in the system of mechanics.*
>
> A. Einstein, 1922

In this way the Principle of equivalence together with Mach's Principle lead to the necessity of reviewing once again our notions of space and time, abandoning the pseudo-Euclidean geometry of Minkowski space-time in favor of a more general geometry able to incorporate the inertial–gravitational phenomena. In General Relativity theory the space-time geometry is not a priori fixed, but it is governed by laws determining it as a function of the matter distribution. The rest of the laws of Physics, like the Principle of inertia and the laws of electromagnetism, will be reformulated in this new geometrical context. Naturally, there is a non-relativistic limit where the theory meets again the Newtonian laws for the particle movement in a gravitational field, and for the gravitational potential as a function of its sources. In order to establish the link with the old theory, we shall bear in mind the relation (8.8) between gravitational potential and metric tensor.

8.4. RIEMANNIAN GEOMETRY

General Relativity postulates that space-time is provided with a *pseudo-Riemannian* geometry determined by the energy and momentum belonging to the matter and the radiation in the universe. The metric tensor plays the role of inertial–gravitational potential.

In a *Riemannian geometry* the distance $\mathrm{d}s$ between two infinitesimally separated points, whose coordinates are x^i and $x^i + \mathrm{d}x^i$ respectively, is a quadratic form of the coordinate differences [9]

$$\mathrm{d}s^2 = g_{ij}(x)\,\mathrm{d}x^i\,\mathrm{d}x^j \tag{8.9}$$

The classical physics' Euclidean space and the Special Relativity's pseudo-Euclidean space-time are particular cases of Riemannian geometries, which are characterized by having privileged coordinates—Cartesian coordinates—that make possible to write $\mathrm{d}s$ in a same and simple way at all the space points (the components of metric tensor have the same values in the whole space). The existence of Cartesian coordinates in these spaces so far concentrated our interest in those transformations linking different Cartesian systems (rotations, Lorentz boosts, and their combinations). However, in an arbitrary geometry there is not such kind of special coordinates. Therefore the transformations of interest have to be extended to all possible coordinate changes. Now, if the quadratic form $\mathrm{d}s^2$ defines a geometrical property of space, then its value will have to be independent of the used coordinates. In order to satisfy this requirement the metric tensor components must transform in a proper way under a *general coordinate change*. Let $x^{i'} = x^{i'}(x^i)$ be an arbitrary coordinate change, then it is $\mathrm{d}x^i = (\partial x^i/\partial x^{i'})\mathrm{d}x^{i'}$ and it results

$$\mathrm{d}s^2 = g_{ij}\,\mathrm{d}x^i\,\mathrm{d}x^j = g_{ij}\,\frac{\partial x^i}{\partial x^{i'}}\,\frac{\partial x^j}{\partial x^{j'}}\,\mathrm{d}x^{i'}\,\mathrm{d}x^{j'}$$

In order that $\mathrm{d}s$ be invariant, the metric tensor components in the new coordinate system must be

$$g_{i'j'} = g_{ij}\,\frac{\partial x^i}{\partial x^{i'}}\,\frac{\partial x^j}{\partial x^{j'}} \tag{8.10}$$

which means that the coefficients for the transformation of tensor components under a general coordinate change are the functions

$$\Lambda^j{}_{j'} = \frac{\partial x^j}{\partial x^{j'}} \tag{8.11}$$

---

[9] We use to say "pseudo-Riemannian" when alluding to spaces like the space-time where the metric does not have a definite signature but one of its eigenvalues has opposite sign. In a pseudo-Riemannian geometry there is a causal structure of "light cones," analogous to that of Minkowski space-time, and $\mathrm{d}s$ is not a distance but the interval between close events.

In particular, the transformation of the components of a contravariant vector

$$V^{j'} = \frac{\partial x^{j'}}{\partial x^j} V^j \tag{8.12}$$

proves that the coordinates x^j of an event do not transform as four-vector components under a general coordinate change (the four-vector character of x^j is only recovered if the coordinate change is restricted to be linear and homogeneous). The object that does retain its condition of being a contravariant four-vector is the four-vector tangent to a curve (the four-velocity, in the case of a particle world line). In fact, if $x^i = x^i(\tau)$ are parametric equations for a curve, where τ is a parameter identifying each point on the curve, then the components of the vector tangent to the curve are $V^i = \mathrm{d}x^i(\tau)/\mathrm{d}\tau$; therefore, under a general coordinate change (which does not affect the parameter, since this must be regarded as a property of the curve), the vector components will transform according to [10]

$$V^{i'} = \frac{\mathrm{d}x^{i'}}{\mathrm{d}\tau} = \frac{\partial x^{i'}}{\partial x^i}\frac{\mathrm{d}x^i}{\mathrm{d}\tau} = \frac{\partial x^{i'}}{\partial x^i} V^i$$

The derivatives $\partial_i O$ of an invariant function O behave as components of a covariant vector:

$$\partial_{i'} O = \frac{\partial x^i}{\partial x^{i'}} \partial_i O = \Lambda^i{}_{i'}\, \partial_i O \tag{8.13}$$

Instead, the partial derivatives of the components of a contravariant vector do not behave as tensor components under general coordinate changes:

$$\begin{aligned} \partial_{k'} V^{i'} &= \partial_{k'} \left(\frac{\partial x^{i'}}{\partial x^i} V^i \right) = \frac{\partial x^k}{\partial x^{k'}} \partial_k \left(\frac{\partial x^{i'}}{\partial x^i} V^i \right) \\ &= \frac{\partial x^k}{\partial x^{k'}} \frac{\partial x^{i'}}{\partial x^i} \partial_k V^i + V^i \frac{\partial x^k}{\partial x^{k'}} \frac{\partial^2 x^{i'}}{\partial x^k\, \partial x^i} \end{aligned}$$

The last term, which vanishes only if the coordinate change is linear (as it happens, for instance, with Lorentz group transformations in Minkowski space-time), prevents that $\partial_j V^i$ can be regarded as tensor components. For a later application, we shall rewrite this last term in a different way; let us differentiate the identity $\delta^{i'}_{k'} = \partial x^{i'} / \partial x^{k'}$:

$$\delta^{i'}_{k'} = \frac{\partial x^k}{\partial x^{k'}} \frac{\partial x^{i'}}{\partial x^k} \qquad \Rightarrow \qquad 0 = \frac{\partial x^k}{\partial x^{k'}} \frac{\partial^2 x^{i'}}{\partial x^k\, \partial x^i} + \frac{\partial x^{j'}}{\partial x^i} \frac{\partial^2 x^k}{\partial x^{j'} \partial x^{k'}} \frac{\partial x^{i'}}{\partial x^k}$$

By exchanging dummy indexes it is then obtained

$$\partial_{k'} V^{i'} = \frac{\partial x^{i'}}{\partial x^i} \left(\frac{\partial x^k}{\partial x^{k'}} \partial_k V^i - V^j \frac{\partial x^{j'}}{\partial x^j} \frac{\partial^2 x^i}{\partial x^{j'} \partial x^{k'}} \right) \tag{8.14}$$

[10] The definitions of vectors and covectors in *differential geometry* are sketched in Complement 8A.

Complement 8A: *Vectors and covectors*

Usually a vector V is represented by means of an arrow characterized by its modulus and the angles it forms with the Cartesian axes. V can be decomposed in an orthonormal basis $\{\mathbf{e}_i\}$ as $V = V^i\,\mathbf{e}_i$. Components V^i change under rotations of the orthonormal basis, but the vector itself does not change: it is a geometric object. The notions of modulus, angle, and orthonormality come from the space metric structure, and the idea of "Cartesian axes" supposes that the metric structure is Euclidean. However, vectors can also be defined in spaces that are not provided with metric structure at all. With this aim, let us consider a curve in a space provided with a coordinate system only, whose parametric equations are $x^i = x^i(\lambda)$. The operator $\mathrm{d}/\mathrm{d}\lambda$ differentiating functions along the curve with respect to the parameter λ is

$$\frac{\mathrm{d}}{\mathrm{d}\lambda} = \frac{\mathrm{d}x^i(\lambda)}{\mathrm{d}\lambda}\,\partial_i \tag{8A.1}$$

Thus the derivative at a point P along each curve going through P is a linear combination of partial derivatives. Inversely, for each linear combination of partial derivatives $V = V^i\,\partial_i$ there is a curve going through P such that $V^i = \mathrm{d}x^i(\lambda)/\mathrm{d}\lambda|_\mathrm{P}$. This means that the set of derivatives along curves going through P is a vector space, since it is closed under vector addition and multiplication by real numbers; the operators $V = \mathrm{d}/\mathrm{d}\lambda|_\mathrm{P} = V^i\,\partial_i$ are the *vectors* and $\{\partial_i\}$ is a *coordinate basis*. The form of the components of $\mathrm{d}/\mathrm{d}\lambda$ in (8A.1) justifies calling this vector space *tangent space* $\mathbf{T}_\mathrm{P}$. The purely geometric character of $\mathbf{T}_\mathrm{P}$ definition is reflected in transformation (8.12), which shows that $V \in \mathbf{T}_\mathrm{P}$ does not depend on the used coordinates: $V^{i'}\,\partial_{i'} = V^i\,\partial_i$. At each point P there is a tangent space; a vector *field* is introduced by choosing a vector at each point.

Covectors or 1-*forms* are linear functionals on the space of vectors, i.e., they are applications $\tilde{\omega}$ such that $\tilde{\omega}(V)$ is a real number independent of the coordinate system. The linearity means that $\tilde{\omega}(aV + bU) = a\tilde{\omega}(V) + b\tilde{\omega}(U)$, where a, b are reals, V, $U \in \mathbf{T}_\mathrm{P}$. Linear combinations of 1-forms are defined as

$$(a\,\tilde{\omega} + b\,\tilde{\eta})(V) \equiv a\,\tilde{\omega}(V) + b\,\tilde{\eta}(V) \tag{8A.2}$$

so achieving that the 1-forms set has properties of a vector space at P, which is called *cotangent space* $\mathbf{T}_\mathrm{P}{}^*$. Using the linearity of 1-forms it results

$$\tilde{\omega}(V) = \tilde{\omega}(V^i\,\partial_i) = V^i\,\tilde{\omega}\,(\partial_i) \equiv V^i\,\omega_i \tag{8A.3}$$

The quantities ω_i transform like the components of a covariant vector: $\omega_{i'} = \tilde{\omega}\,(\partial_{i'}) = \tilde{\omega}\,[(\partial x^i/\partial x^{i'})\,\partial_i] = (\partial x^i/\partial x^{i'})\,\tilde{\omega}\,(\partial_i) = (\partial x^i/\partial x^{i'})\,\omega_i$. In order that the quantities ω_i can be regarded as the components of the 1-form $\tilde{\omega}$, a *dual coordinate basis* $\{\tilde{d}x^i\}$ in the cotangent space $\mathbf{T}_\mathrm{P}{}^*$ must be defined. In fact, if the 1-form $\tilde{d}x^i$ is defined as:

$$\tilde{\mathrm{d}}x^i\,(\partial_j) \equiv \delta^i_j \quad \text{then} \quad \tilde{\omega}\,(\partial_i) = \omega_i \Leftrightarrow \tilde{\omega} = \omega_i\,\tilde{\mathrm{d}}x^i \tag{8A.4}$$

An invariant function O defines the 1-form $\tilde{\mathrm{d}}O \equiv \partial_i\,O|_\mathrm{P}\;\tilde{\mathrm{d}}x^i$ (see Eq. (8.13)), whose application to $V \in \mathbf{T}_\mathrm{P}$ is $\tilde{\mathrm{d}}O(V) = V^i\,\partial_i\,O|_\mathrm{P} = V(O)$ (V acting on O).

In the absence of a metric tensor, the operation of "rising and lowering indexes"—which would establish a relation between vectors and covectors—does not exist. Vectors and covectors are then genuinely different geometric objects.

The derivatives of the components of a covariant vector do not transform as tensor components either:

$$\partial_{k'} C_{i'} = \partial_{k'} \left(\frac{\partial x^i}{\partial x^{i'}} C_i \right) = \frac{\partial x^k}{\partial x^{k'}} \frac{\partial x^i}{\partial x^{i'}} \partial_k C_i + C_i \frac{\partial^2 x^i}{\partial x^{k'} \partial x^{i'}} \tag{8.5}$$

In this case, the term hindering the tensor character—the last one in Eqs. (8.5)—results to be symmetric. Therefore the antisymmetrized partial derivative $\partial_k C_i - \partial_i C_k$ does have tensor character. This property is also true for the antisymmetrized partial derivative of any completely antisymmetric covariant tensor.[11]

8.5. MOTION OF A FREELY GRAVITATING PARTICLE

In this section we shall reformulate the Principle of inertia in the context of a pseudo-Riemannian geometry. As we know, the motion of a freely gravitating particle (exclusively subjected to the inertial–gravitational field) does not depend on the particle properties, as a consequence of the equality between inertial and gravitational masses. This singular quality offers the possibility of *geometrizing* this kind of motion: instead of explaining it by means of a force field, we can describe it as the *natural* motion in a suitable geometry. Since the role of the inertial–gravitational field was assigned to the metric tensor, then a freely gravitating particle has to be considered as genuinely free of forces; therefore it must *inertially* move along those curves that the geometry characterizes as *natural*.

As commonly formulated in classical physics or Special Relativity, the Principle of inertia turns to the natural curves of Euclidean *plane* geometry, i.e. the straight lines. In Euclidean geometry the straight lines are remarkable curves because they have the property of making minimum the length between two given points. Analogously, in Minkowski pseudo-Euclidean geometry the rectilinear timelike world line—the inertial trajectory of a particle—has the property of making maximum the proper time along the curve joining a given pair of causally connected events (this property was shown in §4.5). Both the length in the first case and the proper time in the second one are related to the integral $\int ds$ along the curve.[12] In general, the curve between a pair of given points making stationary (maximum or minimum) the value of the integral $\int ds$ is called *geodesic*. The notion of geodesic is equally defined in a Riemannian geometry, since what characterizes such a geometry is the existence of a notion of distance or interval.

geodesic

[11] This property makes possible to define the differential and integral calculus for completely antisymmetric covariant tensors without resorting to a metric or any other structure except for a well-behaved coordinate system. The completely antisymmetric covariant p-index tensors ($p \leq n$, n is the space dimension) are called *p-forms*.

[12] In General Relativity, like in Special Relativity, the integral $\int c^{-1} ds$ along a timelike curve is identified with the reading of a clock traveling the curve (see §4.4 and 4.5).

Therefore, it is natural that the reformulation of the Principle of inertia turns to the notion of geodesic. To form an idea of the properties of geodesics in a Riemannian geometry, we shall appeal to the Riemannian geometry of a surface embedded in an Euclidean space (in this case the notion of distance on the surface is the one induced by the Euclidean space where the surface is embedded). Figure 8.5 uses as example the spherical surface embedded in a three-dimensional flat space. The sphere geodesics are the great circles. Figure 8.5 shows that two initially "parallel" geodesics (with a notion of parallelism taken from the exterior Euclidean space)—two meridians in this figure—do not remain parallel (in this example the meridians are parallel at the equator but they intersect at the poles). This *geodesic deviation* is proper of the *curved* spaces (as opposed to *flat* space of Euclidean geometry); this is precisely the Riemannian geometry property that is needed to express the relative acceleration between freely gravitating particles in non-uniform gravitational fields, as it was seen in §8.2 (see Figures 8.1 and 8.2). The separation between the initially parallel orbits in Figure 8.1 will be then described as a geodesic deviation in a curved space.

In General Relativity theory the Principle of inertia is stated:

Principle of inertia in a Riemannian geometry

A free particle moves along timelike geodesics ($ds^2 > 0$) *of space-time.*
A light ray propagates along null geodesics ($ds^2 = 0$) *of space-time.*

This entirely natural statement about freely gravitating motions should however pass a primary test: under non-relativistic conditions the classical law for the particle movement in a gravitational field must be recovered. The Principle of equivalence helped to establish a link between the Newtonian gravitational potential and the metric tensor component g_{00} (Eq. (8.8)), which resulted from demanding the gravitational frequency shift. The same relation (8.8) should also explain the particle movements in gravitational fields, in the "weak" potential approximation $|\Phi| << c^2$. To check the fulfillment of this requirement, let us remark that the property defining the geodesics—i.e., the one of making stationary the integral $\int ds$—identifies $\int ds$ with the *action functional* for the

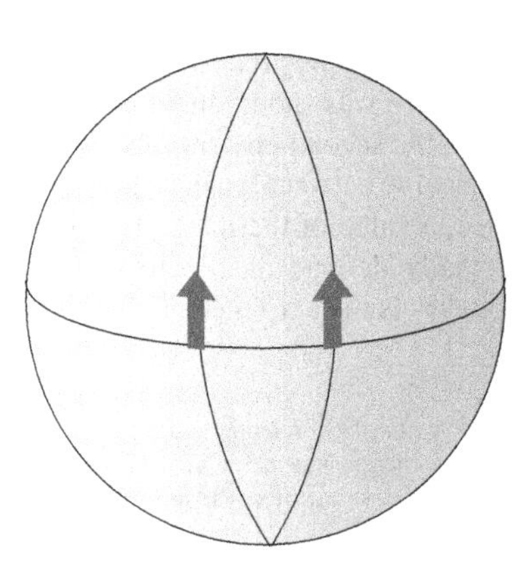

Figure 8.5. Geodesic deviation in a curved surface immersed in a flat space.

freely gravitating particle (see Appendix). In order to provide this quantity with units of action, we shall multiply by $m\,c$:[13]

$$S = -m\,c\int \mathrm{d}s \tag{8.16}$$

The Newtonian movement for a freely gravitating particle should result from this action in the limit of a non-relativistic motion in a "weak" gravitational potential, i.e., when the particle moves at low velocity in a space-time whose geometry is similar to Minkowski space-time geometry. Concretely, the geometry must be such that a coordinate system should exist—at least in some region—where the metric tensor components do not differ much from Minkowskian values of Eq. (7.28) (in particular, $g_{00} \approx 1+2\Phi\,c^{-2}$, $|\Phi| << c^2$, and $g_{0\mu}=0,\ \mu=1,2,3$).[14] The particle movement will be non-relativistic if, in this coordinate system, it is satisfied that $|\mathrm{d}x^\mu| << \mathrm{d}x^0$. Thus the integrand in the action can be approximated as

$$\begin{aligned}
-mc\,\mathrm{d}s &= -mc\sqrt{g_{ij}\mathrm{d}x^i\,\mathrm{d}x^j} \approx -mc\sqrt{(1+2\Phi c^{-2})\mathrm{d}x^{02} - \delta_{\mu\nu}\,\mathrm{d}x^\mu\,\mathrm{d}x^\nu}\\
&= -mc^2\,\mathrm{d}t\left[1+2\Phi c^{-2} - c^{-2}\delta_{\mu\nu}\frac{\mathrm{d}x^\mu}{\mathrm{d}t}\frac{\mathrm{d}x^\nu}{\mathrm{d}t}\right]^{1/2} \cong -mc^2\,\mathrm{d}t\left[1+\frac{\Phi}{c^2}-\frac{1}{2}\frac{u^2}{c^2}\right]\\
&= \mathrm{d}t(-mc^2 - m\Phi + \frac{1}{2}mu^2)
\end{aligned}$$

In this result we can recognize the classical Lagrangian for a particle in a gravitational field (see Appendix), besides the rest energy term which does not take part in the minimization of the action. Therefore, the geodesics of a quasi-Minkowskian geometry, which satisfies the condition (8.8), accurately describe the non-relativistic limit of a particle movement in a gravitational field.

In Special Relativity the inertial world line satisfies the second order differential equations $\mathrm{d}^2x^i/\mathrm{d}\tau^2 = 0$ (i.e., the four-velocity is constant: $\mathrm{d}U^i/\mathrm{d}\tau = 0$). In a Riemannian geometry the geodesic equation can be obtained through a calculus of variations from the action functional (8.16). As known, the curve making stationary a functional like the action (8.16) is that fulfilling the Euler-Lagrange equations (see Appendix). Let us consider any timelike curve joining the initial and final events that play the role of integration limits in the action. If λ is the curve parameter, then $\mathrm{d}s = [g_{ij}\mathrm{d}x^i\mathrm{d}x^j]^{1/2} = [g_{ij}(\mathrm{d}x^i/\mathrm{d}\lambda)(\mathrm{d}x^j/\mathrm{d}\lambda)]^{1/2}\mathrm{d}\lambda$. Thus the Lagrangian for a freely gravitating relativistic particle is

$$L = -mc\sqrt{g_{ij}\frac{\mathrm{d}x^i}{\mathrm{d}\lambda}\frac{\mathrm{d}x^j}{\mathrm{d}\lambda}} \tag{8.17}$$

[13] The minus sign implies that the stationary value of S is a minimum instead of a maximum.
[14] The condition $g_{0\mu}=0$ is always eligible in a region, because it only means that the $x^\mu = constant$ lines are orthogonal to the hypersurfaces $x^0 = constant$.

The Euler-Lagrange equations coming from this Lagrangian result to be

$$\frac{\mathrm{d}}{\mathrm{d}\lambda}\left(\frac{g_{ij}}{\sqrt{g_{ef}\frac{\mathrm{d}x^e}{\mathrm{d}\lambda}\frac{\mathrm{d}x^f}{\mathrm{d}\lambda}}}\frac{\mathrm{d}x^j}{\mathrm{d}\lambda}\right)-\frac{1}{2\sqrt{g_{ef}\frac{\mathrm{d}x^e}{\mathrm{d}\lambda}\frac{\mathrm{d}x^f}{\mathrm{d}\lambda}}}\frac{\partial g_{jk}}{\partial x^i}\frac{\mathrm{d}x^j}{\mathrm{d}\lambda}\frac{\mathrm{d}x^k}{\mathrm{d}\lambda}=0$$

These equations become simpler by using as parameter the interval $\mathrm{d}s$ (or the proper time $\mathrm{d}\tau = c^{-1}\mathrm{d}s$). In fact, in such a case the square root becomes a constant:

$$\frac{\mathrm{d}}{\mathrm{d}\tau}\left(g_{ij}\frac{\mathrm{d}x^j}{\mathrm{d}\tau}\right)-\frac{1}{2}\frac{\partial g_{jk}}{\partial x^i}\frac{\mathrm{d}x^j}{\mathrm{d}\tau}\frac{\mathrm{d}x^k}{\mathrm{d}\tau}=0 \tag{8.18}$$

particle energy–momentum conservation

When written in this way, the motion equations for a freely gravitating particle state that $p_i = mg_{ij}\mathrm{d}x^j/\mathrm{d}\tau$ will be conserved during the motion if the metric tensor components do not depend on the coordinate x^i.

In Eq. (8.18) $\mathrm{d}/\mathrm{d}\tau$ indicates the derivative along the particle world line; then it is valid that $\mathrm{d}g_{ij}/\mathrm{d}\tau = (\partial g_{ij}/\partial x^k)\mathrm{d}x^k/\mathrm{d}\tau$. Therefore

$$g_{ij}\frac{\mathrm{d}^2x^j}{\mathrm{d}\tau^2}+\left(\frac{\partial g_{ij}}{\partial x^k}-\frac{1}{2}\frac{\partial g_{jk}}{\partial x^i}\right)\frac{\mathrm{d}x^j}{\mathrm{d}\tau}\frac{\mathrm{d}x^k}{\mathrm{d}\tau}=0$$

Finally, by recognizing that only the symmetric part of the parenthesis takes part in the contraction with the symmetric tensor $\mathrm{d}x^j/\mathrm{d}\tau\,\mathrm{d}x^k/\mathrm{d}\tau$, and using the inverse metric tensor g^{ij}, it is obtained

equations for a freely gravitating motion

$$\frac{\mathrm{d}^2x^i}{\mathrm{d}\tau^2}+\frac{1}{2}g^{il}\left(\frac{\partial g_{lk}}{\partial x^j}+\frac{\partial g_{lj}}{\partial x^k}-\frac{\partial g_{jk}}{\partial x^l}\right)\frac{\mathrm{d}x^j}{\mathrm{d}\tau}\frac{\mathrm{d}x^k}{\mathrm{d}\tau}=0 \tag{8.19}$$

The functions

$$\Gamma^i_{jk}=\frac{1}{2}g^{il}\left(\frac{\partial g_{lk}}{\partial x^j}+\frac{\partial g_{lj}}{\partial x^k}-\frac{\partial g_{jk}}{\partial x^l}\right) \tag{8.20}$$

are called *Christoffel symbols*.

According to the Principle of equivalence, there are locally inertial frames at each event where the inertial–gravitational field vanishes and the laws of Physics become the ones we know in Special Relativity. As seen in Eq. (8.19), a locally inertial frame at a given event is that associated with coordinates such that Christoffel symbols vanish at the event. In fact, if Christoffel symbols are null then Eq. (8.19) will acquire the form of the Principle of inertia in Special Relativity. On the contrary, the use of different coordinates will introduce inertial–gravitational terms in the description of the freely gravitating motion (inertial or gravitational forces): Christoffel symbols are the geometrization of inertial–gravitational forces per unit of mass.[15]

[15] Christoffel symbols did not appear in the Special Relativity pseudo-Euclidean space-time just because they are null—not only at an event but everywhere—when Cartesian coordinates are used (the Cartesian components of the metric tensor are constant). However, they would appear if other coordinates were used.

The existence of coordinates such that the Christoffel symbols locally vanish is the signal that these symbols cannot be components of a tensor (if a tensor is null in a coordinate system then it will be null in any other coordinate system). This fact by no means invalidates that the geodesic described by Eq. (8.19) is a geometric object (independent of the coordinate system), since the tensor character has to be exhibited not by each separated term but by the entire equation. There is no doubt that the results of the calculus of variations from an invariant action like Eq. (8.16) must have a manifest tensor character. However, as it was mentioned in §8.4, the ordinary derivatives of vector components do not tensorially transform (see Eq. (8.14)). Therefore, the first term in Eq. (8.19)—$\mathrm{d}^2x^i/\mathrm{d}\tau^2 = \mathrm{d}U^i/\mathrm{d}\tau$—which would be the particle four-acceleration in Cartesian coordinates of Minkowski space-time, is not by itself a vector under general coordinate changes. Christoffel symbols in Eq. (8.19) then compensates the improper first term behavior for rendering the ensemble a four-vector; we shall call this ensemble four-acceleration:

$$a^i = \frac{\mathrm{d}U^i}{\mathrm{d}\tau} + \Gamma^i_{jk} U^j U^k \equiv \left(\frac{\mathrm{D}U}{\mathrm{D}\tau}\right)^i \tag{8.21}$$

Thus Eq. (8.19) says that the particle four-acceleration vanishes for a freely gravitating motion.

8.6. COVARIANT DERIVATIVE. MINIMAL COUPLING

Equation (8.21) shows the way of defining the *covariant derivative* of a vector along a curve $x^i = x^i(\tau)$, i.e., a derivative whose result is a vector. The covariant derivative of a field V^i along the curve whose tangent vector is $U^i = \mathrm{d}x^i/\mathrm{d}\tau$ is defined as

$$\begin{aligned}\left(\frac{\mathrm{D}V}{\mathrm{D}\tau}\right)^i &= \frac{\mathrm{d}V^i}{\mathrm{d}\tau} + \Gamma^i_{jk} U^j V^k \\ &= \frac{\mathrm{d}x^j}{\mathrm{d}\tau}\frac{\partial V^i}{\partial x^j} + \Gamma^i_{jk} U^j V^k = U^j(\partial_j V^i + \Gamma^i_{jk} V^k)\end{aligned} \tag{8.22}$$

According to Eq. (8.14), the tensor behavior of differentiation (8.22) will be guaranteed whenever the symbols Γ^i_{jk} transform as

$$\Gamma^{i'}_{j'k'} = \frac{\partial x^{i'}}{\partial x^i}\left(\frac{\partial x^j}{\partial x^{j'}}\frac{\partial x^k}{\partial x^{k'}}\Gamma^i_{jk} + \frac{\partial^2 x^i}{\partial x^{j'}\partial x^{k'}}\right) \tag{8.23}$$

Christoffel symbols (8.20) satisfy this transformation rule.

The covariant derivative (8.22) has the form of a directional derivative: it is the contraction between vector U^i, which gives the direction of differentiation, and a tensor describing the behavior of field V^i in the vicinity of the point at which the derivative is evaluated. Let us introduce the notation

$$V^i{}_{,j} \equiv \partial_j V^i \qquad\qquad V^i{}_{;j} \equiv V^i{}_{,j} + \Gamma^i_{jk} V^k \tag{8.24}$$

Complement 8B: *Derivative of a vector*

Here we shall use the geometric notion of vector that was defined in Complement 8A. The derivative of a vector $V = \mathrm{d}/\mathrm{d}\lambda = V^i \partial_i$ in the direction of another vector $U = \mathrm{d}/\mathrm{d}\mu = U^j \partial_j$ involves the differentiations of functions V^i and vectors ∂_i belonging to the coordinate basis:

$$\frac{\mathrm{D}V}{\mathrm{D}\mu} = \frac{\mathrm{D}}{\mathrm{D}\mu}(V^i \partial_i) = (U^j \partial_j V^i)\partial_i + V^k \frac{\mathrm{D}}{\mathrm{D}\mu}\partial_k \tag{8B.1}$$

Comparing Eqs. (8B.1) and (8.22) it results that the derivative of vector. ∂_k is

$$\frac{\mathrm{D}}{\mathrm{D}\mu}\partial_k = U^j \Gamma^i_{jk} \partial_i \tag{8B.2}$$

Differentiation in an Euclidean space. In an Euclidean space the derivatives of vectors belonging to the Cartesian basis are null (the Cartesian g_{ij} are constant and Christoffel symbols (8.20) vanish). The situation is different when curvilinear coordinates are used. As an example, let us consider polar coordinates in the plane: $x = r\cos\varphi,\ y = r\sin\varphi$. Then

$$\begin{aligned}\partial_r &= \frac{\partial x}{\partial r}\partial_x + \frac{\partial y}{\partial r}\partial_y = \cos\varphi\,\partial_x + \sin\varphi\,\partial_y \\ \partial_\varphi &= \frac{\partial x}{\partial \varphi}\partial_x + \frac{\partial y}{\partial \varphi}\partial_y = -r\sin\varphi\,\partial_x + r\cos\varphi\,\partial_y\end{aligned} \tag{8B.3}$$

Thus, for any vector $\mathrm{d}/\mathrm{d}\mu = U^r\partial_r + U^\varphi\,\partial_\varphi$ it is obtained

$$\frac{\mathrm{D}}{\mathrm{D}\mu}\partial_r = -U^\varphi \sin\varphi\,\partial_x + U^\varphi \cos\varphi\,\partial_y = U^\varphi\, r^{-1}\,\partial_\varphi$$

$$\frac{\mathrm{D}}{\mathrm{D}\mu}\partial_\varphi = U^r(-\sin\varphi\,\partial_x + \cos\varphi\,\partial_y) + U^\varphi(-r\cos\varphi\,\partial_x - r\sin\varphi\,\partial_y) = U^r\, r^{-1}\,\partial_\varphi - U^\varphi\, r\,\partial_r$$

and, according to (8B.2) , it results

$$\Gamma^\varphi_{\varphi r} = r^{-1} = \Gamma^\varphi_{r\varphi} \qquad \Gamma^r_{\varphi\varphi} = -r \qquad \text{(the rest of the symbols vanish)} \tag{8B.4}$$

which are the Christoffel symbols that would result from Eq. (8.20) by using the metric tensor polar components:†

$$\mathrm{d}s^2 = \mathrm{d}x^2 + \mathrm{d}y^2 = \mathrm{d}r^2 + r^2\,\mathrm{d}\varphi^2 \qquad \Rightarrow \qquad g_{rr} = 1, \quad g_{r\varphi} = 0, \quad g_{\varphi\varphi} = r^2 \tag{8B.5}$$

†Cartesian basis $\{\partial_x, \partial_y\}$ is orthonormal, since $g_{xx} = 1 = g_{yy}$, $g_{xy} = 0$. Instead, the polar coordinate basis (8B.3) is not; ∂_φ is not normalized, as it can be seen in (8B.5). On the other hand, the basis of polar versors, $\hat{\mathbf{e}}_r = \partial_r$, $\hat{\mathbf{e}}_\varphi = r^{-1}\,\partial_\varphi$, is orthonormal but it is not a coordinate basis like the ones used in this text. Those bases that cannot be written as partial derivatives in any coordinate system are called *non-holonomic*.

As was already mentioned, functions $V^i{}_{,j}$ do not constitute tensor components. Instead, functions $V^i{}_{;j}$ do transform as tensor components; we call this tensor *covariant derivative of* V^i (without relation to a curve at all). Using this notation, the covariant derivative of V^i along the curve $x^i = x^i(\tau)$ whose tangent vector is $U^i = \mathrm{d}x^i(\tau)/\mathrm{d}\tau$ is written as

$$\left(\frac{\mathrm{D}V}{\mathrm{D}\tau}\right)^i = U^j\, V^i{}_{;j} \tag{8.25}$$

The operation defined in Eq. (8.22) will have the whole features of a differentiation if Leibniz rule is fulfilled whenever the differentiated vector is written as the product of an invariant function and a vector: $V^i = O\,W^i$. For this, the covariant derivative of O is defined to be equal to the ordinary derivative of O (which is likewise invariant):

$$\frac{\mathrm{D}O}{\mathrm{D}\tau} = \frac{\mathrm{d}O}{\mathrm{d}\tau} = U^j\,\partial_j O = U^j\,O_{,j} \tag{8.26}$$

The covariant derivative can be defined for tensors of all types just by requiring the fulfillment of Leibniz rule under tensor product, and its commutativity with the index contraction operation. For instance, if the invariant O in Eq. (8.26) came from the contraction $C_i V^i$, then it would be

$$\frac{\mathrm{d}}{\mathrm{d}\tau}(C_i V^i) = \frac{\mathrm{D}}{\mathrm{D}\tau}(C_i V^i) = \left(\frac{\mathrm{D}C}{\mathrm{D}\tau}\right)_i V^i + C_i\left(\frac{\mathrm{D}V}{\mathrm{D}\tau}\right)^i \tag{8.27}$$

from where we can solve the covariant derivative of the covariant vector C_i:

$$\left(\frac{\mathrm{D}C}{\mathrm{D}\tau}\right)_i = U^j\,C_{i;j} \tag{8.28}$$

where

$$C_{i;j} \equiv C_{i,j} - \Gamma^k_{ji} C_k \tag{8.29}$$

In general, the covariant derivative of any kind of tensor is

$$\begin{aligned} T^{ij\ldots}{}_{ef\ldots;k} \equiv T^{ij\ldots}{}_{ef\ldots,k} \\ + \Gamma^i_{kl}\, T^{lj\ldots}{}_{ef\ldots} + \text{a term for each contravariant index} \\ - \Gamma^l_{ke}\, T^{ij\ldots}{}_{lf\ldots} - \text{a term for each covariant index} \end{aligned} \tag{8.30}$$

Actually, a covariant derivative could be defined without resorting to the Christoffel symbols (8.20) associated with the metric tensor. Even if the space were not provided with a metric structure we only have to endow it with an *affine connection* $\Gamma^i{}_{jk}(x)$, verifying the transformation rule (8.23), to define a directional covariant derivative on it. In this case, the curve that verifies the equation $\mathrm{D}U/\mathrm{D}\tau = 0\,(U^i = \mathrm{d}x^i/\mathrm{d}\tau)$ should not be called geodesic, since it would not be the result of becoming stationary the integral $\int \mathrm{d}s$ along the curve. Anyway this curve will have the property that its tangent vector U^i would remain "equal" to itself along the curve (this notion of "equal" would be defined by the affine connection itself). Such curve could be then called *autoparallel*. It can be said that any affine connection $\Gamma^i{}_{jk}$ defines a notion of *parallelism*: a vector V^i is parallel-transported along a curve if its covariant derivative in the direction of the vector U^i tangent to the curve vanishes. However, if the affine connection is linked to the metric structure as the *metric connection* defined by the Christoffel symbols (8.20) is, then the autoparallel curves will be coincident with geodesics, and the following important properties will result:

- Metric connection is symmetric [16].

$$\Gamma^i_{\ jk} = \Gamma^i_{\ kj} \tag{8.31}$$

The connection symmetry guarantees the commutativity of two covariant derivatives of an invariant function

$$O_{;\,jk} = O_{;\,kj} \tag{8.32}$$

- The covariant derivatives of the metric tensor and its inverse are null

$$g_{ij;\,k} = 0 \qquad g^{ij}{}_{;k} = 0 \tag{8.33}$$

This property makes possible that the operation of index raising and lowering commutes with the covariant derivative, since the metric tensor is "transparent" to the covariant derivative. Besides it establishes the validity of Leibniz rule for the scalar product of vectors $V \cdot W = g_{ij}\, V^i\, W^j$:

$$\frac{\mathrm{D}}{\mathrm{D}\tau}(V \cdot W) = \frac{\mathrm{D}V}{\mathrm{D}\tau} \cdot W + V \cdot \frac{\mathrm{D}W}{\mathrm{D}\tau} \tag{8.34}$$

In particular, if U satisfies the geodesic equation, $\mathrm{D}\,U/\mathrm{D}\tau = 0$, then it will be $\mathrm{D}(U \cdot U)/\mathrm{D}\tau = 0$. This implies that the timelike, spacelike, or null character of a geodesic curve is conserved.

- Contracted Christoffel symbols result to be [17]

$$\Gamma^k_{jk} = \left(\ln \sqrt{|g|}\right)_{,\,j} \qquad \text{where} \quad g = \det(g_{ij}) \tag{8.35}$$

---

[16] Symmetry is lost when non-coordinate or *non-holonomic* bases are used (see Complements 8A and 8B), as a consequence of the lack of tensor character of Christoffel symbols. Non-symmetric connections (in coordinate bases) define a tensor called *torsion*: $T^i{}_{jk} = \Gamma^i{}_{jk} - \Gamma^i{}_{kj}$ (its tensor character is guaranteed by transformation (8.23)). $T^i{}_{jk}$ characterizes the non-commutativity of covariant derivatives of an invariant function: $O_{;\,jk} - O_{;\,kj} = T^i{}_{jk}\, O_{,\,i}$.

[17] Eq. (8.35) is obtained from an equality which is valid for the derivative of the determinant of any matrix: $\partial_k \ln|\det(M_{ij})| = M^{ij}\,\partial_k\, M_{ij}$, where M^{ij} is the inverse of M_{ij}.

As a consequence, the divergences of a vector V^i and an antisymmetric tensor F^{ij} are

$$V^i{}_{;i} = \frac{1}{\sqrt{|g|}}\left(\sqrt{|g|}\,V^i\right)_{,i} \qquad F^{ij}{}_{;i} = \frac{1}{\sqrt{|g|}}\left(\sqrt{|g|}\,F^{ij}\right)_{,i} \tag{8.36}$$

A locally inertial frame at an event O is a coordinate system $\{X^a\}$ such that $\Gamma^a{}_{bc}(x_O) = 0$ and $g_{ab}(x_O) = \mathrm{diag}(1, -1, -1, -1)$. According to Eq. (8.33), in a locally inertial frame at O it results $g_{ab,c}(x_O) = 0 = g^{ab}{}_{,c}\,(x_O)$. In order to construct such a frame, notice that the coordinate change

locally inertial frame

$$x^i \to x^i + \frac{1}{2}\Gamma^i{}_{jk}(x_O)(x^j - x_O{}^j)(x^k - x_O{}^k) + O[(x - x_O)^3]$$

cancels the $\Gamma^i{}_{jk}$'s at O (use the inverse transformation of Eq. (8.23) to prove it). The construction of coordinates $\{X^a\}$ is completed by performing a linear transformation, which keeps the $\Gamma^i{}_{jk}$'s null (see Eq. (8.23)), to leave the metric at O to the form diag $(1, -1, -1, -1)$. The locally inertial frames retain their character under Lorentz transformations. Complement 8C shows the construction of the locally inertial frame of *Riemann normal coordinates.*

If the metric tensor at an event O has the Minkowskian form in coordinates $\{X^a\}$, then the four-volume at the event O will adopt the Special Relativity usual form:

$$\text{four-volume} = \mathrm{d}X^0\,\mathrm{d}X^1\,\mathrm{d}X^2\,\mathrm{d}X^3$$

Therefore, in another coordinate system the four-volume is

$$\text{four-volume} = \frac{\partial(X^0, X^1, X^2, X^3)}{\partial(x^0, x^1, x^2, x^3)}\,\mathrm{d}x^0\,\mathrm{d}x^1\,\mathrm{d}x^2\,\mathrm{d}x^3$$

where the first factor is the Jacobian $J = \det(\partial X^a/\partial x^i)$ for the coordinate change $\{X^a\}\leftrightarrow\{x^i\}$. Taking determinant in (8.10) and using that det $(g_{ab}) = -1$:

$$g_{ij} = \frac{\partial X^a}{\partial x^i}\frac{\partial X^b}{\partial x^j}\,g_{ab} \qquad \Rightarrow \qquad |g| = \det\left(\frac{\partial X^a}{\partial x^i}\right)^2 \times 1$$

Thus $J = |g|^{1/2}$ ($J > 0$ if the coordinate change does not modify the volume orientation). Then,

$$\text{four-volume} = \sqrt{|g|}\,\mathrm{d}x^0\,\mathrm{d}x^1\,\mathrm{d}x^2\,\mathrm{d}x^3 \tag{8.37}$$

Complement 8C: *Riemann normal coordinates*

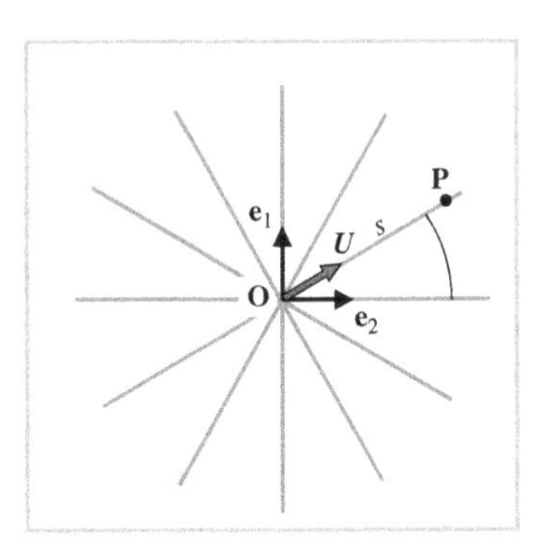

The figure shows the construction of Cartesian coordinates in a flat space by using concepts that can be transferred to a Riemannian geometry, with the aim of providing this last geometry with a *normal* coordinate system in which a symmetric connection became null at a point O. We must draw the autoparallel curves (in this case, straight lines) going through point O, which will play the role of coordinate origin. We choose an orthonormal basis of vectors $\{\mathbf{e}_m\}$ at O. Each autoparallel curve can be identified with its tangent unitary vector U at the origin O. This vector U is expanded in the basis $\{\mathbf{e}_m\}$ as $U = U^m\, \mathbf{e}_m$.

Thus, the Cartesian coordinates of point P can be written as

$$X^m = s\, U^m \tag{8C.1}$$

where s is the distance between origin O and point P (in fact, $U^1 = \cos\,\theta$, $U^2 = \sin\,\theta$).

The procedure of drawing the autoparallel curves going through a given point O, and choosing a vector orthonormal basis $\{\mathbf{e}_m\}$ at O, can be applied in any Riemannian geometry. Then, for each point P there exists the length $s = \int ds$ (or, as appropriate, the proper time $s\,c^{-1}$) of the autoparallel curve joining O and P (this autoparallel curve is unique in a vicinity of the origin), and the components U^m of the respective unitary tangent vector at O. Equation (8C.1) then defines *Riemann normal coordinates* X^m with origin at O for each point P. Equation (8C.1) also says that the parametric equations for autoparallel curves going through O are linear in the parameter s when they are written for normal coordinates:$X^m(s) = s\, U^m$. So, the components of the tangent vector U result to be constant in the coordinate basis $\{\partial/\partial X^m\}$: $U^m(s) = dX^m(s)/ds = U^m$ (see Complement 8A). The autoparallel equation is $DU/Ds = 0$, thus yielding

$$\Gamma^m_{np}\,(\mathrm{P})\; U^n\; U^p = 0 \tag{8C.2}$$

Equation (8C.2) is fulfilled at any point P belonging to the autoparallel curve associated with U, but this does not imply the vanishing of connection $\Gamma^m{}_{np}$ (P). In particular, Eq. (8C.2) is fulfilled at O. Since any direction for U can be chosen at O, the only way that Eq. (8C.2) can be satisfied at O is through the vanishing of the symmetric connection: $\Gamma^m{}_{np}\,(\mathrm{O}) = 0$.

Finally, since the components of U in the coordinate basis $\{\partial/\partial X^m\}$ are $dX^m/ds = U^m$, it is clear that coordinate basis $\{\partial/\partial X^m\}$ and orthonormal basis $\{\mathbf{e}_m\}$ are coincident at O. As it happens in flat space, the orthonormality of the coordinate basis provides the standard (pseudo) Euclidean form to the metric tensor components g_{mn} at O, what completes the proof that Riemann normal coordinate system is a locally inertial frame at O:

$$\Gamma^m_{np}\,(\mathrm{O}) = 0 \qquad g_{mn}\,(\mathrm{O}) = \pm\,\mathrm{diag}(\pm 1, 1, 1, 1) \tag{8C.3}$$

It can also be proved that Riemann tensor (8.42) in coordinate basis $\{\tilde{d}X^m\}$ verifies that

$$g_{mn,\,pq}\,(\mathrm{O}) = -\frac{1}{3}\,[R_{mpnq}\,(\mathrm{O}) + R_{mqnp}\,(\mathrm{O})] \tag{8C.4}$$

In general it is $|g'|^{1/2} = |\det(\Lambda^{l}{}_{l'})|\,|g|^{1/2}$. Multiplying this relation and Eq. (7.59) it results that $|g|^{1/2}\,\varepsilon_{ijkl}$ transforms as a pseudotensor under general coordinate change (see Note 12 in Chapter 7). The addition of factor $|g|^{1/2}$ to the Levi-Civita symbol gives to expression (7.58) the status of n-volume in any coordinate system. The hypersurface element (7.61) must also include factor $|g|^{1/2}$ in order to covariantly transform under general coordinate change.

The covariant derivative is the proper tool to give the Principle of equivalence a mathematical content. As it was seen, the Principle of equivalence states that the laws of Physics in a Riemannian geometry are such that they reduce to the Special Relativity form in a locally inertial frame. In their covariant formulation, the laws of Physics in Special Relativity are built with tensor magnitudes under Lorentz group transformations and ordinary Cartesian derivatives. Correspondingly, geometric (independent of the coordinate system) objects, namely tensors under general coordinate changes and covariant derivatives, will be used in Riemannian geometry to represent physical magnitudes and write the laws governing them. The laws of Physics will then have tensor character under general coordinate changes (*general covariance*), then manifesting that their statements are independent of the used coordinate system. But at each event there exists a locally inertial frame where Christoffel symbols locally vanish, the covariant derivatives locally become ordinary derivatives, and the metric tensor components locally adopt the form diag $(1, -1, -1, -1)$. In order that these properties of the locally inertial frames can reduce the laws of Physics to the form they have in Special Relativity, the Principle of equivalence is implemented in the following way:

Principle of equivalence and minimal coupling

> *The physical laws in the covariant formulation of Special Relativity translate to a Riemannian geometry by substituting partial derivatives by covariant derivatives.*

This statement prescribes the way in which the inertial–gravitational field, through Christoffel symbols, couples to the remaining fields in Nature by entering in the laws governing them.[18] This type of coupling is called *minimal* because it rules out any other possible modification of physical laws, as the addition of terms including tensors associated with the curvature—like Riemann tensor and others that we shall see in the next section—that vanish in Minkowski space-time. The geodesic equation, which governs the freely gravitating motion according to the Principle of inertia, exemplifies the former statement. As a second example, let us apply the statement for writing the conservation of the energy–momentum tensor of an isolated system in a Riemannian geometry (see Eq. (7.64)):

$$T^{ik}{}_{;k} = 0 \qquad (8.38)$$

[18] Since the covariant derivatives of vectors do not commute, while partial derivatives do commute, the prescription is ambiguous for laws containing second derivatives because it does not fix the ordering of the derivatives. Maxwell's laws do not suffer this problem because they can be written as *first* order differential equations for the field tensor F_{ij}.

8.7. RIEMANN TENSOR. EINSTEIN EQUATIONS

Newton's law of universal gravitation implies that the gravitational potential satisfies a second order differential equation:

$$\nabla^2 \Phi = 4\pi G \rho_\mathrm{m} \tag{8.39}$$

where ρ_m is the mass density which plays the role of source for the gravitational field, and $G = 6.67 \times 10^{-11}\,\mathrm{m}^3\,\mathrm{kg}^{-1}\,\mathrm{s}^{-2}$ is the universal gravitation constant. Since the gravitational potential is a feature of the metric tensor, then the tensor laws governing the space-time geometry have to be differential equations in second derivatives of metric tensor. There is a tensor built with second derivatives of metric tensor which takes part in the study of geodesic deviation (the relative acceleration between close geodesics). The geodesic deviation, that characterizes the curved spaces, is the geometric correlate of tidal forces between freely gravitating particles in non-uniform gravitational fields. According to Eq. (8.5) tidal forces are second derivatives of the gravitational potential; their geometrization leads to a tensor depending on second derivatives of the metric (see Complement 8D). This tensor is the *Riemann tensor* [19]

$$R^i{}_{jkl} \equiv \Gamma^i{}_{lj,k} - \Gamma^i{}_{kj,l} + \Gamma^i_{ke}\,\Gamma^e_{lj} - \Gamma^i_{le}\,\Gamma^e_{kj} \tag{8.40}$$

Each affine connection $\Gamma^i{}_{jk}$ defines a Riemann tensor. In the case of the metric connection—which is given by the Christoffel symbols (8.20)—the Riemann tensor becomes a magnitude that depends on second and first derivatives of the metric. The quantities $R^i{}_{jkl}$ defined in Eq. (8.40) behave as tensor components because of the transformation (8.23) of elements $\Gamma^i{}_{jk}$. As a tensor associated with the affine connection, the Riemann tensor expresses the non-commutativity of covariant derivatives of a vector, whose direct consequence is that the vector parallel transport between two given points depends on the path chosen to join the points.[20] In Euclidean plane geometry there is neither geodesic deviation nor the parallel transport depends on the path; for this reason, we call *flat* any connection whose Riemann tensor vanishes.[21]

The Riemann tensor satisfies the *Bianchi identities*:[22]

$$R^i{}_{jkl;e} + R^i{}_{jle;k} + R^i{}_{jek;l} = 0 \tag{8.41}$$

---

[19] Riemann tensor is antisymmetric in the last pair of indexes: $R^i{}_{lkj} = -R^i{}_{ljk}$.

[20] Riemann tensor characterizes the non-commutativity of the covariant derivatives of a vector: $V^i{}_{;jk} - V^i{}_{;kj} = T^l{}_{jk}\,V^i{}_{;l} + R^i{}_{lkj}\,V^l$.

[21] The element of arc for a cylinder of radius a is $d\ell^2 = a^2\,d\varphi^2 + dz^2$, and its Riemann tensor is null. So the cylinder is an example of flat geometry. Except if they are regarded as surfaces embedded in a space of a larger dimension, the cylinder and the plane are *locally* indistinguishable, since they share the same geometry (by substituting the angle φ with the Cartesian coordinate $y = a\varphi$ it results $dl^2 = dy^2 + dz^2$). The only differences between the cylinder and the plane are *global* (the cylinder has closed geodesics), and go outside the scope of the local description provided by Riemann tensor.

[22] Equation (8.41) is valid for any torsionless connection.

When Christoffel symbols (8.20) are used, the Riemann tensor takes the form

$$R_{ijkl} = \frac{1}{2}(g_{il,jk} + g_{jk,il} - g_{ik,jl} - g_{jl,ik}) + g_{ef}\left(\Gamma^e_{jk}\Gamma^f_{il} - \Gamma^e_{jl}\Gamma^f_{ik}\right) \tag{8.42}$$

where $R_{ijkl} = g_{ie} R^e{}_{jkl}$.

Ricci tensor is defined as

$$R_{jl} \equiv R^k{}_{jkl} \tag{8.43}$$

Ricci tensor for the symmetric connection (8.20) is symmetric and equal to [23]

$$R_{ij} = \frac{1}{\sqrt{|g|}}\left(\sqrt{|g|}\,\Gamma^k_{ij}\right)_{,k} - \left(\ln\sqrt{|g|}\right)_{,ij} - \Gamma^k_{li}\Gamma^l_{jk} \tag{8.44}$$

Finally, the *curvature scalar* is the invariant

$$R \equiv g^{ij} R_{ij} \tag{8.45}$$

In order to build equations governing the metric tensor, Einstein took into account that the mass density acting as a source in Eq. (8.39) is not by itself a magnitude with a defined geometric behavior. In Relativity the relativistic energy density, which replaces the mass density in the conservation laws, is merely one of the components of the energy–momentum tensor T^{ij}; therefore this tensor has to take the role of source in the equations governing the metric tensor. The other side of the equations should be built from the Ricci tensor, since it is a tensor of the same type than T^{ij} that contains second derivatives of the metric. After 2 years of diverse attempts, toward the end of 1915 Einstein proposed to equalize the energy-momentum tensor T^{ij} to a symmetric tensor which had, like T^{ij}, zero covariant divergence (1915b). Einstein equations express the relation between geometry and matter in the following way:[24]

$$R_{ij} - \frac{1}{2} g_{ij} R = \frac{8\pi G}{c^4} T_{ij} \tag{8.46}$$

Einstein equations

[23] The torsionless connections satisfy $R^i{}_{lkj} + R^i{}_{kjl} + R^i{}_{jlk} = 0$. The metric connection besides fulfills $R_{ilkj} = R_{kjil}$, what provides the symmetry to Ricci tensor.

[24] *Einstein tensor* $G_{ij} \equiv R_{ij} - (1/2)\, g_{ij} R$ is the sole two index symmetric tensor built from first and second derivatives of the metric which results to be linear in the last ones, is divergenceless and vanishes in a flat geometry. In four or less dimensions this uniqueness is retained even if the requirement of linearity on the second derivatives is released (Lovelock, 1971). It can be proved that $g^{jk} G_{ij;k} \equiv 0$ by starting from Bianchi identities. For a direct calculation, use a locally inertial frame with origin at the point where the involved magnitudes are evaluated. Thus the expressions will become simpler as a consequence of the vanishing of Christoffel symbols and the first derivatives of the metric tensor at the coordinate origin. The result will be valid in any coordinate system, since the vanishing of a tensor does not depend on the used coordinates.

Complement 8D: *Geodesic deviation*

In §8.2 we studied the relative acceleration between two close freely falling movements, in the framework of Newtonian gravity. Equation (8.5) shows that the relative acceleration depends on second derivatives of the gravitational potential. In General Relativity the geometric correlate of this relative acceleration is the *geodesic deviation.* If $x^i(\tau)$ is a geodesic or autoparallel curve (in what follows the affine connection is not necessarily metric but it must be symmetric), a close curve $x^i(\tau)+\delta x^i(\tau)$ will be geodesic as well if the variation $\delta x^i(\tau)$ is such that

$$\delta\left[\frac{d^2 x^i}{d\tau^2}+\Gamma^i_{\ jk}\frac{dx^j}{d\tau}\frac{dx^k}{d\tau}\right]=0 \tag{8D.1}$$

In this way the geodesic equation, which is fulfilled by the first curve, will also be satisfied by the close curve under study. The former condition means that

$$\frac{d^2\delta x^i}{d\tau^2}+\Gamma^i_{\ kj,l}\,\delta x^l\frac{dx^k}{d\tau}\frac{dx^j}{d\tau}+2\Gamma^i_{\ jk}\frac{dx^j}{d\tau}\frac{d\delta x^k}{d\tau}=0 \tag{8D.2}$$

The displacement $\delta x^i(\tau)$ is a four-vector; its derivative along the geodesic is

$$\frac{D\delta x^i}{D\tau}=\frac{d\delta x^i}{d\tau}+\Gamma^i_{\ jl}\frac{dx^j}{d\tau}\delta x^l \tag{8D.3}$$

Differentiating Eq. (8D.3) along the geodesic one obtains

$$\frac{D^2\delta x^i}{D\tau^2}=\frac{d}{d\tau}\left[\frac{d\delta x^i}{d\tau}+\Gamma^i_{jl}\frac{dx^j}{d\tau}\delta x^l\right]+\Gamma^i_{ke}\frac{dx^k}{d\tau}\left[\frac{d\delta x^e}{d\tau}+\Gamma^e_{jl}\frac{dx^j}{d\tau}\delta x^l\right]=$$

$$\frac{d^2\delta x^i}{d\tau^2}+\Gamma^i_{lj,k}\frac{dx^k}{d\tau}\frac{dx^j}{d\tau}\delta x^l+\Gamma^i_{le}\frac{d^2x^e}{d\tau^2}\delta x^l+2\Gamma^i_{jk}\frac{dx^j}{d\tau}\frac{d\delta x^k}{d\tau}+\Gamma^i_{ke}\Gamma^e_{lj}\frac{dx^k}{d\tau}\frac{dx^j}{d\tau}\delta x^l$$

The second derivative of $x^e(\tau)$ can be substituted by $-\Gamma^e_{\ kj}\,dx^k/d\tau\;dx^j/d\tau$, since $x^e(\tau)$ is a geodesic. Using the Eq. (8D.2) and applying the Riemann tensor definition (8.40), then the relative acceleration between close geodesics or geodesic deviation results to be

$$\frac{D^2\delta x^i}{D\tau^2}=R^i_{\ jkl}\frac{dx^j}{d\tau}\frac{dx^k}{d\tau}\delta x^l \tag{8D.4}$$

As we are going to show now, constant $8\pi G c^{-4}$ is required for recovering the gravitation law (8.39) in the non-relativistic limit of weak field.

Einstein equations (8.46) can be written in a slightly different way: contracting indexes in Eq. (8.46), and noting that the trace $g^i_{\ i}=g^{ij}g_{ji}=\delta^i_{\ i}$ is equal to the space-time dimension $n=4$, it results

$$-R=8\pi G c^{-4}\,T \tag{8.47}$$

where $T = T^i{}_i$ is the trace of the energy–momentum tensor. Replacing the result in Eq. (8.46) it is obtained

$$R_{ij} = \frac{8\pi G}{c^4}(T_{ij} - \frac{1}{2} g_{ij} T) \qquad (8.48)$$

linearization of Einstein equations

Einstein equations (8.46) are involved non-linear differential equations for the components of metric tensor. Nevertheless, they can be linearized if the field is "weak," acquiring a more familiar aspect. A weak field is a geometry allowing a coordinate system such that the metric tensor components are written as $g_{ij} = \eta_{ij} + h_{ij}$, where $\eta_{ij} = \mathrm{diag}(1, -1, -1, -1)$ and $|h_{ij}| << 1$. In Eq. (8.42) the products of Christoffel symbols are quadratic in derivatives of the h_{ij}'s; therefore they do not take part in an approximation that is linear in the h_{ij}'s. Then

$$R_{ij} = g^{kl} R_{kilj} \approx \frac{1}{2} \eta^{kl} (h_{kj,il} + h_{il,kj} - h_{kl,ij} - h_{ij,kl}) \qquad (8.49)$$

The linearization of equations (8.46) should lead to Newton's law of universal gravitation for the potential Φ associated with h_{00} (according to Eq. (8.8), it is $h_{00} \approx 2\Phi c^{-2}$). Let us then consider the component 00 of the linearized Einstein equations, for an inertial–gravitational field independent of the time coordinate: $g_{ij,0} = 0$. According to Eq. (8.49), the linearized form of R_{00} is

$$R_{00} \approx -\frac{1}{2} \eta^{\mu\nu} h_{00,\nu\mu} = \frac{1}{2} \nabla^2 h_{00}$$

On the other hand, non-relativistic matter satisfies that $p_o << \rho_o$, $T \approx \rho \approx T_{00}$ (see §7.6, in particular Eqs. (7.72–7.74). Here ρ is the relativistic energy density: $\rho \approx \rho_m c^2$. After replacing in Eq. (8.48) it results

$$\nabla^2 h_{00} \approx \frac{8\pi G}{c^2} \rho_m$$

which is precisely the law of universal gravitation with $h_{00} \approx 2\Phi c^{-2}$.

The linearized Riemann tensor is invariant under the transformation

$$h_{ij} \to h_{ij} - \xi_{i,j} - \xi_{j,i} \qquad (8.50)$$

that resembles the gauge transformation (7.87) for the electromagnetic potential. This gauge invariance is then present in linearized Einstein equations. The quantities $|\xi_{i,j}|$ in Eq. (8.50) have to be much smaller than the unity in order to preserve the character of metric perturbation for h_{ij}. The gauge transformation (8.50) can be regarded as a first order result for the transformation (8.10) of metric tensor components under the infinitesimal coordinate change

$$x^i \to x^i + \eta^{ij} \xi_j(x^i) \qquad |\xi_{i,j}| << 1 \qquad (8.51)$$

The gauge freedom for the perturbation h_{ij} can be exploited to choose those coordinate systems where the linearized equations have their simpler form. Thus, it is possible to choose the four functions ξ_j that force the h_{ij}'s to accomplish conditions similar to Lorentz gauge (7.88):

Lorentz gauge

$$\eta^{kl}\, h_{kj,l} = \frac{1}{2}\, \eta^{kl}\, h_{kl,j} \tag{8.52}$$

By replacing Eq. (8.52) in Eq. (8.49)

$$R_{ij} \approx -\frac{1}{2}\, \eta^{kl}\, h_{ij,kl} = -\frac{1}{2}\,\Box h_{ij} \tag{8.53}$$

The linearized Einstein equations (8.46) and the gauge condition (8.52) become clearer when written with the tensor

$$\bar{h}_{ij} \equiv h_{ij} - \frac{1}{2}\eta_{ij}\,\eta^{kl}\, h_{kl} \quad \text{or} \quad h_{ij} = \bar{h}_{ij} - \frac{1}{2}\eta_{ij}\,\eta^{kl}\,\bar{h}_{kl} \tag{8.54}$$

so yielding

$$\Box \bar{h}_{ij} \approx -\frac{16\pi G}{c^4} T_{ij} \qquad \eta^{kl}\,\bar{h}_{kj,l} = 0 \tag{8.55}$$

Newtonian limit

The Newtonian limit of the theory is reached when the only considered source for the gravitational field in Eq. (8.55) is a rest energy ρ independent of x^0. In such a case it is $T_{ij} \approx \mathrm{diag}\,(\rho, 0, 0, 0)$ and Eq. (8.55) are solved by $\bar{h}_{ij} = \mathrm{diag}\,(4\,\Phi\, c^{-2},\ 0,\ 0,\ 0)$, where Φ is the Newtonian potential. Therefore it is $h_{ij} \approx 2\,\Phi\, c^{-2}$ diag (1, 1, 1, 1), so the metric corresponding to the Newtonian limit, in a coordinate system where the Lorentz gauge is satisfied, is

$$\mathrm{d}s^2 \approx \left(1 + \frac{2\Phi}{c^2}\right) c^2\, \mathrm{d}t^2 - \left(1 - \frac{2\Phi}{c^2}\right) (\mathrm{d}x^2 + \mathrm{d}y^2 + \mathrm{d}z^2) \tag{8.56}$$

Since this metric does not depend on x^0, the freely gravitating particles conserve p_0 (see Eq. (8.18)). If the particles are non-relativistic ($u << c$), then the proper time along their world lines can be approximated by $\mathrm{d}\tau^2 = c^{-2}\,\mathrm{d}s^2 \approx \mathrm{d}t^2\,(1 + 2\Phi\, c^{-2} - u^2\, c^{-2})$; so the conservation of $c\, p_0$ corresponds to the conservation of the classical mechanical energy: $c\, p_0 = mc^2\ g_{00}\ \mathrm{d}t/\mathrm{d}\tau \approx mc^2(1 + 2\Phi\, c^{-2})\,(1 + 2\Phi\, c^{-2} - u^2\, c^{-2})^{-1/2} \approx mc^2 + (1/2)\ m\ u^2 + m\,\Phi$. [25]

gravitational waves

According to Eq. (8.48) "vacuum" Einstein equations, i.e., the equations for vanishing energy–momentum tensor, are $R_{ij} = 0$. Equation (8.53) shows that the vacuum equations for h_{ij} are

$$\Box h_{ij} = 0 \tag{8.57}$$

---

[25] The Newtonian limit for the geodesic deviation (8D.4) between non-relativistic particles is $\mathrm{d}^2\delta\, x^\alpha/\mathrm{d}t^2 \approx R^\alpha{}_{00\beta}\, c^2\, \delta x^\beta \approx (1/2)\,\eta^{\alpha\gamma}\, h_{00,\gamma\beta}\, c^2\, \delta x^\beta \approx \eta^{\alpha\gamma}\,\Phi_{,\gamma\beta}\,\delta x^\beta$, in agreement with Eq. (8.5).

indicating that the perturbation h_{ij} is a *gravitational wave* which propagates at the speed of light in Minkowski space-time.[26] As an example we shall study a plane gravitational wave: $h_{ij} = H_{ij}\,\psi(k_k\, x^k)$. Equation (8.57) is fulfilled for any null wave four-vector k_k and any symmetric and constant H_{ij}. But it is still necessary to verify the Lorentz gauge (8.52), which is indispensable to the validity of Eq. (8.57):

$$H_{kj}\, k^k = \frac{1}{2}\,\eta^{kl}\, H_{kl}\, k_j \tag{8.58}$$

As it happens in electromagnetism (see §7.7), Lorentz gauge does not completely fix the gauge freedom of the theory, since there exist gauge transformations (8.50) that do not alter Lorentz gauge. In fact, once Lorentz gauge is chosen, the still remaining gauge transformations are those such that $\Box\xi_i = 0$.[27] In order to exemplify this residual gauge freedom, let us consider a monochromatic plane wave that propagates along the x direction: $k_j = (k, -k, 0, 0)$, $h_{ij} = H_{ij} \cos k(x - ct)$. As above mentioned, the gauge transformation generated by $\xi_i = \Xi_i\, k \sin k(x - ct)$ does not alter Lorentz gauge. This functional form of ξ_i is such that the resulting gauge transformation (8.50), for the wave h_{ij} under consideration, becomes mere algebraic relations among amplitudes Ξ_i and H_{ij}. This makes possible to use the four constants Ξ_i to cancel four components of H_{ij}. However, because ξ_i only depends on t, x, the transversal sector of H_{ij}, $(H_{22},\ H_{23},\ H_{33})$, is not affected by the gauge transformation generated by ξ_i, so it cannot be cancelled. Instead, wave components H_{i0} can be cancelled. In fact, the gauge transformation (8.50) for $j = 0$ becomes

$$H_{00} \to H_{00} + 2\,\Xi_0 \qquad H_{10} \to H_{10} + \Xi_1 - \Xi_0$$

$$H_{20} \to H_{20} + \Xi_2 \qquad H_{30} \to H_{30} + \Xi_3$$

which show how to choose the amplitudes Ξ_i to cancel the components H_{i0}. After this procedure, the resulting wave is replaced in Lorentz gauge (8.58) in order to obtain four restrictions for the $H_{\alpha\beta}$'s. Thus, it results that

$$0 = H_{11} + H_{22} + H_{33} \qquad H_{11} = 0 \qquad H_{12} = 0 \qquad H_{13} = 0$$

This means that the genuine degrees of freedom reduce only to two, which are represented by the wave transversal part, $H_{22},\ H_{23},\ H_{33}$, subjected to the

[26] In Eqs. (8.55) and (8.57) the symbol $\Box$ indicates the operator $\eta^{kl}\,\partial_k\,\partial_l$, which does not have invariant character under general coordinate changes. In order to built a second order operator invariant under general coordinate changes, replace V^i in Eq. (8.36) with the gradient of an invariant O: $V^i = g^{ij}\,\partial_j\, O$. Thus it results that the Laplacian (Riemannian geometry) or the d'Alembertian (pseudo-Riemannian geometry) of O is $g^{ij}\, O_{;\,ij} = |g|^{-1/2}\,\partial_i |\,g|^{1/2}\, g^{ij}\,\partial_j\, O$.
[27] Under a gauge transformation it is $\eta^{kl}\,\bar{h}_{kj,l} \to \eta^{kl}\,\bar{h}_{kj,l} - \Box\xi_j$. Therefore, Lorentz gauge $\eta^{kl}\,\bar{h}_{kj,l} = 0$ is reached if $\Box\xi_j = \eta^{kl}\,\bar{h}_{kj,l}$. This equation does not completely determine ξ_j because any covector with zero d'Alembertian can be added to ξ_j.

condition of zero trace, $H_{22}+H_{33}=0$. This way of fixing the gauge is called *TT- gauge* (transverse-traceless).

The particularly simple form adopted by the plane wave in this gauge leads to a result that seems strange at first sight: $\Gamma^{i}{}_{00}=0$ for all value of i. This means that the timelike world lines $x^0=\tau,\ x^\alpha=constant$ ($\alpha=1,2,3$) are geodesic for the metric of this gravitational wave. Said in this way, it could be believed that particles initially at rest do not undergo any change due to the gravitational wave passing. However, the wave effects should not be read in terms of the chosen coordinates but in terms of the involved physical magnitudes. Instead of paying attention to the coordinates, it should be remarked that the metric oscillation modifies the *distances* between particles. As seen, only those distances that are perpendicular to the propagation direction will undergo changes. To study this effect it is convenient to distinguish two independent linear polarizations: (a) $H_{23}=0,\ H_{22}=-H_{33}\neq 0$; (b) $H_{23}\neq 0,\ H_{22}=0=H_{33}$. A 45° rotation in the $(x^2,\ x^3)$-plane transforms each one of these polarizations in the other; then we can reduce our study to the understanding of the effect produced by the first polarization. In this case, the distances between particles placed along a line parallel to the x^2 axis will increase at the same time that the distances between particles arranged along the direction of the x^3 axis will decrease (and vice versa): the oscillations of distances are in counter-phase because it is $H_{22}=-H_{33}$. Owing to the wave passing, a ring of particles in the $(x^2,\ x^3)$-plane deforms to an ellipse whose axes oscillate in counter-phase along the directions $x^2,\ x^3$. The same effect is produced by the other linear polarization, but the ellipse axes are rotated in 45°. The change of distances due to the wave passing is reflected in the change of the travel time of a light ray making round trips between two particles.

degrees of freedom and automatic conservation

Certainly, the number of genuine degrees of freedom of the gravitational field is two, not only in the linearized version of the theory but also for the exact Einstein equations (8.46). In principle, there are ten independent Einstein equations to determine the ten independent components of the symmetric tensor g_{ij}. However, it would be disappointing that Einstein equations completely determine the metric tensor, because such situation would imply the determination of the coordinate system as well. Einstein equations should determine the components of metric tensor, but not beyond the freedom associated with coordinate changes. In other words, Einstein equations should determine the space-time *geometry* without getting involved with the coordinate system at all. The metric tensor resulting from Einstein equations should then keep a *gauge freedom* associated with the choice of the coordinate system. Remarkably this is what really happens, as a consequence of the fact that the left side in Eq. (8.46)—the *Einstein tensor* $G_{ij}\equiv R_{ij}-(1/2)\,g_{ij}\,R$—verifies the *automatic conservation*: $G^{ik}{}_{;k}\equiv 0$ (see Note 24). These equations are contracted Bianchi identities that play a role similar to that of Eq. (7.83) in electromagnetism. In fact, Eq. (7.83) are the covariant form of Maxwell's equations without sources (1B.1), which suggest writing the fields as derivatives of the potentials A^i (see Eq. (5.18)). This means that the *generalized* Bianchi identities (7.83) of electromagnetism entail the relation (7.78) and the gauge freedom (7.87) generated by ξ. The existence of generalized Bianchi identities is characteristic of any theory whose action has gauge invariance. As it was shown in the linear approximation, gauge transformations have, in this case, four "generators" ξ_i. The free choice of coordinates makes possible to fix four components of the metric tensor, so leaving six components to describe the geometry. However the gauge freedom has not been completely fixed: as it happens in electromagnetism, each generator ξ_i alludes not to one but two spurious degrees of freedom; thus, the ten independent components of metric tensor g_{ij} contain only two genuine degrees of freedom.

To confirm this conclusion, let us first remark that $g_{00,00}$ does not appear in Riemann tensor (8.42); so it does not take part in Einstein equations.[28] This indicates that g_{00} is not a genuine dynamical variable and can be fixed by means of one of the four gauge transformations. We shall use the three remaining gauge transformations to choose the three components $g_{\mu 0}$ $(\mu = 1, 2, 3)$ to be zero.[29] Therefore, Einstein equations will describe the dynamics of the six components $g_{\mu\nu}$ being left. But, in this gauge the four equations associated with Einstein tensor components G_{i0} do not contain temporal second derivatives $g_{\mu\nu,00}$. In fact, there is no presence of $g_{\mu\nu,00}$ in Einstein tensor components $G_{\mu 0} = R_{\mu 0}$ (note that $\Gamma^0{}_{\mu 0} = (1/2)\, g^{00}\, g_{00,\mu}$ in (8.44)); besides, the contributions $(1/2)\, g^{\mu\nu}\, g_{\mu\nu,00}$ appearing in $G_{00} = R_{00} - (1/2)\, g_{00}\, g^{ij} R_{ij} = (1/2)\,(R_{00} - g_{00}\, g^{\mu\nu} R_{\mu\nu})$ cancel out.[30] Therefore, the four Einstein equations associated with components G_{i0} are not dynamical equations in this gauge, because they do not contain temporal second derivatives. Thus they are four equations that constrain the possible initial values for $g_{\mu\nu}$ and $g_{\mu\nu,0}$, therefore reducing the number of degrees of freedom contained in the six components $g_{\mu\nu}$ to only two.

The automatic conservation $G^{ik}{}_{;k} \equiv 0$, which expresses the existence of gauge freedom, also means that Einstein equations *impose* the conservation of the energy–momentum tensor T^{ik}; otherwise, the equations would be inconsistent. Therefore, Einstein equations not only describe the dynamics of the geometry but also contain essential features of the behavior of its sources.

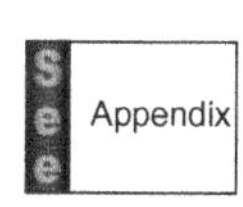

[28] Instead $h_{00,00}$ does appear in Eqs. (8.55) and (8.57) as a consequence of Lorentz gauge (8.52). Compare R_{ij} before and after the Lorentz gauge (Eqs. (8.49) and (8.53)).

[29] The vanishing of components $g_{\mu 0}$ $(\mu = 1, 2, 3)$—and then, the $g^{\mu 0}$'s as well—is reached when the coordinate lines $x^\mu = constant$ are orthogonal to the hypersurfaces $x^0 = constant$.

[30] $g^{00} = g_{00}{}^{-1}$ if $g_{\mu 0} = 0$.

9

Results of General Relativity

9.1. SCHWARZSCHILD SOLUTION. BLACK HOLE

Although Einstein equations are involved non-linear equations, the searching for solutions can be facilitated by proposing a high symmetry. For instance, a spherical spatial symmetry makes possible to state the existence of coordinates $(r,\ \theta,\ \varphi)$ such that the element of arc length on each sphere is written in the usual way: $\mathrm{d}l^2 = r^2\ (\mathrm{d}\theta^2 + \sin^2\theta\,\mathrm{d}\varphi^2)$. Once the form of the sector $(\theta,\ \varphi)$ in the metric tensor is thus anticipated, the equations to be solved become markedly simplified. In 1916, K. Schwarzschild obtained the following spherically symmetric solution to vacuum Einstein equations $(R_{ij} = 0)$:

$$\mathrm{d}s^2 = \left(1 - \frac{2MG}{c^2\,r}\right) c^2\,\mathrm{d}t^2 - \frac{\mathrm{d}r^2}{1 - \dfrac{2MG}{c^2\,r}} - r^2\,(\mathrm{d}\theta^2 + \sin^2\theta\,\mathrm{d}\varphi^2) \tag{9.1}$$

where M is an integration constant. In particular if $M = 0$ the interval (9.1) becomes the Minkowski space-time interval written in spherical coordinates. Thus, in the region where $r >> MGc^{-2}$ the geometry of Eq. (9.1) can be regarded as a perturbation of the flat geometry. According to Eq. (8.8) the gravitational potential is $\Phi = -MG/r$ (in spite of the fact that the interval (9.1) is not written in Lorentz gauge, leading to the form (8.56)). Then the vacuum solution (9.1) corresponds to the *outer* geometry of a spherically symmetric body, M being its Newtonian gravitational mass. The geometry inside the body is different, since the energy–momentum tensor and the state equation of matter will take part in the inner region. The inner and outer solutions must join with continuity at the body surface.

We shall study the solution in all range of the coordinate r, as if the field source were a point-like body. It is easily noticeable that the metric tensor components have two singularities: one of them at $r = 0$ and the other at the *Schwarzschild radius*

$$r_S \equiv \frac{2\,M\,G}{c^2} \tag{9.2}$$

Besides, in the region $r < r_S$ the coordinate r is timelike, since $g_{rr} > 0$, while the coordinate t is spacelike ($g_{tt} < 0$). The singularity at $r = r_S$ is nothing but a singularity of the employed coordinate system; in fact, it is easy to show that a suitable coordinate change eliminates the singularity at $r = r_S$ (see §9.4). On the contrary, the singularity at $r = 0$ is an authentic geometric singularity (independent of the coordinates), as proved by the invariant $R^{ijkl}\,R_{ijkl}$ which takes the value $12\,r_S{}^2\,r^{-6}$ in Schwarzschild geometry, thus being divergent at $r = 0$. Nevertheless, the coordinates used in (9.1) are the most convenient to study the weak field region ($r >> r_S$).[1]

Let us consider a static source at $r_{src} > r_s$ emitting luminous signals of frequency ν_{src}. Since the metric tensor components do not depend on t, we can use Eq. (8.7) to compute the frequency measured by a detector fixed at the position r_{obs}:

$$\frac{\nu_{obs}}{\nu_{src}} = \sqrt{\frac{g_{00}(r_{src})}{g_{00}(r_{obs})}} = \left(\frac{1 - \dfrac{r_S}{r_{src}}}{1 - \dfrac{r_S}{r_{obs}}}\right)^{\frac{1}{2}} \tag{9.3}$$

So the observed frequency undergoes a gravitational redshift, and vanishes when the source is at the horizon ($r_{src} = r_S$).

Let us now study the causal structure or light-cone structure of the geometry described by the interval (9.1). The light cones are generated by light rays, which are null world lines of the geometry under consideration. Let us focus on light rays having radial trajectories ($d\theta = 0 = d\varphi$). In such a case the interval (9.1) is null if

$$\left(1 - \frac{r_S}{r}\right) c^2\,dt^2 - \frac{dr^2}{1 - \dfrac{r_S}{r}} = 0$$

i.e.,

$$c\,\frac{dt}{dr} = \pm\frac{1}{1 - \dfrac{r_S}{r}} \tag{9.4}$$

Of course, not any null world line is a geodesic. But the radial world lines satisfying Eq. (9.4) are the only radial null world lines; so they must correspond to radial null geodesics of the spherically symmetric geometry (9.1).

causal structure

Figure 9.1 shows the curves verifying Eq. (9.4), together with the future light-cone structure. Notice that the choice of "future" is merely conventional in

---

[1] In 1923, Birkhoff proved that Schwarzschild solution is the only spherically symmetric vacuum solution. Therefore the interval (9.1) can also be applied inside a spherically symmetric hollow shell. But in such a case there is no reason for the existence of a geometric singularity at the center of symmetry, what forces to choose the integration constant M equal to zero. The space-time inside the shell has Minkowski geometry.

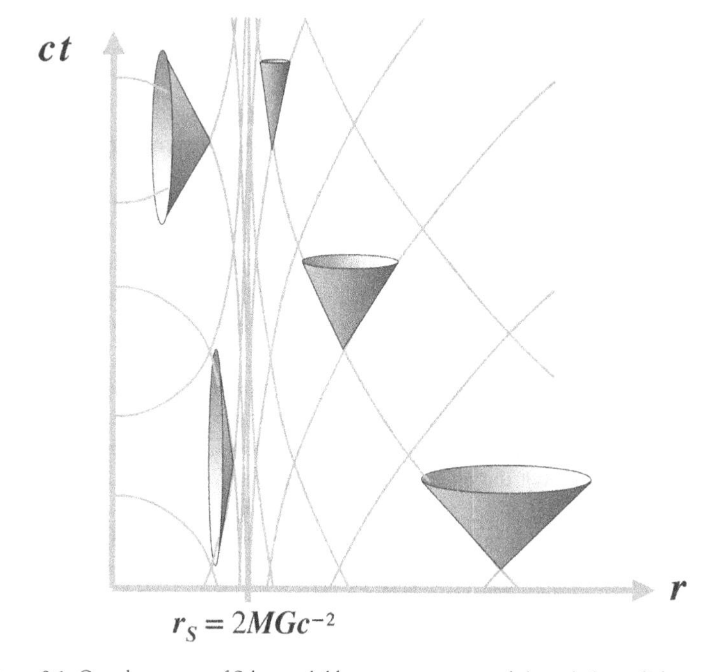

Figure 9.1. Causal structure of Schwarzschild geometry represented through future light cones.

the region $r > r_S$, since the interval (9.1) does not change under the inversion of the temporal coordinate t. Instead, in the region $r < r_S$ the temporal coordinate is r, and the interval does perceive the change $r \to -r$; in this region each one of the two possible choices of future leads to different causal structures. The causal structure in Figure 9.1 (future cones pointing toward the singularity $r = 0$) is the one corresponding to objects falling toward the singularity. The most outstanding characteristic of this causal structure is the fact that the world line of any particle or light ray inside the region $r < r_S$ unavoidably ends at the singularity $r = 0$ (remember that particles travel inside the light cones). Nothing, not even light, can evade this fate. An observer in region $r > r_S$ will not receive any information about the events that happened in region $r < r_S$ since neither a particle nor a light ray can go through the Schwarzschild radius to escape the region $r < r_S$. The surface defined by the Schwarzschild radius is called *event horizon*, since it hides those events that are "behind" it. The region inside the Schwarzschild radius is black for an external observer, because nothing, not even light, emerges from it. In order that a spherically symmetric object can exhibit its event horizon, its mass has to be compacted in a sphere whose radius is smaller than r_S (remember that geometry (9.1) is a solution for *vacuum* Einstein equations). In such a case the object is perceived as a *black hole*. The Schwarzschild radius for an object of mass equal to the solar mass is 2.95 km.

event horizon and black hole

Figure 9.1 also shows that a particle that falls toward the black hole reaches the event horizon when $t \to \infty$. Although this result seems to indicate that the event horizon is never reached, we must remember that the employed coordinate system is singular at the horizon. It will then be better to turn to the particle invariant proper time to avoid hasty conclusions; in fact, in §9.2 we shall prove that the particle reaches the horizon in a finite proper time.

When the particle reaches the horizon, any light or signal emitted from the particle will be infinitely redshifted to an external static observer (in this case, gravitational redshift is composed with Doppler redshift). After going through the horizon, the particle will irreversibly head toward the singularity; neither force could avoid this fate, since the light cones unavoidably lead to the singularity. Although the horizon had already been crossed, the singularity will never be visible; the structure of future cones prevents the existence of light rays coming from the singularity.

spatial geometry

In region $r > r_S$ the spatial geometry in interval (9.1), i.e., the element of arc length corresponding to a hypersurface $t = constant$, can be pictorially represented as the geometry of a surface embedded in an Euclidean space. In fact, we shall show that the element of arc length is analogous to the Euclidean distance on a paraboloid of revolution. Let us then consider the paraboloid generated by the revolution of the parabola $r = r_S + z^2/(4\ r_S)$, where r_S must be regarded as a parabola parameter. The relation between the differentials $\mathrm{d}r$ and $\mathrm{d}z$ on the parabola is $\mathrm{d}r^2 = (z/2\ r_S)^2\mathrm{d}z^2 = (r\ r_S^{-1} - 1)\,\mathrm{d}z^2$. Then the Euclidean distance on the parabola is

$$\mathrm{d}r^2 + \mathrm{d}z^2 = \left(1 + \frac{1}{r\,r_S^{-1} - 1}\right)\mathrm{d}r^2 = \frac{\mathrm{d}r^2}{1 - \frac{r_S}{r}}$$

which is coincident with the radial part of interval (9.1). This analogy makes possible to represent the spatial geometry of region $r > r_S$ in the way shown in Figure 9.2.

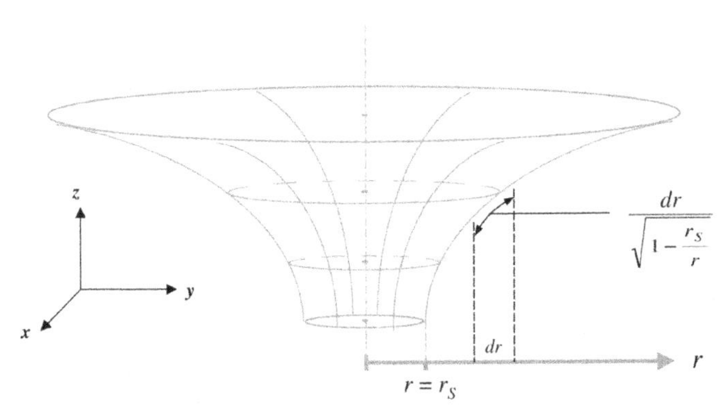

Figure 9.2. The spatial geometry of region $r > r_S$ is analogous to the geometry on a paraboloid of revolution embedded in an Euclidean geometry. The bidimensional surface in the graphics corresponds to an equatorial surface ($\theta = \pi/2$) of Schwarzschild geometry at a given time.

9.2. INERTIAL MOVEMENT IN SCHWARZSCHILD GEOMETRY

Let us study the movement of a freely gravitating particle in Schwarzschild geometry. Instead of solving the geodesic equation (8.19), we shall use the conservation theorem resulting from Eq. (8.18): since the metric tensor components in Eq. (9.1) do not depend on the coordinates t, φ, then p_t and p_φ are conserved. We shall call E and L—alluding to the energy and the angular momentum—the corresponding constants of motion:

$$p_t = g_{tt}\, p^t = g_{tt}\, m \frac{\mathrm{d}t}{\mathrm{d}\tau} = c^2 \left(1 - \frac{r_S}{r}\right) m \frac{\mathrm{d}t}{\mathrm{d}\tau} = E \tag{9.5}$$

$$p_\varphi = g_{\varphi\varphi}\, p^\varphi = g_{\varphi\varphi}\, m \frac{\mathrm{d}\varphi}{\mathrm{d}\tau} = r^2 \sin^2\theta\, m \frac{\mathrm{d}\varphi}{\mathrm{d}\tau} = L \tag{9.6}$$

Besides we can invoke the spherical symmetry to state that the freely gravitating motion is always contained in an equatorial surface. This surface can be characterized with $\theta = \pi/2$, so it results $p^\theta = m\,\mathrm{d}\theta/\mathrm{d}\tau = 0$. By definition it is $p \cdot p = m^2\, U \cdot U = m^2 c^2$, then

$$m^2 c^2 = g_{ij}\, p^i p^j = g_{tt}{}^{-1} E^2 + g_{rr}\, p^{r\,2} + g_{\varphi\varphi}{}^{-1} L^2$$

thus, by replacing $p^r = m\,\mathrm{d}r/\mathrm{d}\tau$, an equation for the radial component of the motion as a function of the proper time is obtained:

$$\frac{1}{2} m \left(\frac{\mathrm{d}r}{\mathrm{d}\tau}\right)^2 + \frac{L^2}{2\,m\,r^2} - \frac{G\,m\,M}{r} - \frac{G\,M\,L^2}{m\,r^3 c^2} = \frac{E^2 - m^2 c^4}{2\,m\,c^2} \equiv e \tag{9.7}$$

This conservation equation differs from the respective Newtonian equation in the presence of a term that is proportional to r^{-3}, and because the proper time appears instead of the coordinate time.

For a radial trajectory ($L = 0$) it results

$$\frac{\mathrm{d}r}{\mathrm{d}\tau} = \pm\sqrt{2\,\overline{e} + \frac{2\,G\,M}{r}}$$

radial motion

where $\overline{e} \equiv e/m$ is an energy per unit of mass. The previous equation can be integrated to show that the proper time for joining a pair of allowed positions (i.e., such that $\overline{e} + (G\,M/r) \geq 0$) is always finite:

$$\Delta\tau = -\int_{r_1}^{r_2} \frac{\mathrm{d}r}{\sqrt{2\,\overline{e} + \dfrac{2\,G\,M}{r}}} \qquad r_1 > r_2 \tag{9.8}$$

In fact $\Delta\tau$ coincides with the (finite) time of the Newtonian theory for a particle having mechanical energy e. It is clear that $\Delta\tau$ does not even diverge when

$r_2 \to 0$ (the integrand goes to zero in that limit). In particular, $\Delta\tau|_{r_1}^{0}$ has a simple form when $e = 0$: $\Delta\tau|_{r_1}^{0} (e = 0) = [2\, r_1{}^3/(9GM)]^{1/2}$.

orbital motion

Let us now study the orbital movements ($L \neq 0$). The effective potential in Eq. (9.7), which includes the centrifugal potential, the Newtonian gravitational potential and the relativistic term, i.e.

$$V^{\text{effective}} = \frac{\overline{L}^2}{2\,r^2} - \frac{G\,M}{r} - \frac{G\,M\,\overline{L}^2}{r^3 c^2} \tag{9.9}$$

is shown in Figure 9.3 for several values of the angular momentum per unit of mass $\overline{L} \equiv L/m$, together with the Newtonian effective potential. When $\overline{L}^2\, c^2 > 12\, G^2\, M^2$, the relativistic potential has a maximum and a minimum that are located at

$$\frac{\overline{L}^2}{2\,G\,M}\left(1 \pm \sqrt{1 - \frac{12\,G^2\,M^2}{\overline{L}^2\,c^2}}\right)$$

The maximum of potential—which does not exist in the non-relativistic potential—allows an unstable circular orbit. If the energy value $\overline{e}$ coincides with the minimum of potential, then the motion becomes a stable circular ($r = constant$) orbit. The radius of the stable circular orbit is

$$r_{\text{circ. orb.}} = \frac{\overline{L}^2}{2\,G\,M}\left(1 + \sqrt{1 - \frac{12\,G^2\,M^2}{\overline{L}^2\,c^2}}\right) \tag{9.10}$$

The Newtonian value results in the limit $c \to \infty$

$$r_{\text{Newtonian circ. orb.}} = \frac{\overline{L}^2}{G\,M} \tag{9.11}$$

So long as L decreases, the radius of the stable circular orbit diminishes.

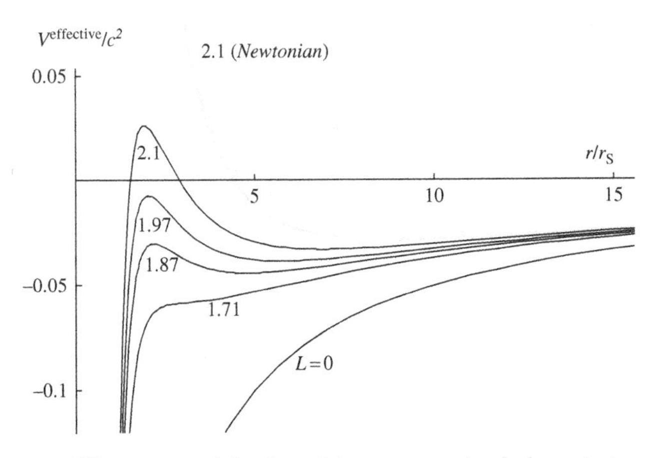

Figure 9.3. Effective potential for the radial component of a freely gravitating motion in Schwarzschild geometry, compared with the Newtonian potential. The numbers are values for $\overline{L}/(r_S\, c)$.

Therefore the smallest orbit corresponds to the critical value $\overline{L}^2 c^2 = 12\, G^2 M^2$, and results to be

$$r_{\text{circ. orb. minimum}} = \frac{6\,G\,M}{c^2} = 3\,r_S \tag{9.12}$$

"last" stable circular orbit

The equation governing the trajectory $r = r(\varphi)$ on the equatorial surface $\theta = \pi/2$ is obtained by replacing $\mathrm{d}\tau = r^2\,\overline{L}^{-1}\,\mathrm{d}\varphi$ in Eq. (9.7) (see Eq. (9.6)):

$$\frac{\overline{L}^2}{2\,r^4}\left(\frac{\mathrm{d}r}{\mathrm{d}\varphi}\right)^2 + \frac{\overline{L}^2}{2\,r^2} - \frac{G\,M}{r} - \frac{G\,M\,\overline{L}^2}{r^3 c^2} = \overline{e}$$

The most convenient variable to examine this equation is $u \equiv r^{-1}$. Thus $\mathrm{d}r/\mathrm{d}\varphi = -u^{-2}\mathrm{d}u/\mathrm{d}\varphi$ and the trajectory equation turns out to be

$$\left(\frac{\mathrm{d}u}{\mathrm{d}\varphi}\right)^2 = \frac{2\,\overline{e}}{\overline{L}^2} - u^2 + \frac{2\,G\,M\,u}{\overline{L}^2} + \frac{2\,G\,M\,u^3}{c^2} \tag{9.13}$$

The non-relativistic equation excludes the last term; thus the form of Eq. (9.13) is not useful to look for approximations not merely leading to the Newtonian movement. Nevertheless, we could think in some kind of motion allowing the substitution of the last term for an approximate expression. We shall then study elliptical orbits of small eccentricity ($\overline{e}$ is close to the minimum of potential), in the conditions $G\,M << \overline{L}\,c$ where the radius of the circular orbit approaches the Newtonian value $u_N = G\,M/\overline{L}^2$ (then $u_N\,\overline{L} << c$). By using the binomial expansion we can write u^3 as

$$u^3 = (u - u_N)^3 + 3\,u^2\,u_N - 3\,u\,u_N{}^2 + u_N{}^3$$

Replacing in Eq. (9.13) and completing the square:

$$\left(\frac{\mathrm{d}u}{\mathrm{d}\varphi}\right)^2 + \omega^2 (u - u_N\,A)^2 = \frac{2\,\overline{e}}{\overline{L}^2} + u_N{}^2\,\omega^2 A^2 + \frac{2\,u_N\,\overline{L}^2}{c^2}\,[u_N{}^3 + (u - u_N)^3] \tag{9.14}$$

where ω and A are constants that equal 1 in the non-relativistic theory:

$$\omega^2 = 1 - \frac{6\,u_N{}^2\,\overline{L}^2}{c^2} \qquad \omega^2 A = 1 - \frac{3\,u_N{}^2\,\overline{L}^2}{c^2} \tag{9.15}$$

In order that the eccentricity of the orbit be small, the energy $\overline{e}$ must be near enough to the minimum value of the effective potential ($\overline{e} \sim V(u_N) \sim -\,G^2 M^2/(2\,\overline{L}^2) = -u_N{}^2\,\overline{L}^2/2$). In such a case it is $|u - u_N| << u_N$, and the term $(u - u_N)^3$ in Eq. (9.14) can be neglected. Thus, Eq. (9.14) acquires the form

$$\left(\frac{\mathrm{d}u}{\mathrm{d}\varphi}\right)^2 + \omega^2\,(u - u_N\,A)^2 \cong B^2, \tag{9.16}$$

where $B^2 = 2\,\overline{e}\,\overline{L}^{-2} + u_N^2 \omega^2 A^2 + 2\,u_N^4\,\overline{L}^2 c^{-2}$. The solution of Eq. (9.16) is

$$u\,(\varphi) = u_N\,A + \omega^{-1}\,B\,\cos\,(\omega\,\varphi + \alpha_o) \tag{9.17}$$

α_0 being an integration constant. The solution (9.17) indicates that u oscillates with angular frequency ω, around a value $u_N A$ similar to the Newtonian value u_N:

$$u_N A \approx u_N \left(1 + \frac{3\, u_N{}^2 \overline{L}^2}{c^2}\right) \tag{9.18}$$

The oscillation amplitude $\omega^{-1}B$ is small, since $\omega \sim 1$ and $B << u_N$ (remember that $2\,\overline{e}/\overline{L}^2 \sim -u_N^2$).

The exact non-relativistic solution is obtained in the limit $c \to \infty$. In particular, the non relativistic value of ω is 1. Therefore the radius r of classical solution performs a complete oscillation each time the angle φ varies in 2π; this means that non-relativistic orbits are closed. Equation (9.17) with $\omega = 1$ is simply the ellipse equation in polar coordinates. On the contrary, the relativistic orbits are not closed, because a complete radius oscillation requires that $\omega\Delta\varphi = 2\pi$, where $\omega \neq 1$. Since $\omega < 1$, it results that the radius oscillation is completed once the angle φ has made more than a turn:

$$\Delta\varphi = 2\pi\,\omega^{-1} \approx 2\pi + \frac{6\pi\, u_N{}^2 \overline{L}^2}{c^2} = 2\pi + \frac{6\pi\, u_N\, G\, M}{c^2}$$

Figure 9.4 shows the excess angle, whose value is

$$\delta\varphi \approx \frac{6\pi\, u_N\, G\, M}{c^2} = 3\,\pi\, u_N\, r_S \tag{9.19}$$

Mercury's perihelion shift

In 1915 Einstein used this result to explain Mercury's perihelion shift (1915a). Mercury's perihelion shifts forward about 574 arcseconds per century. This behavior was ascribed to gravitational perturbations on the orbit due to the other planets. However, the Newtonian computation of these perturbations results in 531 arcseconds per century. Owing to its proximity to the Sun, Mercury is the planet that receives the greatest influence from the Schwarzschild geometry

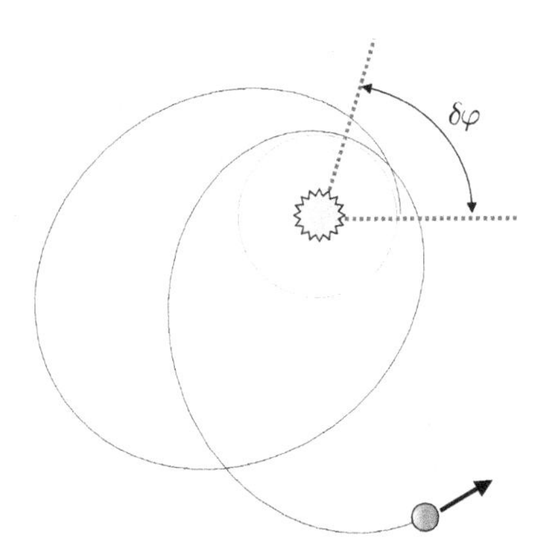

Figure 9.4. Perihelion shift of a relativistic orbit.

generated by the solar mass. By substituting $u_N = G\, M/\overline{L}^2$ in Eq. (9.19) with the value $1.802 \times 10^{-11}\,\mathrm{m}^{-1}$ coming from Mercury's angular momentum and the solar mass, then the perihelion shift of the relativistic orbit (without the perturbations due to the other planets) is of 5×10^{-7} radians per orbit. Since Mercury travels 415.2 orbits per century, then the perihelion shift sums 0.00021 radians after a century, i.e. 43 arcseconds per century. This relativistic effect must be added to the 531 arcseconds per century coming from the perturbations of other planets, to obtain a perfect agreement with the observed value.

9.3. LIGHT DEFLECTION IN SCHWARZSCHILD GEOMETRY

The conservation of p_i when the metric tensor components do not depend on the coordinate x^i (see Eq. (8.18)) is satisfied whatever (timelike, spacelike, or null) the geodesic character be. Therefore, null geodesics in Schwarzschild geometry have also two constants of motion: $p_t \equiv E$ and $p_\varphi \equiv L$. In order to study light ray trajectories in Schwarzschild geometry we can then start from Eq. (9.13). The null geodesic character requires the cancellation of mass m; so $\overline{L}$ diverges, but $2\,\overline{e}\,\overline{L}^{-2} \to E^2\,L^{-2}\,c^{-2} \equiv b^2$ (see the definition of e in Eq. (9.7)). Equation (9.13) for null geodesics becomes

$$\left(\frac{\mathrm{d}u}{\mathrm{d}\varphi}\right)^2 = b^2 - u^2 + r_S\, u^3 \tag{9.20}$$

which is analogous to the conservation equation for a classical "energy" b^2 in the "potential" $V = u^2 - r_S\, u^3$ that is displayed in Figure 9.5.

If $b^2 > 4\,r_S{}^{-2}/27$, then a light ray coming from the infinite ($u = 0$) will end at the singularity. If $b^2 = 4\,r_S{}^{-2}/27$, then the light ray is captured in an unstable circular orbit of radius $3\,r_S/2$. If $b^2 < 4\,r_S{}^{-2}/27$, the light ray will reach its maximum approach to the center of the gravitational field when $\mathrm{d}u/\mathrm{d}\varphi$ will vanish, i.e., for the value $u_{\max}$ such that

$$b^2 = u_{\max}^2\,(1 - r_S\, u_{\max}) \tag{9.21}$$

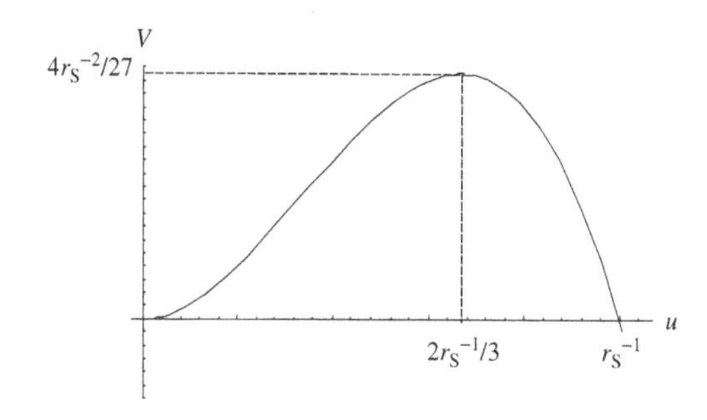

Figure 9.5. Effective potential for a light ray in Schwarzschild geometry.

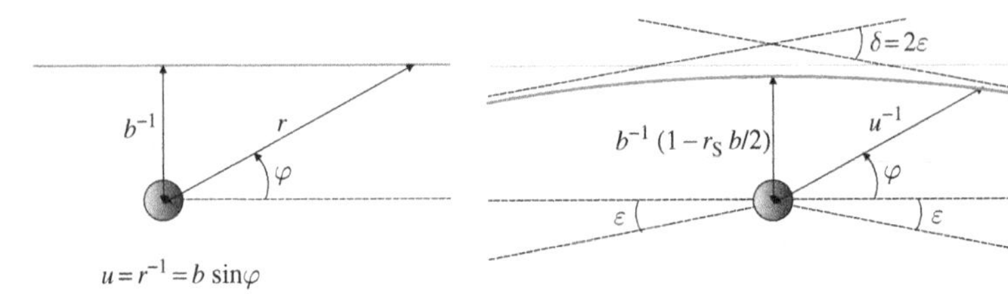

Figure 9.6. Equation for a rectilinear trajectory (left). Parameters characterizing the deflected ray trajectory (right).

In the non-relativistic limit ($r_S \rightarrow 0$), Eq. (9.20) has the solution

$$u_{nr} = b \sin(\varphi + \alpha_o), \tag{9.22}$$

where the integration constant α_o can be chosen to be zero by using an appropriate origin for the polar angle. As Figure 9.6 shows, Eq. (9.22) is nothing but the equation for a straight line with $u_{max} = b$, written in polar coordinates.

We shall work out the relativistic equation (9.20) in the region $r_S u << 1$ (in particular, $r_S b << 1$). In such a case, the solution will be a perturbation of the non-relativistic solution (9.22). The perturbation must be symmetric with respect to the maximum approach angle $\varphi = \pi/2$, and should modify u_{max} in agreement with Eq. (9.21):

$$b^2 = u_{max}^2(1 - r_S u_{max}) \cong u_{max}^2(1 - r_S b) \Rightarrow u_{max} \cong b\left(1 + \frac{r_S b}{2}\right) \tag{9.23}$$

Equation (9.20) is easier to manage after differentiating it:

$$\frac{d^2u}{d\varphi^2} = -u + \frac{3}{2} r_S u^2 \tag{9.24}$$

This procedure adds an integration constant, which will be fixed through the boundary condition (9.23). Since $r_S b << 1$, we shall substitute function u in the last term of Eq. (9.24) with its approximate expression (9.22):

$$\frac{d^2u}{d\varphi^2} + u \cong \frac{3}{2} r_S b^2 \sin^2\varphi \tag{9.25}$$

This non-homogeneous linear equation is easily solved. The integration constants are chosen in such a way that $\varphi = \pi/2$ is the maximum approach angle, and condition (9.23) is accomplished. The result is[2]

$$u \cong b \sin\varphi + r_S b^2 - \frac{r_S b^2}{2} \sin^2\varphi \tag{9.26}$$

[2] The consistency between the light deflection predicted by the Principle of equivalence and the result of General Relativity is analyzed in R. F., American Journal of Physics **71** (2003), 168–170.

which describes the deflected trajectory in Figure 9.6. When $u \to 0\,(r \to \infty)$, the polar angle goes to $-\varepsilon$ or $\pi + \varepsilon$, with $\varepsilon << 1$. Thus $\sin(-\varepsilon) \approx -\varepsilon$, and Eq. (9.26) will say that $\varepsilon \approx r_S\, b$. Then, the ray undergoes a deflection

$$\delta = 2\,\varepsilon \approx 2\,r_S\, b \tag{9.27}$$

For a ray going close to the solar limb it is $b \cong (solar\ \ radius)^{-1} = (696,000\,\mathrm{km})^{-1}$, so the deflection results of 8.48×10^{-6} radians, i.e., 1.75 arcseconds.

9.4. KRUSKAL-SZEKERES COORDINATES

As mentioned in §9.1, coordinates $(r,\, t)$ are the most suitable to work in the weak field region of Schwarzschild geometry ($r >> r_S$), but they are singular at the vicinity of the event horizon. In particular, the spatial and temporal coordinates exchange roles at the horizon. Differing from the geometric singularity at $r = 0$, the singularity at the horizon can be avoided through a coordinate change. In 1960, M.D. Kruskal and G. Szekeres proposed the following change $(r,\, t) \to (u,\, v)$:

- *if* $r > r_S$

$$u = \frac{r_S}{2} \left| \frac{r}{r_S} - 1 \right|^{1/2} \exp\left[\frac{r}{2\,r_S}\right] \cosh\left[\frac{c\,t}{2\,r_S}\right] \tag{9.28a}$$

$$v = \frac{r_S}{2} \left| \frac{r}{r_S} - 1 \right|^{1/2} \exp\left[\frac{r}{2\,r_S}\right] \sinh\left[\frac{c\,t}{2\,r_S}\right] \tag{9.28b}$$

- *if* $r < r_S$

$$u = \frac{r_S}{2} \left| \frac{r}{r_S} - 1 \right|^{1/2} \exp\left[\frac{r}{2\,r_S}\right] \sinh\left[\frac{c\,t}{2\,r_S}\right] \tag{9.29a}$$

$$v = \frac{r_S}{2} \left| \frac{r}{r_S} - 1 \right|^{1/2} \exp\left[\frac{r}{2\,r_S}\right] \cosh\left[\frac{c\,t}{2\,r_S}\right] \tag{9.29b}$$

This coordinate transformation is singular at $r = r_S$, where u and v vanish for all finite value of t. Naturally, the coordinate transformation itself has to be singular in order to heal the singularity of coordinates $(r,\, t)$. Coordinate r can be expressed as an implicit function of u and v by subtracting the squared former equations:

$$\frac{r_S{}^2}{4}\left(\frac{r}{r_S}-1\right)\exp\left[\frac{r}{r_S}\right]=u^2-v^2 \tag{9.30}$$

while the coordinate t can be solved through the ratio of u and v definitions:

$$\text{th}\left[\frac{ct}{2\,r_S}\right]=\begin{cases}\dfrac{v}{u} & \text{if } r>r_S\\[2ex] \dfrac{u}{v} & \text{if } r<r_S\end{cases} \tag{9.31}$$

In the black hole outer region—which will be called *region* I—coordinate t vanishes on the line $v=0$. Instead, inside the black hole—*region* II—t is zero on the line $u=0$. Since t is a temporal coordinate in I but is spatial in II, this behavior of the surface $t=0$ anticipates that coordinate v will be temporal in both regions.

According to Eq. (9.30), the lines $r=constant$ are hyperbolas in coordinates $(u,\ v)$, while Eq. (9.31) says that lines $t=constant$ are straight lines going through the coordinate origin of the $(u,\ v)$ plane. In Eq. (9.30) the event horizon corresponds to the hyperbola asymptotes: if $r=r_S$ it will result in $u=\pm v$. In Eq. (9.31), $u=\pm v$ means $t=\pm\infty$. Another characteristic feature of the transformation (9.28–9.29) is that $u+v\geq 0$ for all the values of r and t. These properties of the coordinate transformation are displayed in Figure 9.7.

Differentiating Eq. (9.30–9.31) and replacing them in (9.1), the Schwarzschild geometry interval results to be expressed in Kruskal–Szekeres coordinates as

$$ds^2=16\frac{r_S}{r}\exp\left[-\frac{r}{r_S}\right](dv^2-du^2)-r^2\,(d\theta^2+\sin^2\theta\,d\varphi^2), \tag{9.32}$$

where r must be regarded as the function $r(u,v)$ given in Eq. (9.30).

light cones

In coordinates $(u,\ v)$, a radial (θ, φ are constant) world line will be null if $v=\pm u+constant$ (in fact, $ds^2=0\Leftrightarrow dv=\pm du$). This means that the radial light rays are 45° straight lines. The light cones in $(u,\ v)$ coordinates look then equal at all the events. Figure 9.7 includes some future light cones (remember that v is always temporal). Figure 9.7 again shows the main feature of the causal structure of Schwarzschild geometry: once the particle went through the *future event horizon* ($r=r_S$, $t=+\infty$) to enter in region II, its unavoidable fate is the singularity $r=0$. Neither particle nor a light ray can escape from region II.

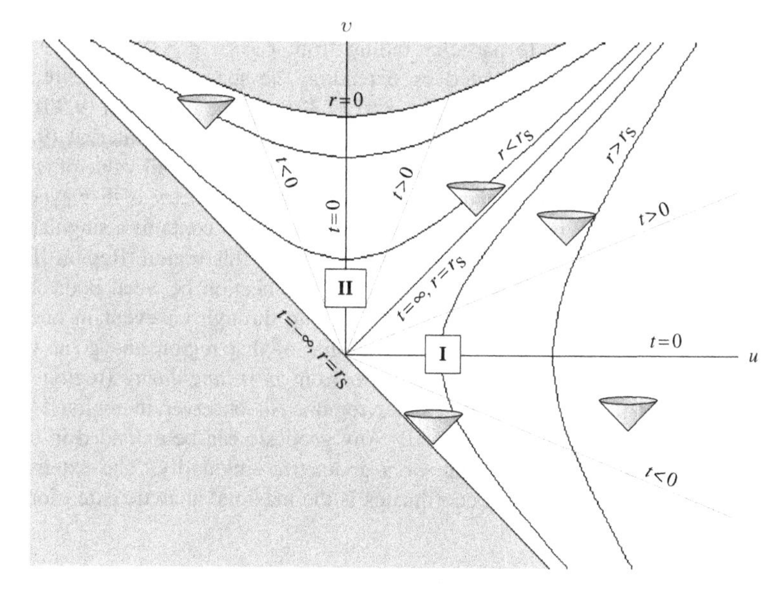

Figure 9.7. Light cones and coordinate lines (r,t) in terms of Kruskal–Szekeres coordinates (u,v).

Schwarzschild geometry extended

The space-time described in Figures 9.1 and 9.7 suffers from two enigmatic features. On one hand the choice of future cones in region II was not the sole possible choice (see §9.1); Schwarzschild solution equally allows the opposite choice, which leads to a different causal structure. On the other hand, the geodesics in region I having $\bar{e} < 0$ (see Eq. (9.8))—which describe particles going up to maximum r and then falling towards the black hole—cannot be extended toward the past beyond the *past event horizon* ($r = r_S$, $t = -\infty$), in spite of the fact that the particle proper time is still finite and there is not at that horizon geometric singularity at all. As Figure 9.8 shows, the only geodesics

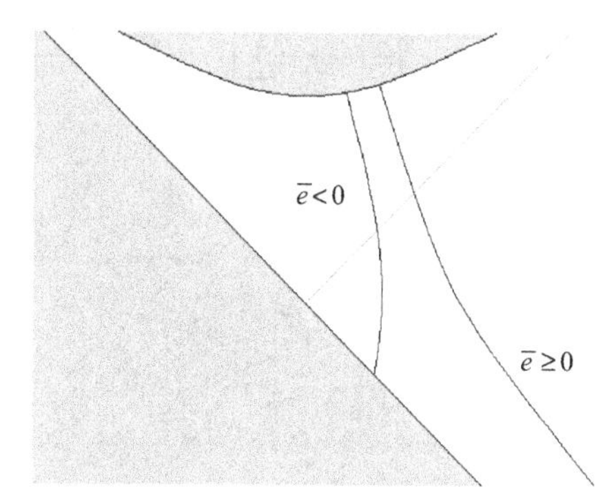

Figure 9.8. Particles falling from the infinity ($\bar{e} \geq 0$), or going through a maximum radius ($\bar{e} < 0$).

that can be infinitely extended toward the past in terms of their proper time are those corresponding to particles falling from $r = \infty\,(\bar{e} \geq 0$ in Eq. (9.8)). Both questions can be elucidated by extending the space-time to the region $u + v < 0$, keeping the form of the interval (9.32) and the definition (9.30) for the function $r(u, v)$. This extension is admissible because the interval (9.32) solves vacuum Einstein equations not only if $u + v \geq 0$, but for all value of (u, v) such that $r(u, v) > 0$. The so-extended Schwarzschild geometry is displayed in Figure 9.9. Region IV is a replica of the inner region II; it contains a singularity, but this singularity is in the past of all the events in this region. Region III is external, like region I, but there is not causal connection between both outer regions. The light rays (45° straight lines) going through an event in one of the outer regions either come from the infinite of that region and go toward the *future singularity* (regi6n II), or come from *past singularity* (region IV) and go toward the infinite of that outer region. No observer in region I can know about the existence of region III. Any geodesic can be extended in both directions up to reach the infinite or a geometric singularity. The extension provided by Kruskal–Szekeres coordinates is the maximal analytic extension of Schwarzschild geometry.

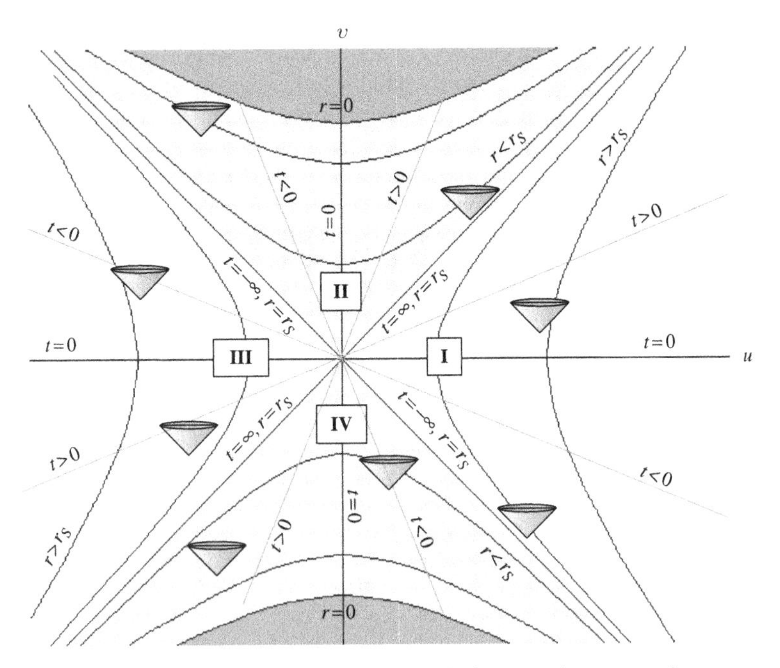

Figure 9.9. The maximal analytic extension of Schwarzschild geometry has two singularities–one of them in the past and the other in the future–and two outer regions–I and III–non-causally connected.

9.5. COSMOLOGICAL MODELS

As a second example of highly symmetric geometry, let us consider a space-time accepting a coordinate system where the metric acquires the form investigated by A.A. Friedmann in 1922:

$$\mathrm{d}s^2 = c^2\mathrm{d}t^2 - a(t)^2 \left[\frac{\mathrm{d}r^2}{1-K\,r^2} + r^2(\mathrm{d}\theta^2 + \sin^2\theta\,\mathrm{d}\varphi^2) \right] \tag{9.33}$$

On each hypersurface $t = constant$ (the "space"), the geometric properties are described by the element of arc length enclosed between square brackets. This spatial element of arc length, which is manifestly spherically symmetric, has a uniform scalar curvature equal to $6K$, thus telling that the spatial geometry is homogeneous. Function $a(t)$ plays the role of a global scale factor that modifies the spatial distances as time t elapses, without affecting the space symmetries. The world lines $x^\mu = (r, \theta, \varphi) = constant$ are timelike geodesics of this geometry—as it can be easily verified by noticing that $\Gamma^\mu_{00} = 0$—and orthogonal to the hypersurfaces $t = constant$ since $g_{\mu 0} = 0$. The proper time along these geodesics coincides with the coordinate time t because $g_{00} = 1$.

Apart from being homogeneous, the hypersurfaces $t = constant$ are isotropic, not only at the coordinate origin—as it is evident from the spherical symmetry of the interval (9.33)—but at all the points. In order to exhibit this property, it is sufficient to examine the cases $K = 0$, 1, or -1 (any other factor $K \neq 0$ can be absorbed through the change $|K|r^2 \to r^2$, $|K|^{-1}a^2 \to a^2$).[3]

flat space

In the case $K = 0$ the spatial geometry characterized by the square bracket in Eq. (9.33) is nothing but the Euclidean geometry written in spherical coordinates. Thus the interval (9.33) for $K = 0$ describes a flat space whose distances are magnified or reduced by the scale factor $a(t)$. Naturally, the flat geometry is isotropic with respect to any point.

closed space

If $K = 1$ then r is such that $0 \leq r \leq 1$. So we can perform the coordinate change $r = \sin\chi$; therefore $\mathrm{d}\chi^2 = \mathrm{d}r^2/(1-r^2)$. In this way it becomes evident that the spatial geometry results to be a spherical geometry. In fact, on a surface $\varphi = constant$ the spatial metric yields $\mathrm{d}\sigma^2 = a(t)^2\,(\mathrm{d}\chi^2 + \sin^2\chi\,\mathrm{d}\theta^2)$, which is analogous to the geometry of a bidimensional sphere of radius $a(t)$ embedded in an Euclidean space, where χ is the azimuthal angle (extended to the range $0 \leq \chi \leq \pi$). Let us now restore the polar angle φ for getting a three-dimensional spherical geometry. Certainly, a spherical geometry is isotropic with respect to any point. The sphere isotropy is based on the invariance of the sphere under the group of rotations, whose transformations move points on the sphere in any direction. The spherical geometry indicates that the space is closed if $K = 1$, and its 3-volume is $a(t)^3 \int \sin^2\chi\,\sin\theta\,\mathrm{d}\chi\,\mathrm{d}\theta\,\mathrm{d}\varphi = 2\,\pi^2\,a(t)^3$.

open space

If $K = -1$ then the coordinate change $r = \sinh\chi$ leads the spatial geometry to the form $\mathrm{d}\sigma^2 = a(t)^2[\mathrm{d}\chi^2 + \sinh^2\chi\,(\mathrm{d}\theta^2 + \sin^2\theta\,\mathrm{d}\varphi^2)]$. Although

[3] If K is non-dimensional then r is non-dimensional as well, but the scale factor has units of length.

less familiar, $-d\sigma^2$ is the interval for a hyperboloid $w^2 - x^2 - y^2 - z^2 = a(t)^2$ embedded in a Minkowski space having temporal coordinate w. In fact, this is the conclusion after defining coordinates $\chi\,\theta\,\varphi$ on the hyperboloid as $w = a \cosh \chi$, $(x, y, z) = a \sinh \chi\,(\sin\theta\cos\varphi, \sin\theta\sin\varphi, \cos\theta)$. Combining boosts and rotations, the Lorentz group transformations in Minkowski space $w\,x\,y\,z$ move points on the hyperboloid in any direction, but the hyperboloid remains invariant. This property guarantees the isotropy of $d\sigma^2$ at any point. The isotropy at all points implies the space homogeneity.

Figures 9.10–9.12 summarize the characteristics of the three—flat, closed, and open—isotropic and homogeneous geometries.

$$
\begin{aligned}
K = 0:&\quad ds^2 = c^2dt^2 - a(t)^2\,[d\chi^2 + \chi^2\,(d\theta^2 + \sin^2\theta\, d\varphi^2)], && 0 \le \chi < \infty\\
K = 1:&\quad ds^2 = c^2dt^2 - a(t)^2\,[d\chi^2 + \sin^2\chi\,(d\theta^2 + \sin^2\theta\, d\varphi^2)], && 0 \le \chi \le \pi\\
K = -1:&\quad ds^2 = c^2dt^2 - a(t)^2\,[d\chi^2 + \sinh^2\chi\,(d\theta^2 + \sin^2\theta\, d\varphi^2)], && 0 \le \chi < \infty
\end{aligned}
$$

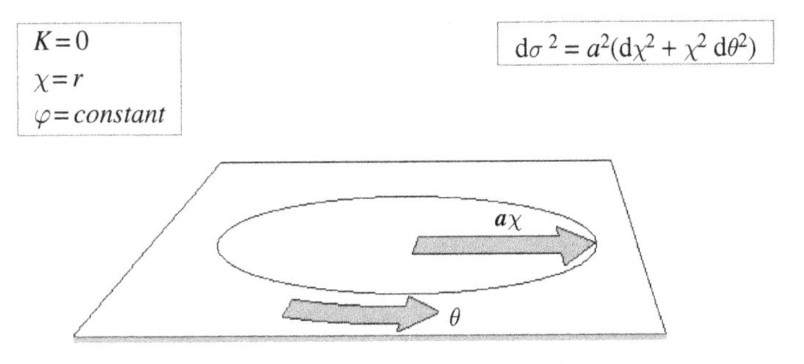

Figure 9.10. Flat isotropic and homogeneous geometry.

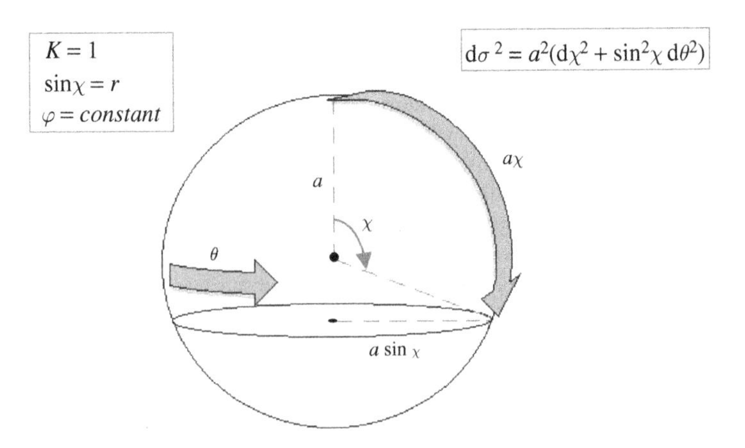

Figure 9.11. Closed isotropic and homogeneous geometry.

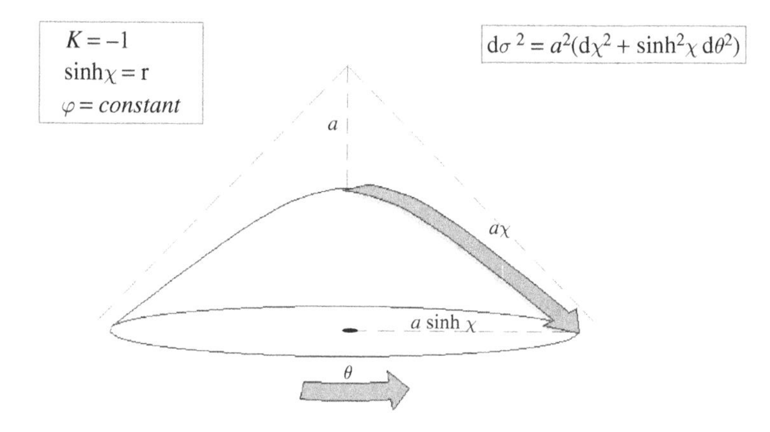

Figure 9.12. Open isotropic and homogeneous geometry.

The spatial geometry of Friedmann models with $K = \pm 1$ does not satisfy that the ratio between the arc length and the distance to the arc center equals the subtended angle; such result is specific of the flat case ($K = 0$). The form of $d\sigma^2$ says that the length of the arc subtended by the angle $\Delta\theta$ is $a(t)\, r\, \Delta\theta$, while the distance (the radius) between the arc and the (arbitrary) point chosen as coordinate origin is $a(t)\chi$. Therefore the quotient between the arc length and the angle is smaller than the radius if $K = 1$ ($r = \sin\chi$) and larger than the radius if $K = -1$ ($r = \sinh\chi$).

In 1935 and 1936, H.P. Robertson and A.G. Walker proved that the interval (9.33) is the most general interval compatible with the requirements of homogeneity and isotropy. Friedmann-Robertson-Walker metrics are used for cosmological models, which describe the behavior of the universe at large scales. This use supposes the acceptation of the *cosmological principle* that claims the homogeneity and isotropy of the universe, a justifiable hypothesis at scales larger than 100 Mpc.[4] In an isotropic and homogeneous cosmological model the scale factor $a(t)$ is determined by Einstein equations, which are greatly simplified due to the high symmetry anticipated in the searched solution. The source in Einstein equations, i.e., the matter distribution in the universe, must be consistent with the hypotheses of homogeneity and isotropy. For this reason, the large-scale matter behavior will be described as a perfect fluid of uniform density and pressure, whose elements move along the geodesics $x^\mu = constant$. So, the fluid is at rest in the coordinate system used to write the interval (9.33) (otherwise the fluid motion would destroy the isotropy since it would select a privileged direction at each location); in spite of this, distances between "particles"—the galaxies—can

---

[4] 1 Mpc = 10^6 pc. The *parsec* (pc) is the distance at which the radius of the Earth's orbit subtends an angle of 1 arcsecond, and is equal to 3.09×10^{16} m or 3.26 light years. A characteristic size for a galaxy is 1–50 kpc. Galaxy clusters—systems of gravitationally bound galaxies—typically have sizes of 2–10 Mpc. Galaxy clusters gather in superclusters of filamentary structure that surround voids at scales of 100 Mpc, as shown by 2dF-SDSS luminous matter surveys (Colless, 2001; Cannon et al., 2006).

vary since they are subjected to the time evolution of the scale factor $a(t)$. Owing to its particular relation with the fluid world lines, the coordinate system in Eq. (9.33) is comoving. To solve Einstein equations and get the scale factor $a(t)$, we should know the state equation of the fluid, which relates the pressure to the energy density.

cosmological frequency shift

Before solving Einstein equations we shall study a phenomenon that will appear in case that the scale factor $a(t)$ would vary in time. Thus, we shall see what kind of observation, in the context of General Relativity, would reveal that the universe is not immutable but changes in time. Let us consider a set of light pulses emitted from a galaxy with a period T_{src}. We want to know the observed period T_{obs} at another galaxy. Both galaxies take part in the fluid representing the average behavior of matter; then, by neglecting local peculiar motions, we can state that their spatial coordinates χ_{src} and χ_{obs} are fixed in the comoving system (the isotropy releases us from considering other coordinates than the radial one; in particular, we could choose the coordinate origin at one of these galaxies). Whatever the K value is, a radial light ray verifies that

$$ds^2 = c^2\,dt^2 - a(t)^2\,d\chi^2 = 0 \tag{9.34}$$

which can be integrated to yield

$$\int_{t_{emission}}^{t_{reception}} \frac{c\,dt}{a(t)} = |\Delta\chi| \tag{9.35}$$

This result is repeated for the travel of each pulse (remember that $|\Delta\chi|$ does not change). By applying the result to two successive pulses, emitted at $t_{emission}$ and $t_{emission} + T_{src}$, and received at $t_{reception}$ and $t_{reception} + T_{obs}$ (coordinate time t is coincident with the proper time used in each galaxy to measure the respective periods), it is obtained

$$\int_{t_{emission}}^{t_{reception}} \frac{c\,dt}{a(t)} = \int_{t_{emission}+T_{src}}^{t_{reception}+T_{obs}} \frac{c\,dt}{a(t)} \Rightarrow \int_{t_{emission}}^{t_{emission}+T_{src}} \frac{c\,dt}{a(t)} = \int_{t_{reception}}^{t_{reception}+T_{obs}} \frac{c\,dt}{a(t)}$$

In the last expression, the measures of the integration intervals are equal to the periods. Since the universe does not undergo significant changes in such a short time, the scale factor can be considered as constant in each integral. Then

$$\frac{T_{src}}{a(t_{emission})} = \frac{T_{obs}}{a(t_{reception})} \tag{9.36}$$

If the scale factor increases, then the observed period will be larger than the source period. This means that a redshift of cosmological origin will be observed. This frequency shift could be recognized in the emission and absorption lines of galaxies, which would thus confirm the universe evolution. The cosmological redshift is a consequence that the distance between galaxies increases when the

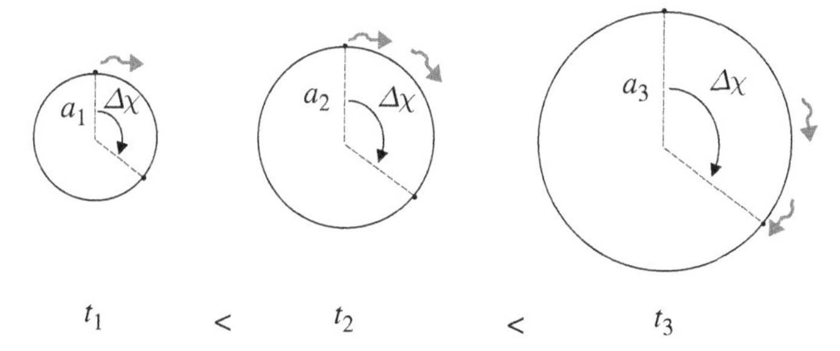

Figure 9.13. The growing of the scale factor a forces the second pulse to travel a distance larger than the one covered by the first pulse. The consequent delay produces an increasing of the period (redshift).

scale factor grows; the second pulse then travels a distance larger than the one covered by the first pulse and undergoes a "delay," thus increasing the observed period. This situation is represented in Figure 9.13.

The relative frequency shift

$$z = \frac{\nu_{\text{src}} - \nu_{\text{obs}}}{\nu_{\text{obs}}} = \frac{\nu_{\text{src}}}{\nu_{\text{obs}}} - 1 = \frac{a(t_{\text{reception}})}{a(t_{\text{emission}})} - 1 \tag{9.37}$$

can be expanded in Taylor series around the reception time t_r as a function of the emission time t_e. If $z << 1$, it is valid that

$$1 + z = \frac{a(t_r)}{a(t_e)} \approx 1 - \frac{\dot{a}(t_r)}{a(t_r)}(t_e - t_r) = 1 + H_r(t_r - t_e) \tag{9.38}$$

where we define

$$H(t) = \frac{\dot{a}(t)}{a(t)} \tag{9.39}$$

and $H_r = H(t_r)$. The travel time $\Delta t = t_r - t_e$ can be written as a function of distance σ between emitter and receiver,

$$\sigma(t) = \int d\sigma = \int a(t) d\chi = a(t)\,|\Delta\chi| \tag{9.40}$$

In fact, the relation between distance and travel time is estimated from Eq. (9.35) at the same approximation order used in Eq. (9.38):

$$c\,\Delta t \approx a(t_r)\,|\Delta\chi| = \sigma(t_r) \tag{9.41}$$

Therefore

$$z \approx c^{-1} H_r\, \sigma_r \tag{9.42}$$

so the frequency shift results to be proportional to the distance σ_r. The proportionality "constant" H is called *Hubble constant*, in honor of Edwin P. Hubble who discovered the phenomenon in 1929. The redshift observed by Hubble indicates that the universe expands. The value of $H(t_{\text{today}})$ is about $70\,\text{km}\,\text{s}^{-1}\,\text{Mpc}^{-1}$, with a significant margin of error due to the poor determination of astronomical distances.

Hubble constant also appears in the velocity at which the galaxies separate as a consequence of the expansion. According to Eq. (9.40) it is

$$\dot{\sigma}(t) = \dot{a}(t)\ |\Delta\chi| = H(t)\,\sigma(t) \tag{9.43}$$

which allows us to rewrite Eq. (9.42) as

$$z \approx \frac{\dot{\sigma}_r}{c}$$

which coincides with the relation between frequency shift and relative velocity in Doppler effect (for the same approximation order).

Distance σ_r does not agree with the *luminous distance* d_L used in Astronomy, since d_L emerges from the relation between energy flux F (energy per unit of time and surface) and luminosity L (energy per unit of time) characteristic of an Euclidean geometry:

$$L = 4\,\pi\,{d_L}^2\,F \tag{9.44}$$

In this way, the distance to an object is evaluated from the energy flux F registered by the receiver, together with an estimation of the object luminosity L.[5] However, this relation does not take into account that the universe expansion redshifts the frequency (energy) of photons, so weakening the flux in a factor $\nu_{obs}/\nu_{src} = (1+z)^{-1}$. This shift was explained as a delay between successive pulses caused by the expansion (see Figure 9.13); besides this delay weakens the number of photons received per unit of time by the same factor $T_{src}/T_{obs} = (1+z)^{-1}$. Finally the energy per unit of time, weakened by the factor $(1+z)^{-2}$, spreads on a sphere whose surface is $4\pi {a_r}^2 {r_{src}}^2$, which is the 2-volume of a surface $t = constant$, $r = constant$ with the metric (9.33):

$$F = \frac{L}{4\pi\,{a_r}^2\,{r_{src}}^2\,(1+z)^2} \tag{9.45}$$

[5] The *cepheids* are variable stars used as standard candels to establish distances, because their luminosities can be determined from their periods. The relation period–luminosity is obtained from close cepheids whose distances can be geometrically measured (for instance, by using the parallax), and their luminosities can be thus known. Since the furthest cepheids that Hubble space telescope can resolve are located at less than 80 millions of light years or 25 Mpc ($z \sim 0.005$), then the calibration of larger distances requires another type of standard candel or other techniques.

i.e., $r_{\rm src}$ is the radial position of the source with respect to the observer. From Eq. (9.44) and (9.45) the luminous distance results to be

$$d_{\rm L} = (1+z)\,a_{\rm r}\,r_{\rm src} = (1+z)\,a_{\rm r}\,f(\chi_{\rm src}) \tag{9.46}$$

Function $f(\chi)$ is χ, $\sin\chi$, or $\sinh\chi$, depending on the universe is flat, closed, or open. In all cases it is $f(\chi) = \chi + O(\chi^3)$. In this way, if $z << 1$ (i.e., if $\chi_{\rm src} << 1$) it will be valid that $d_{\rm L} \cong a_{\rm r} r_{\rm src} \cong a_{\rm r}\chi_{\rm src} = \sigma_{\rm r}$; so $\sigma_{\rm r}$ can be substituted with luminous distance $d_{\rm L}$ in Eq. (9.42). For large values of z we should come back to Eq. (9.35) to obtain a precise dependence of $\chi_{\rm src}$ and z. But, to reach this result we should know the entire evolution of the universe from the emission to the reception of the signal. On the contrary, Eq. (9.42) just depends on the value of the Hubble constant at the reception time. The substitution of $\chi_{\rm src}\ (z)$ in Eq. (9.46) will lead to the exact relation between luminous distance and frequency shift.

To obtain the expression for $\chi_{\rm src}\ (z)$ we shall perform a change of variable in the integral (9.35). Each time t' of the signal journey can be associated with the value of z' that would correspond if the signal were emitted at that time and location. By differentiating the Eq. (9.37) it results

$$dz' = -\frac{a_{\rm r}}{a(t')^2}\,\dot{a}(t')\,dt' = -a_r\,H(t')\,\frac{dt'}{a(t')}$$

thus, replacing in Eq. (9.35), and using that $z'(t_{\rm e}) = z$, $z'(t_{\rm r}) = 0$:

$$\chi_{\rm src}(z) = \int_{t_{\rm e}}^{t_{\rm r}} \frac{c\,dt'}{a(t')} = \int_0^z \frac{c\,dz'}{a_{\rm r}\,H(z')} \tag{9.47}$$

In the next section we shall find out the behavior of $H(z')$ along the history of the universe (see Eq. (9.62)). Here we shall improve the kinematic approach for small z. We can expand $H(z')$ around $z = 0$: $H(z') = H_{\rm r} + (dH/dz')_{\rm r} z' + \cdots$, the derivative of H being

$$\frac{dH}{dz'} = \frac{dH}{dt'}\frac{dt'}{dz'} = \left(\frac{\ddot{a}(t')}{a(t')} - \frac{\dot{a}(t')^2}{a(t')^2}\right)\left(-\frac{a(t')}{a_{\rm r}\,H(t')}\right) = \frac{a(t')}{a_{\rm r}}\,H(t')\,(1+q(t'))$$

where we have defined the *deceleration parameter*

$$q(t) = -\frac{\ddot{a}(t)\,a(t)}{\dot{a}(t)^2} \tag{9.48}$$

Replacing $H(z') = H_{\rm r} + H_{\rm r}\ (1+q_{\rm r})\ z' + \cdots$ in the integral (9.47), it results that $\chi_{\rm src}(z) = c(a_{\rm r}H_{\rm r})^{-1}\,(z - {}^1\!/\!_2(1+q_{\rm r})z^2 + \cdots)$. At this approximation order it still remains valid that $f(\chi) \cong \chi$. Then the luminous distance in Eq. (9.46) is expressed as a function of the redshift in the way

$$c^{-1}\,H_{\rm r}\,d_{\rm L} = z + \frac{1}{2}(1-q_{\rm r})\,z^2 + \cdots \tag{9.49}$$

9.6. EVOLUTION OF THE UNIVERSE

We shall solve the equations governing geometry and matter. Since the high symmetry imposed to the searched type of geometry has left only a function $a(t)$ to be determined, then just one Einstein equation for geometry will be needed, together with an equation describing the dynamics of matter and a state equation. The average behavior of matter will be characterized by the energy–momentum tensor of a homogeneous perfect fluid. In the comoving frame $T^{i}_{\ j}$ has the form diag $(\rho_o, -p_o, -p_o, -p_o)$ studied in §7.6 (see Eq. (7.72)). The isotropy of the spatial sector guarantees that $T^{i}_{\ j}$ keeps its form under changes of spatial coordinates. The three equations above mentioned will then be used to find out the scale factor $a(t)$, the energy density $\rho_o(t)$ and the pressure $p_o(t)$.

In §8.7 it was emphasized that not all Einstein equations are of second order in temporal derivatives of the metric, but there are first order equations that constrain the initial values in the second order equations. In particular, Einstein equation "00" is an initial value equation for the scale factor $a(t)$. In fact, let us compute G^0_0 with the help of Christoffel symbols[6]

$$\Gamma^{k}_{\ 00} = 0, \quad \Gamma^{0}_{\ \mu 0} = 0, \Gamma^{\mu}_{\ \nu 0} = c^{-1}\, a^{-1}\, \dot{a}\, \delta^{\mu}_{\nu}, \quad \Gamma^{0}_{\ \mu\nu} = -c^{-1}\, a^{-1}\, \dot{a}\, g_{\mu\nu} \qquad (9.50)$$

Then, noticing that $g^i_j = g^{ik} g_{kj} = \delta^i_j$, it is

$$R_{00} = -3\, c^{-2} \frac{\ddot{a}}{a}, \quad R = -\frac{6\, c^{-2}}{a^2}\, (a\, \ddot{a} + \dot{a}^2 + K c^2), \quad G^0_0 = \frac{3\, c^{-2}\, \dot{a}^2}{a^2} + \frac{3K}{a^2} \qquad (9.51)$$

so Einstein equation "$^0_{\ 0}$" results to be

initial value equation

$$\frac{\dot{a}^2}{a^2} + \frac{K\, c^2}{a^2} = \frac{8\, \pi\, G}{3\, c^2}\, \rho_o \qquad (9.52)$$

The dynamics of matter is contained in the conservation equations $T^{ij}_{\ \ ;j} = 0$. For $i = 0$ it results

energy conservation

$$\frac{\mathrm{d}}{\mathrm{d}t}\, (\rho_o\, a^3) = -p_o\, \frac{\mathrm{d}}{\mathrm{d}t}\, a^3 \qquad (9.53)$$

which is the relation between the energy variation in a fluid volume and the work done in the adiabatic expansion of a perfect fluid. Equations (9.52) and (9.53) must be joined to the fluid state equation $p_o = p_o(\rho_o)$.

---

[6] Although the comoving frame is based on geodesic coordinate lines, it is not, as seen, a locally inertial frame. The comoving frame only guarantees that geodesics $x^\mu = constant$ ($\mu = 1, 2, 3$) locally look as straight lines, since $\Gamma^k_{00} = 0$ ($k = 0, 1, 2, 3$). But in a locally inertial frame *all* geodesic lines locally look as straight lines.

The rest of Einstein equations are either trivial or second order equations constrained by Eqs. (9.52) and (9.53). In fact, Einstein equations for the three diagonal spatial components of G^i_j result to be

$$2\frac{\ddot{a}}{a}+\frac{\dot{a}^2}{a^2}+\frac{K\,c^2}{a^2}=-\frac{8\,\pi\,G}{c^2}\,p_o \tag{9.54}$$

which could also be obtained by differentiating Eq. (9.52) and using Eq. (9.53). Hence, Eq. (9.52) governs Eq. (9.54) by constraining its initial values. In these equations, the last term on the left is essentially the (3-dimensional) curvature $^{(3)}R = 6K/a^2$ associated with the element of spatial arc length $d\sigma(t)$ in the interval (9.33).

In order to describe the behavior of the fluid, we shall use in Eq. (9.53) a type of state equation that is suitable for embracing different contributions to the matter and energy contents of the universe:

state equation

$$p_o(t) = \varpi\,\rho_o(t) \tag{9.55}$$

For dust (matter composed by non-colliding particles) it is $\varpi = 0$ (see Eq. (7.70)). For gases made up of photons or other massless particles and, in an approximated way, ultra-relativistic particles (i.e., particles whose kinetic energy is much larger than their rest energy), it is $\varpi = 1/3$ (see §7.7). Integrating the Eq. (9.53) it is obtained

$$\rho_o(t)\,a(t)^{3(1+\varpi)} = constant \tag{9.56}$$

Equation (9.56) for dust matter ($\varpi = 0$) says that the universe expansion does not modify the energy—essentially, the number of particles—contained in a coordinate volume (defined by given values of $\Delta\chi$, $\Delta\theta$, $\Delta\varphi$). Instead, the photon-gas energy ($\varpi = 1/3$) contained in a coordinate volume diminishes when the universe expands, as a consequence of the frequency redshift.

According to Eqs. (9.52) and (9.54), function $\ddot{a}(t)$ results to be negative,

$$2\frac{\ddot{a}}{a}=-\frac{8\,\pi\,G}{c^2}\left(p_o+\frac{\rho_o}{3}\right) \tag{9.57}$$

for any value $\varpi > -1/3$. This means that if the universe only contained conventional forms of matter and energy, then its expansion would slow down as a consequence of the gravitational attraction among its parts. Going back through the history of the universe we should thus observe the scale factor diminishing more quickly each time (see Figure 9.14) and, according to Eq. (9.56), increasing energy densities. This behavior would lead to a null scale factor at a finite past time. At that time, distances would go to zero and densities (and temperatures) would become infinite. This singular beginning of the evolution of the universe is known with the name of *Big-Bang*. According to Figure 9.14, the present value of the Hubble constant would give an upper bound for the age of the universe:

$$\text{Age of the universe} < \frac{a(t_{\text{today}})}{\dot{a}(t_{\text{today}})} = H(t_{\text{today}})^{-1} \sim 14\times 10^9\ \text{years}$$

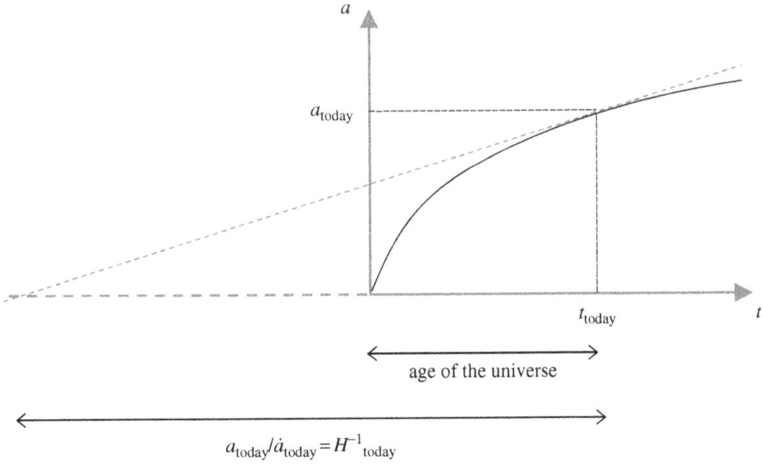

Figure 9.14. The present value of H^{-1} gives an upper bound for the age of the universe in a decelerated expansion.

According to Eq. (9.52) a spatially flat universe ($K = 0$) requires a critical relation between density ρ_o and Hubble constant $H = \dot{a}/a$:

$$\rho_{o\,critical}(t) = \frac{3\,c^2}{8\pi\,G}\,H(t)^2 \tag{9.58}$$

The current value of Hubble constant means that the critical density corresponding to this epoch is of the order of $\rho_{o\,critical}\,(t_{today})c^{-2} \sim 10^{-26}\,\text{kg}\,\text{m}^{-3}$. It is convenient to measure the different contributions to the matter and energy of the universe as fractions of the critical density. For this, we define

$$\Omega(t) = \frac{\rho_o(t)}{\rho_{o\,critical}(t)} \tag{9.59}$$

In the present epoch of the universe, the averaged luminous matter density is $\rho_o^{\text{lum mat}}(t_{today})c^{-2} \sim 4 \times 10^{-29}\,\text{kg}\,\text{m}^{-3}$ (i.e., $\Omega^{\text{lum mat}}(t_{today}) \sim 0.004$). The existence of dark matter in galactic halos, indirectly detected through the gravitational effects on the movement of luminous matter (Rubin, 1980), increases the matter density in one order of magnitude. Effects of dark matter has also been observed around galaxy clusters and superclusters that would raise the matter density to the value $\Omega^{\text{mat}}(t_{today}) \sim 0.3$. Thus, some 99% of the matter in the universe would be dark matter. The nature of this dark matter is not known. A fraction of dark matter should be baryonic matter (like ordinary matter) in the form of white dwarfs, brown dwarfs (proto-stars that did not reach to start the thermonuclear reaction), planetoids, black holes, etc. But dark matter cannot be completely baryonic, since in such a case the observed deuterium abundance would be incompatible with the one synthesized as the universe cooled (see

dark matter

below *nucleosynthesis*). Some 85 % of $\Omega^{\text{mat}}(t_{\text{today}})$ should be another type of matter. If neutrinos (leptonic matter) had mass—as it seems to be the case, according to the observed "oscillations" among the three types of neutrinos—they would take part in non-baryonic dark matter. Other candidates for dark matter have a more speculative character.

In the present epoch matter dominates on the radiation. The density of photons—most of them coming from the cosmic background radiation of 2.725 K (see below)—is $\rho_o^{\text{phot}}\, c^{-2} = 4.66 \times 10^{-31}\,\text{kg m}^{-3}$ ($\Omega^{\text{phot}}(t_{\text{today}}) \sim 5 \times 10^{-5}$). However, the current matter domination reverses in the past. In fact, the ordinary matter in the present universe can be described as dust, since $p_o^{\text{mat}} << \rho_o^{\text{mat}}$, and Eq. (9.56) says that dust density ($\varpi = 0$) increases towards the past as a^{-3}. But radiation density ($\varpi = 1/3$) does it more quickly: $\rho_o^{\text{rad}} \propto a^{-4}$. Therefore, previous to the matter-dominated era the universe was governed by radiation.[7]

In an adiabatic evolution, the ordinary matter temperature rises as the volume diminishes. On the other hand the radiation density is proportional to the fourth power of its temperature (i.e., $T^{\text{rad}} \propto a^{-1}$).[8] This means that the temperatures of matter and radiation increases toward the past. At times when the temperature was very elevated, the matter was highly ionized; there was abundance of free electrons that interacted with photons and scattered them. This interaction thermalized the electromagnetic radiation, thus giving it the spectrum characteristic of a black body.[9] When the expansion cooled the universe below 3000 K, the photon energies turned to be insufficient to avoid the stabilization of the atomic structures of hydrogen and helium. Once the matter organized in the form of atoms, the interaction between radiation and matter ended.[10] Before this *decoupling* of matter and radiation, the universe was opaque, like the interior of stars. After the decoupling, the universe became transparent to the electromagnetic radiation. If this model of universe is correct, an electromagnetic radiation with black body spectrum— the relic of the state of the universe before the decoupling of matter and radiation—should be observed coming from all directions. In fact, each time we observe a far event, we are watching the past: since light propagates with a finite speed, the observation of an object does not tell us how the object is now, but how it was when the light was emitted. Hence, today we should receive the black body radiation emitted in other places at the decoupling time. The present-day universe should be filled with this black body radiation, the relic of the opaque universe prior to decoupling. The accidental

decoupling of matter and radiation

cosmic background radiation

[7] The separated treatment of matter and radiation in Eq. (9.53) supposes that the interaction between both fluids is negligible. The pressure of ordinary matter can always be neglected; fluid pressure is only important in the early universe, when radiation dominates.

[8] Stefan–Boltzmann law states: $\rho_o^{\text{photons}} = (8\pi^5/15)k_B^4\, h^{-3}\, c^{-3}\, T^4 = 7.565 \times 10^{-16} T^4\,\text{J m}^{-3}\,\text{K}^{-4}$. Masless neutrinos fulfill a similar law.

[9] This thermalization implies an interaction between matter and photons, since the photon gas transfers energy to matter to keep their temperatures equalized (otherwise, matter would cool more quickly than photons). Even so, the fluxes $T^{0\alpha}$ are negligible if compared with T^{00}, except at times close to the Big-Bang.

[10] Rayleigh scattering of photons due to hydrogen atoms can be neglected.

discovery of the *cosmic background radiation* in 1965 by A.A. Penzias and R.W. Wilson (Nobel Prizes in Physics, 1978), and the immediate identification of its origin by Dicke, Peebles, Roll and Wilkinson, was a great success for the Big-Bang model. This is a highly isotropic microwave radiation with black body spectrum at the temperature of $T^{\mathrm{phot}}{}_{(\mathrm{today})} = 2.725\,\mathrm{K}$.[11] If at the photon decoupling time the temperature was about 3000 K,[12] then $T^{\mathrm{phot}}{}_{(\mathrm{decoupling})}/T^{\mathrm{phot}}{}_{(\mathrm{today})} = a_{(\mathrm{today})}/a_{(\mathrm{decoupling})} = 1 + z \sim 1000$. Therefore, the cosmic background radiation is redshifted by a factor $z \sim 1000$. Being opaque to the electromagnetic radiation, the universe prior to the decoupling time is not visible in this type of radiation. Nevertheless, the black body electromagnetic radiation emitted at the decoupling time contains information about the previous history of the universe. The present data about matter and radiation distributions allows inferring that the decoupling happened shortly after the radiation-dominated era ended. After the decoupling, the ordinary matter evolves independently of photons, and it begins the process of its condensation in stars, galaxies and galaxy clusters, by starting from preexisting small inhomogeneities in its distribution (whose imprints can be tracked in the cosmic background radiation).

nucleosynthesis

Besides describing the expansion of the universe and enlightening the origin of the cosmic background radiation, the *standard Big-Bang model* also explains the hydrogen–helium abundance ratio in the universe. Primordial *nucleosynthesis*—the process where free protons and neutrons join together to compose light nuclides—happened when the universe cooled below 10^9 K. Since proton mass is 1.3 MeV smaller than neutron mass, neutron decays into a proton (β decay) by emitting an electron and an anti-neutrino. The high temperatures of the early universe also allow inverse processes, because the particles have enough kinetic energy to overcome the mass difference between proton and neutron. Thus, an equilibrium relation resulted, which depended on the temperature and the mass difference. When the universe expanded to cool below 10^{10} K ($kT \sim 1$ MeV), the kinetic energy of particles became insufficient to produce neutrons from protons. The neutrons then present could have ended their lives by decaying all of them into protons; however, at the temperature of 10^9 K the photons lacked the necessary energy to avoid that neutrons and protons join together for composing deuterons and helium nuclei. Thanks to these stable structures, the ratio proton–neutron was frozen. Primordial nucleosynthesis mostly produced light nuclides, mainly hydrogen and helium, and a few of deuterium and lithium (the heavy nuclides are synthesized in stars and supernovae). In 1948, G. Gamow calculated that, at the time of nucleosynthesis, there were seven protons per

---

[11] The data from COBE probe (Smoot et al., 1992; Mather et al., 1994) showed that temperature fluctuates with the direction less than a part in 10^5, within the range of angular scales larger than 7° covered by the maximum resolution of the instrument (J. C. Mather and G. F. Smoot won the Nobel Prize in Physics 2006 for this work). Some 99.97% of the electromagnetic radiation in the universe comes from the cosmic background radiation.

[12] Although this temperature corresponds to a characteristic energy $kT \sim 0.26$ eV that is smaller than the hydrogen-ionizing energy of 13.6 eV, the very large ratio of photons to protons assured a sufficient number of hydrogen-ionizing photons in the tail end of the black body energy distribution.

neutron, so the ratio between "primordial" helium (also the stars can produce helium) and hydrogen had to be 1–3, which agrees with the observed ratio.

In the epoch prior to nucleosynthesis, higher temperatures allowed the energies needed to create massive particles, like positron–electron pairs, other leptons (muons and tauons), pions, etc. At times when $kT \sim 300\,\text{MeV}$ or larger, the quarks that constitute the hadrons (baryons and mesons) were free. But, as long as the expansion cooled the universe, pions annihilated into photons. Then the annihilation of muons and tauons came. At $T \sim 3\times 10^{10}\,\text{K}$ ($kT \sim 3\,\text{MeV}$) the equilibrium among electrons, positrons, and neutrinos ended (the annihilation of positron–electron pairs became more important than their creation by neutrinos); neutrinos decoupled from matter and began to travel freely. At $T \sim 5\times 10^{9}\,\text{K}$ ($kT \sim 0.5\,\text{MeV} \sim m_e c^2$) positron–electron pairs annihilated into photons. This process reheated the photon gas; so photon temperature became higher than neutrino temperature. Neutrinos present-day temperature is 1.9 K.[13] Since neutrino masses would be very small (1 eV or less), they were ultrarelativistic during the radiation-dominated era. At the end of nucleosynthesis the universe mostly contained photons, neutrinos, electrons, protons, ^{4}He nuclei, and non-baryonic dark matter, and was radiation dominated. Radiation domination ended at the time t_{rm} when $\rho_o^{\text{rad}}(t_{\text{rm}}) = \rho_o^{\text{mat}}(t_{\text{rm}})$, shortly before the photon decoupling time.[14]

In the standard Big-Bang model the universe goes through a radiation-dominated era that is followed by a matter-dominated era. Equations (9.52) and (9.56) govern this evolution. It is useful to write the constants in these equations as functions of the values a_{o}, H_{o}, and Ω_{o} corresponding to a given time t_{o}. The initial value equation (9.52) thus reads

$$K\, c^2 = a(t)^2\, H(t)^2 (-1 + \sum_i \Omega_i(t)) = a_{\text{o}}^2\, H_{\text{o}}^2 (-1 + \sum_i \Omega_{\text{o}_i}) \tag{9.60}$$

where the sum is made on the constituents of the universe. Equation (9.60) says that the universe is closed, flat, or open according to $\sum \Omega_i$ is larger, equal, or smaller than 1. On the other hand, the conservation equation (9.56) can be rewritten as

$$\Omega_i(t)\, H(t)^2 a(t)^{3(1+\varpi_i)} = \Omega_{\text{o}_i}\, H_{\text{o}}^2\, a_{\text{o}}^{3(1+\varpi_i)} \tag{9.61}$$

Replacing these results in Eq. (9.52), it is obtained

$$H(t)^2 = H_{\text{o}}^2 \left(\frac{a_{\text{o}}}{a(t)}\right)^2 \left[\sum_i \Omega_{\text{o}_i} \left(\frac{a_{\text{o}}}{a(t)}\right)^{1+3\varpi_i} + \left(1 - \sum_i \Omega_{\text{o}_i}\right)\right] \tag{9.62}$$

[13] This temperature should be taken as a mere particle counter parameter in the case of massive neutrinos. The relation between photon and neutrino temperatures results from the entropy conservation. Neutrinos contribution to radiation is comparable to photons contribution ($\Omega^{\text{neutrinos}} = 0.68\,\Omega^{\text{phot}}$, for three neutrino species).

[14] According to Eq. (9.56) $1 + z(t) = a(t_{\text{today}})/a(t) = [\rho_o^{\text{mat}}(t_{\text{today}})/\rho_o^{\text{mat}}(t)][\rho_o^{\text{rad}}(t)/\rho_o^{\text{rad}}(t_{\text{today}})]$. Then $1 + z(t_{\text{rm}}) = \rho_o^{\text{mat}}(t_{\text{today}})/\rho_o^{\text{rad}}(t_{\text{today}}) = 1.28\times 10^{30} \rho_o^{\text{mat}}(t_{\text{today}})\ c^{-2}\,\text{kg}^{-1}\,\text{m}^3$ (assuming three neutrino species taking part in the radiation).

Choosing $t_o = t_{today}$, Eq. (9.62) gives the value of Hubble constant at the time of emission of a signal that is received today with a redshift $1+z = a_{today}/a(t)$.[15] Equations (9.61) and (9.62) can be combined to obtain the constituent densities $\Omega_i(z)$ at the time of emission of such signal:

$$\Omega_i(z) = \frac{\Omega_{today_i}(1+z)^{1+3\varpi_i}}{\sum_j \Omega_{today_j}(1+z)^{1+3\varpi_j} + (1 - \sum_j \Omega_{today_j})} \tag{9.63}$$

The evolution of the scale factor $a(t)$ results from Eq. (9.62), which can be regarded as an "energy conservation" equation in a potential proportional to $-\sum \Omega_o (a_o/a)^{1+3\varpi}$:

$$\frac{\dot{a}(t)^2}{a_o^2} - H_o^2 \sum_i \Omega_{o_i} \left(\frac{a_o}{a(t)}\right)^{1+3\varpi_i} = H_o^2(1 - \sum_i \Omega_{o_i}) \tag{9.64}$$

Unless a non-conventional constituent exists, all the potential terms are negative, increasing (if $\varpi_i > -1/3$), and goes to zero for a going to infinite. This means that the conventional constituents slow down the expansion of the universe; however they could only be able to stop it ($\dot{a} = 0$) if $\sum \Omega_i > 1$ (closed universe). In such a case, the scale factor will reach a maximum value given by the relation $\sum \Omega_{o\ i}(a_o/a_{max})^{1+3\varpi_i} = -1 + \sum \Omega_{o\ i}$; afterwards an epoch of contraction will come, where the universe collapses in a *big-crunch*. If $\sum \Omega_i < 1$ (open universe) the expansion will continue eternally. If $\sum \Omega_i = 1$ (flat universe) the expansion will continue but it will asymptotically stop when a goes to infinite.

In the radiation-dominated era the matter contribution can be neglected, at least far from the transition to the following era. Integrating Eq. (9.64) for an only constituent Ω^{rad} with $\varpi = 1/3$, it results

radiation-dominated universe

$$\frac{a(t)}{a_o} = \left[2\Omega_o^{rad^{1/2}} H_o t + (1 - \Omega_o^{rad}) H_o^{\ 2} t^2\right]^{1/2} \tag{9.65}$$

where $t = 0$ is the Big-Bang time. The elapsed time till the scale factor reaches the value a_o is

$$t_o = \frac{1}{H_o(1 + \Omega_o^{rad1/2})} \tag{9.66}$$

In the closed case ($\Omega^{rad} > 1$) the universe reaches the maximum expansion, $a_{max}/a_o = [\Omega_o^{rad}/(\Omega_o^{\ rad} - 1)]^{1/2}$, at $t_{max} = H_o^{-1}\Omega_o^{rad1/2}/(\Omega_o^{rad} - 1)$.

The radiation-dominated era is followed by a matter-dominated era. Since $p_o^{\ mat}(t) << \rho_o^{mat}(t)$, we shall describe this era as a universe filled with dust matter ($\varpi = 0$). As long as it is possible to neglect other constituents, Eq. (9.64) for only a constituent with $\varpi = 0$ becomes

matter-dominated universe

[15] This expression has to be used in Eq. (9.47) to calculate the luminous distance (9.46). See Eq. (9.87).

$$\frac{\dot{a}(t)^2}{a_o^2} - \frac{\Omega_o^{mat}\ H_o^2 a_o}{a(t)} = H_o^2(1-\Omega_o^{mat}) \tag{9.67}$$

This equation is easily integrated if the universe is flat. We shall call $t = 0$ the time when the scale factor gets null; then

- $\Omega^{mat} = 1$ (flat universe):[16]

$$\frac{a(t)}{a_o} = \left(\frac{3}{2}\ H_o\ t\right)^{2/3} \tag{9.68}$$

In other cases, the integration of Eq. (9.67) leads to a function $t(a)$ that does not allow to solve $a(t)$. It is then better to express the solution in a parametric way; for this, we shall use the parameter $\eta(t)$ defined as

$$d\eta = \frac{H_o\ a_o}{a(t)}\sqrt{|1-\Omega_o^{mat}|}dt$$

Therefore it is $\dot{a} = a'\ d\eta/dt$, and Eq. (9.67) becomes

$$\left(\frac{a'}{a_o}\right)^2 + K\left[\frac{a}{a_o} + \frac{\Omega_o^{mat}}{2(1-\Omega_o^{mat})}\right]^2 = -K\left[\frac{\Omega_o^{mat}}{2(1-\Omega_o^{mat})}\right]^2$$

where $K = \pm 1$ is the sign of $\Omega_o{}^{mat} - 1$. The solutions are

- $\Omega^{mat} < 1$ (open universe):

$$\begin{aligned} \frac{a(\eta)}{a_o} &= \frac{\Omega_o^{mat}}{2(1-\Omega_o^{mat})}(\cosh\ \eta - 1) \\ \Omega^{mat}(\eta) &= 2(\cosh\ \eta + 1)^{-1} \\ t &= \frac{\Omega_o^{mat}}{2H_o(1-\Omega_o^{mat})^{3/2}}(\sinh\ \eta - \eta) \end{aligned} \tag{9.69}$$

- $\Omega^{mat} > 1$ (closed universe):

$$\begin{aligned} \frac{a(\eta)}{a_o} &= \frac{\Omega_o^{mat}}{2(\Omega_o^{mat} - 1)}(1-\cos\eta) \\ \Omega^{mat}(\eta) &= 2\ (1+\cos\ \eta)^{-1} \\ t &= \frac{\Omega_o^{mat}}{2H_o(\Omega_o^{mat} - 1)^{3/2}}(\eta - \sin\eta) \end{aligned} \tag{9.70}$$

---

[16] $a(t)/a_o = [3(1+\varpi)H_o\ t/2]^{2/[3(1+\varpi)]}$ is the solution of Eq. (9.64) for a flat universe with only one constituent. The time elapsed from $a = 0$ till $a = a_o$ is $t_o = 2/[3(1+\varpi)H_o]$.

In the closed case, the scale factor reaches the maximum value $a_{max}/a_o = \Omega_o^{mat}/(\Omega_o^{mat} - 1)$ when $\eta = \pi$.
Combining Eqs. (9.57) (with $p_o = 0$), (9.58), and (9.59), it is obtained

$$\Omega^{mat} = -\frac{2\,a\,\ddot{a}}{\dot{a}^2} = 2\,q \tag{9.71}$$

Until the end of the twentieth century it was believed that the present-day universe was dominated by a sole constituent in the form of dust matter. For such universe, Eq. (9.71) shows that the status of open ($\Omega^{mat} < 1$) or closed ($\Omega^{mat} > 1$) can be elucidated through the (positive) deceleration parameter q. In 1999, Gurvits, Kellermann, and Frey based on the relation between angular sizes and redshifts of compact radio-sources (quasars or other active galactic nuclei) in the range $0.01 < z < 4.7$, and certain suppositions about their constitution and evolution, to conclude that a present-day dust filled universe is compatible with $q(t_{today})$ ranging between 0 and 0.5. In the same year, however, redshift measurements for type Ia supernovae in the range $0.18 < z < 0.83$ have convincingly showed that the present-day expansion is accelerated ($q < 0$). This data would reveal the existence of a "cosmological constant" or some type of "dark energy" (a non-conventional constituent with $\varpi < -1/3$) accompanying the ordinary matter, whose contribution in the potential of Eq. (9.64) would provide the acceleration. The new constituent will enter Eq. (9.71) as well, hence changing the link between deceleration parameter and the open or closed status of the universe.

9.7. NON-MACHIAN SOLUTIONS. COSMOLOGICAL CONSTANT

Mach's Principle—the idea that locally inertial frames are determined by the mass distribution in the universe—played a major role in the construction of General Relativity (§8.1). However, the theory just partially realizes this idea. It is true that the energy–momentum distribution influences the inertial behavior of bodies, since it acts as the source in Einstein equations to determine the space-time geometry. Nevertheless, Einstein equations also allow vacuum solutions (for null energy–momentum tensor). Some of these vacuum solutions can be anyway regarded as the geometry resulting from a singular matter distribution.

> *In a consistent theory of relativity there can be no inertia* relatively to "space," *but only an inertia of masses* relatively to one another. *If, therefore, I have a mass at a sufficient distance from all other masses in the universe, its inertia must fall to zero.*
>
> A. Einstein, 1917b

Schwarzschild solution does not realize this Einstein's statement, because inertia does not disappear at the infinite (i.e., far from the object that produce the curvature of space-time) but the inertial trajectories are determined by an asymptotically Minkowskian metric.

Other vacuum solutions, as Minkowski space-time, are free of singularities. In Minkowski space-time the inertial trajectories—the straight timelike world lines—are not determined by matter distribution at all. Minkowskian solution is only determined by the boundary conditions required to solve Einstein equations. In Minkowski space-time there are inertial forces in non-inertial frames, since the Christoffel symbols only vanish in Cartesian coordinates. These inertial forces are bereft of matter to which ascribe their origin, as Mach's principle demands. Even worst, Minkowski space-time is not the only vacuum solution free of singularities (for instance, see the exact plane wave studied by Bondi, Pirani, and Robinson (1959)), which means that the theory accepts different inertial structures without apparent physical origin.[17]

Einstein (1917b) attempted to modify the theory to exclude the vacuum solutions because of their non-Machian character. With this aim, he introduced the conserved term $\Lambda\, g_{ij}$ into his equations, which does not alter the automatic conservation. Thus, Eq. (8.46) becomes

$$R_{ij} - \frac{1}{2}\, g_{ij}\, R - \Lambda\, g_{ij} = \frac{8\,\pi\, G}{c^4} T_{ij} \tag{9.72}$$

Λ is called *cosmological constant.* The new term amounts an isotropic and homogeneous energy–momentum tensor with $\rho_{o\Lambda} = -p_{o\Lambda} = c^4 (8\pi G)^{-1}\Lambda$. It should be remarked that this cosmological term does not disappear in the Newtonian limit. Therefore, it should be satisfied that $\rho_o >> |\rho_{o\Lambda}|$ for any system that is properly described by Newtonian gravity. This relation makes possible to find an upper bound for the cosmological constant by considering the least dense systems accepting a Newtonian description: certain small galaxy clusters with $\rho_o\, c^{-2} \sim 10^{-26}\,\mathrm{kg\,m^{-3}}$. From here it results that $|\Lambda| < 10^{-52}\,\mathrm{m^{-2}}$. The cosmological constant prevents that Minkowski space-time be a vacuum solution of Eqs. (9.72).

Einstein universe

Einstein was pleased seeing that the modified equations with a positive cosmological constant accepted a solution independent of t for dust matter ($p_o = 0$) homogeneously distributed:

$$\mathrm{d}s^2 = c^2 \mathrm{d}t^2 - \frac{\mathrm{d}r^2}{1-\Lambda\, r^2} - r^2(\mathrm{d}\theta^2 + \sin^2\theta\, \mathrm{d}\varphi^2) \tag{9.73}$$

---

[17] Some authors suggest that the boundary conditions should be regarded as gravitational degrees of freedom, in an extended conception of Mach's Principle where not only the matter energy–momentum but also the energy–momentum belonging to the gravitational field determine the locally inertial frames. In such case the absence of matter is not equal to vacuum. However, it should be remarked that the concept of gravitational energy is not well defined in the theory.

This solution describes an isotropic and homogeneous closed universe with constant scale factor. The spatial curvature and matter density for *Einstein universe* are

$$ {}^{(3)}R = \frac{6\,K}{a^2_{\text{Einstein}}} = 6\,\Lambda \qquad \frac{4\,\pi\,G}{c^4}\,\rho_{o_{\text{Einstein}}} = \Lambda \tag{9.74} $$

Einstein universe harmonized with the idea of a static universe prevailing at that time. On the other hand, a closed universe has not boundaries, thus there are not boundary conditions determining the solution. This characteristic suggested that the theory had succeeded in accomplishing Mach's Principle. However W. de Sitter (1917) soon got a vacuum solution for the new equations:

de Sitter solution

$$ ds^2 = (1 - \frac{\Lambda}{3}\,\xi^2)\,c^2 dT^2 - \frac{d\xi^2}{1 - \frac{\Lambda}{3}\,\xi^2} - \xi^2 (d\theta^2 + \sin^2\theta\; d\varphi^2) \tag{9.75} $$

Einstein thought that this vacuum solution contained some type of singularity. After he was convinced that de Sitter solution was free of singularities, his sympathy for Mach's Principle begun to decline.

In 1925, G.E. Lemaître found a coordinate change leading the interval (9.8) (with $\Lambda > 0$) to the form (9.33) of the isotropic and homogeneous cosmological models. The spatial curvature and the scale factor of *de Sitter universe* in the form found out by Lemaître are[18]

de Sitter universe

$$ K = 0, \qquad a(t')_{de\,Sitter} \propto \exp\left[(\Lambda\, c^2/3)^{1/2}\, t'\right] \tag{9.76} $$

When Einstein was told that his static universe was unstable (see below), he recommended forgetting the cosmological constant. At the end of his life Einstein had completely abandoned Mach's Principle. He considered illusory the pretension that matter determinates the metric tensor through the energy–momentum tensor, since the energy–momentum tensor definition itself presupposes the existence of a metric.

General Relativity with or without cosmological constant also accepts solutions of cosmological type—with matter homogeneously distributed—where matter rotates with respect to the directions of freely gravitating gyroscopes: a

---

[18] de Sitter solution describes a geometry with constant scalar curvature: $R = -4\,\Lambda$. In this geometry it is possible to "cut" an isotropic and homogeneous space in different ways. So, if $\Lambda > 0$ another coordinate change makes possible to regard de Sitter geometry as a universe with $K = 1$ and $a(t) = (\Lambda/3)^{-1/2} \cosh\,[(\Lambda c^2/3)^{1/2} t]$. It can be proved that, while these last coordinates cover all the space-time, the former ones –associated with t'—cover just a half. If $\Lambda < 0$ de Sitter solution can be also regarded as a universe with $K = -1$ and $a(t) = (-\Lambda/3)^{-1/2} \cos[(-\Lambda c^2/3)^{1/2}\, t]$ (*anti-de Sitter* universe). If $\Lambda = 0$ de Sitter solution becomes Minkowski space-time, where it is possible to cut an isotropic and homogeneous space with $K = 0$ and $a(t) = constant$ by using the usual coordinates; however there also exist coordinates covering just the future light cone of the origin, which cut an isotropic and homogeneous space with $K = -1$ and $a(t') = c\,t'$ (*Milne universe*).

gyroscope accompanying a galaxy with its spin pointing at another galaxy will lose this condition with time (Gödel, 1949; Ozsváth and Schücking, 1962, 1969; Matzner, Shepley, and Warren, 1970).[19] It could be said that matter in these universes is animated of an absolute rotation. Therefore General Relativity does not solve the Newton's vessel problem (or that of Einstein's spherical bodies, §8.1): it admits solutions where there are centrifugal forces in the absence of matter, or where torque-free gyroscopes undergo a rotation relative to the matter of the universe.

The adding of the cosmological constant to the isotropic and homogeneous models in the previous section changes the Eqs. (9.52) and (9.54):

$$\frac{\dot{a}^2}{a^2} + \frac{K\,c^2}{a^2} - \frac{\Lambda\,c^2}{3} = \frac{8\,\pi\,G}{3\,c^2}\,\rho_o \tag{9.77}$$

$$2\frac{\ddot{a}}{a} + \frac{\dot{a}^2}{a^2} + \frac{K\,c^2}{a^2} - \Lambda\,c^2 = -\frac{8\pi\,G}{c^2}\,p_o \tag{9.78}$$

and subtracting them,

$$\frac{\ddot{a}}{a} = \frac{\Lambda\,c^2}{3} - \frac{4\pi G}{c^2}\,(p_o + \frac{\rho_o}{3}) \tag{9.79}$$

Thus a positive cosmological constant plays the role of a repulsive force that accelerates the expansion. Since $\rho_o^{\text{mat}} \propto a^{-3}$ and $\rho_o^{\text{rad}} \propto a^{-4}$, Eqs. (9.77–9.78) show that the cosmological constant has no influence at the early epoch ($a \to 0$) of a universe filled with ordinary matter and radiation. On the other hand, if nowadays Λ dominated in Eq. (9.79) then it would be $\ddot{a}(t_{\text{today}}) > 0$, so the form of the expansion in Figure 9.14 and the relation between Hubble constant and the age of the universe would be modified.

As already remarked, the cosmological constant in Einstein equations amounts a perfect fluid with $\varpi_\Lambda = -1$ and density Ω^Λ equal to

$$\Omega^\Lambda(t) = \frac{8\pi G\,\rho_{o\Lambda}(t)\,c^{-2}}{3\,H(t)^2} = \frac{\Lambda\,c^2}{3\,H(t)^2} \tag{9.80}$$

Neglecting the contribution of radiation, Eq. (9.60) turns out to be

$$K\,c^2 = a^2\,H^2\,(-1\,+\,\Omega^{\text{mat}} + \Omega^\Lambda) \tag{9.81}$$

which says that the universe is closed or open according to $\Omega^{\text{mat}} + \Omega^\Lambda$ is larger or smaller than 1. If the matter of universe is dust ($p_o = 0$), it results from Eqs. (9.78) and (9.81) that the deceleration parameter q is[20]

$$q = \frac{\Omega^{\text{mat}}}{2} - \Omega^\Lambda \tag{9.82}$$

---

[19] The direction of torqueless gyroscope is governed by Fermi–Walker transport (see Eq. (7.102), where $\mathrm{d}/\mathrm{d}\tau$ must be substituted with the covariant derivative $\mathrm{D}/\mathrm{D}\tau$). Fermi–Walker transport along a geodesic coincides with parallel transport because the four-acceleration is zero.

[20] It is easily proved that, in general, it is $2q = \Sigma\,\Omega_i + 3\Sigma\,\varpi_i\,\Omega_i$.

Equations (9.81) and (9.82) show that the question about the open or closed status of the universe is not more directly linked with the deceleration parameter value, as it would happen if $\Lambda = 0$.

The evolution of the universe is studied with Eq. (9.64); for a universe dominated by dust matter ($\varpi_{\text{mat}} = 0$) and cosmological constant ($\varpi_\Lambda = -1$), the equation becomes

universe dominated by matter and cosmological constant

$$\frac{\dot{a}^2}{a_o^2} - H_o^2\left[\Omega_o^\Lambda \frac{a^2}{a_o^2} + \Omega_o^{\text{mat}} \frac{a_o}{a}\right] = H_o^2(1 - \Omega_o^\Lambda - \Omega_o^{\text{mat}}) \tag{9.83}$$

If $\Lambda < 0$, the "potential" in Eq. (9.83) is increasing and goes to infinite. In such case there exists a turning point whatever the right side value is: the expansion stops, the universe then contracts going toward a big-crunch ($\Lambda < 0$ amounts an attractive force). Instead, if $\Lambda > 0$ the effective potential displays a maximum giving rise to an unstable equilibrium solution that corresponds to Einstein universe. The potential maximum is at $(a/a_o)^3 = \Omega_o^{\text{mat}}/(2\Omega_o^\Lambda)$, where the potential value is $-3\,H_o^2\,(\Omega_o^{\text{mat}\;2}\,\Omega_o^\Lambda/4)^{1/3}$. As can be noticed in Figure 9.15, if the right side in Eq. (9.83) exceeds the potential maximum then the universe will expand forever. In other case, either the universe does not come from a Big-Bang and expands forever, or the universe comes from a Big-Bang and eventually recollapses. This alternative happens if

$$\left(1 - \Omega^{\text{mat}} - \Omega^\Lambda\right)^3 < -\frac{27}{4}\Omega^{\text{mat}^2}\Omega^\Lambda < 0 \tag{9.84}$$

Einstein universe is critical for this expression, since it turns the inequality into an equality. In order to verify this, we should replace the left side with Eq. (9.81), and use the results (9.74) together with Eq. (9.80).

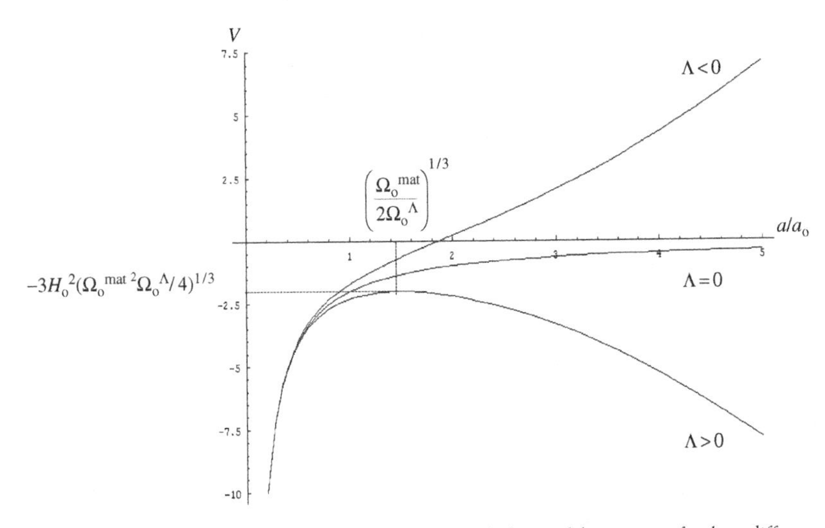

Figure 9.15. "Potentials" governing the evolution of the scale factor of the universe, for three different cosmological constants.

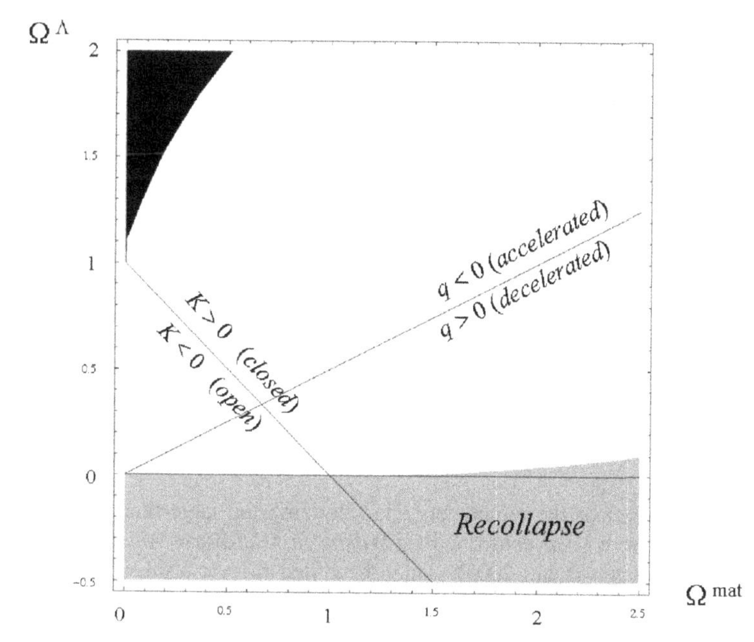

Figure 9.16. Universes filled with dust matter in the space of parameters Ω^{mat} and Ω^{Λ}. The gray region corresponds to recollapsing universes. The dark region corresponds to universes whose returning point is on the right side of the "potential" (they do not come from a Big-Bang).

Figure 9.16 displays Eq. (9.81), (9.82), and (9.84). Notice that if $\Lambda = 0$ then Eqs. (9.81) and (9.84) say that universe is closed and recollapses whenever $\Omega^{\text{mat}} > 1$, as was the case in the previous section. Instead, in the presence of a cosmological constant, the closed universe ($\Omega^{\text{mat}} + \Omega^{\Lambda} > 1$) not necessarily recollapses (except for $\Lambda < 0$).

Redshift data for type Ia supernovae—taken as standard candels—in the range $0.18 < z < 0.83$ support the relation $\Omega^{\Lambda}(t_{\text{today}}) \sim 1.3\Omega^{\text{mat}}(t_{\text{today}}) + 0.3$ (and $\Omega^{\Lambda}(t_{\text{today}}) < 1.5$) leading to a negative value for the deceleration parameter in Eq. (9.82) (Riess et al., 1998; Perlmutter et al., 1999). This means that the expansion of the universe accelerates. On the other hand the analysis of fluctuations of the cosmic background radiation favors a spatially flat universe; according to Eq. (9.81), this amounts to the relation $\Omega^{\Lambda}(t_{\text{today}}) + \Omega^{\text{mat}}(t_{\text{today}}) \sim 1$ (Balbi et al., 2000; de Bernardis et al., 2000). Combining both results it is obtained $\Omega^{\text{mat}}(t_{\text{today}}) \sim 0.3$ and $\Omega^{\Lambda}(t_{\text{today}}) \sim 0.7$; thus $q(t_{\text{today}}) \sim -0.6$.[21] The values of $\Omega^{\text{rad}}(t_{\text{today}})$, $\Omega^{\text{mat}}(t_{\text{today}})$ and $\Omega^{\Lambda}(t_{\text{today}})$ can be substituted in Eq. (9.63) to get the behaviors of these densities throughout the expansion of the universe (see Figure 9.17).

---

[21] The accelerated expansion could be due either to the cosmological constant or to the domination of energy sources with $\rho_o + p_o/3 < 0$ in Eq. (9.79) (for instance, with $\varpi < -1/3$ in the state equation (9.55)). Such possible contributions to the energy–momentum tensor are embraced with the name of *dark energy*.

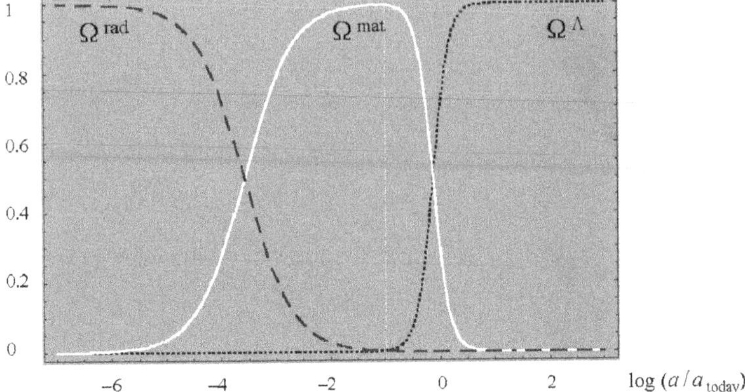

Figure 9.17. Behavior of the constituents as the universe expands.

The values of the parameters characterizing the large-scale behavior of the universe have been confirmed by *Wilkinson Microwave Anisotropy Probe* (WMAP) (Bennett et al., 2003), which benefited from a resolution of 0.3° to make the best map of the fluctuations of the cosmic background radiation. The study of these fluctuations, which contain crucial information about the origin, structure, and fate of the universe, will be enriched with the data to be provided by ESA-Planck probe.

9.8. PROBLEMS OF THE STANDARD BIG-BANG MODEL

particle horizon

Although the standard Big-Bang model makes a suitable description of the observed properties of the universe, it leaves open uncomfortable questions from a theoretical point of view. Since the age of the universe is finite, then the universe has *particle horizons*. Because light travels with a finite velocity, each observer can just see those regions of the universe that are close enough to allow a light ray make the travel in the time of existence of the universe. This also means that the observer can only be causally related with such part of the universe. The particle horizon is characteristic of each observer, and grows as the universe gets older. To study this issue we shall use the *conformal time* $\eta(t)$ defined as

$$d\eta = \frac{c\,dt}{a(t)} = \frac{c\,da}{a^2 H(a)} \tag{9.85}$$

According to Eq. (9.34), the radial trajectories of light rays verify that $d\eta = \pm d\chi$, i. e., the light ray world lines are 45° straight lines in the plane η–χ. Then, the particle horizon of an observer placed at the (arbitrary) coordinate origin has radial coordinate equal to $\chi_{\text{horizon}} = \eta_{\text{today}}$ (see Figure 9.18); so the horizon radius measures $\sigma_{\text{horizon}} = a_{\text{today}}\,\eta_{\text{today}}$. The conformal time η_{today} elapsed from

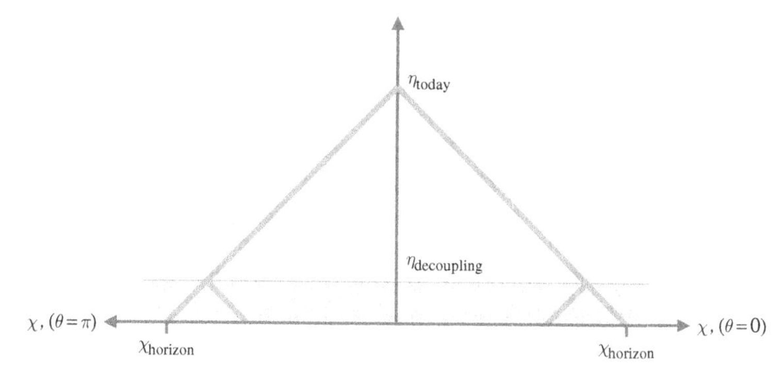

Figure 9.18. Owing to the finite age of the universe, each observer can only see the part of the universe contained within its particle horizon. The radiation emitted at the decoupling time comes from regions of the universe that, in general, were not causally connected since their particle horizons did not intersect.

the Big-Bang up to now will result from integrating Eq. (9.85). If $\int c\,a^{-2}H^{-1}\,da$ were convergent for $a \to 0$ then η_{today} would be finite. In such a case the radius of the particle horizon would be finite, and there would exist regions of the universe from which we would still be causally disconnected.

Using for $H(a)$ the expression (9.62), and calling $u = a/a_{today}$, it results

$$\sigma_{horizon} = a_{today}\,\eta_{today} = \int_0^1 \frac{c\,H_{today}^{-1}\,du}{u\left[\sum_i \Omega_{today_i}\,u^{-(1+3\varpi_i)} + 1 - \sum_i \Omega_{today_i}\right]^{1/2}} \tag{9.86}$$

In order that this integral be finite, it is sufficient that some of the constituents of the universe verifies that $\varpi_i > -1/3$. In particular, the horizon radius for a flat universe having a sole constituent $\Omega = 1$ is

$$\sigma_{horizon}(t_{today}) = \frac{2}{1+3\varpi}\frac{c}{H_{today}} = \frac{3(1+\varpi)}{1+3\varpi}c\,t_{today},$$

where $c\,H_{today}^{-1} \sim 4300$ Mpc (the age of the universe t_{today} was obtained in Note 16).

The particle horizon is also the distance $\sigma_{src}(t_{today}) = a_{today}\chi_{src}$ to an infinitely redshifted source, since z diverges when $a(t_{emission})$ vanishes (see Eq. (9.37)). In fact, the radial position $\chi_{src}(z)$ displaying a given redshift z can be computed by substituting in Eq. (9.47) the expression for $H(z)$ obtained in Eq. (9.62); then, calling $u = a/a_{today} = (1+z')^{-1}$ in Eq. (9.47), it results

$$\chi_{src}(z) = \int_{\frac{1}{1+z}}^1 \frac{c\,H_{today}^{-1}\,du}{a_{today}u\left[\sum_i \Omega_{today_i}\,u^{-(1+3\varpi_i)} + 1 - \sum_i \Omega_{today_i}\right]^{1/2}} \tag{9.87}$$

When $z \longrightarrow \infty$, the distance $a_{today}\chi_{src}$ goes to the radius of the particle horizon (9.86).

The existence of finite particle horizons leads to one of the problems of the standard Big-Bang model. Figure 9.18 shows that the cosmic background radiation coming from opposite directions was emitted at the decoupling time (some 4×10^5 years after the Big-Bang) from places that would have no possibility of exchanging information before the emission, since each of them was outside the particle horizon of the other. As we are going to see, it would be sufficient that two sky directions differed in more than 1° for their lack of causal relation at the decoupling time. But this raises the question of how could the cosmic background radiation temperatures get so similar for different directions, if they were not causally related?

The cosmic background radiation received today is redshifted by a factor $z \sim 1000$. This radiation was emitted from sources placed at the radial position $\chi_{src}(z_{decoupling} \sim 1000)$, which can be calculated with Eq. (9.87). At the decoupling time, the distance between two of these sources separated by an angle $\Delta\theta$ was $\sigma_{\Delta\theta}(t_{decoupling}) = a_{decoupling} f(\chi_{src}(z_{decoupling}))\Delta\theta$, where $f(\chi)$ is χ, $\sin\chi$ or $\sinh\chi$, according to whether the universe is flat, closed or open. In order that two sources emitting cosmic background radiation can be causally connected, the distance $\sigma_{\Delta\theta}(t_{decoupling})$ should be smaller than the radius of the particle horizon at the decoupling time (see Figure 9.19). Then,

$$\Delta\theta < \frac{\sigma_{horizon}(t_{decoupling})}{a_{decoupling}\, f(\chi_{src}(z_{decoupling}))}$$

Using the current values of the constituents, $\Omega^{rad}(t_{today}) \sim 10^{-4}$, $\Omega^{mat}(t_{today}) \sim 0.3$, $\Omega^{\Lambda}(t_{today}) \sim 0.7$, it results that $\chi_{src}(1000) \sim 3\,c\ (a_{today}\,H_{today})^{-1}$ (the value given by the integral (9.87) shows that the sources are at a distance 2% shorter than the present particle horizon). On the other hand, the value

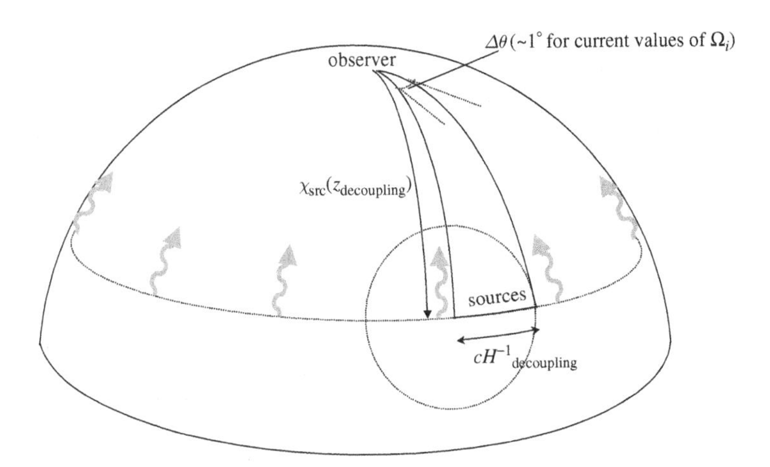

Figure 9.19. Maximum angular separation of two sources that were causally connected at the decoupling time.

of $\sigma_{\text{horizon}}(t_{\text{decoupling}})$ can be computed as in Eq. (9.86), using the values $\Omega^{\text{rad}}(t_{\text{decoupling}}) \sim 0.25$, $\Omega^{\text{mat}}(t_{\text{decoupling}}) \sim 0.75$, $\Omega^{\Lambda}(t_{\text{decoupling}}) \sim 10^{-9}$, which are obtained from Eq. (9.63); the result is $\sigma_{\text{horizon}}(t_{\text{decoupling}}) \sim 1.3\, c\, H^{-1}_{\text{decoupling}}$. In this flat universe it is $f(\chi) = \chi$, thus

$$\Delta\theta < \frac{1.3\, a_{\text{today}} H_{\text{today}}}{3\, a_{\text{decoupling}} H_{\text{decoupling}}} = \frac{0.4}{\left[\sum_i \Omega_{\text{today}\ i}\, 1000^{1+3\varpi_i}\right]^{1/2}}$$

where the right side is a consequence of Eq. (9.62). Thus we obtain that $\Delta\theta < 1°$ in order that two sky directions can be causally connected. Therefore, the only explanation for the isotropy of the cosmic background radiation, within the context of the standard Big-Bang model, is that this isotropy is a feature of the initial condition of the universe. However it is very unappealing to ascribe to the initial conditions those features that the model is unable to explain. For this reason, people have tried to modify the standard Big-Bang model in order that all the sky directions result to be causally connected. In the *inflationary cosmologies* the standard model is preceded by a stage governed by a nearly homogeneous scalar field. If the field is in a state where its kinetic energy is negligible, then its energy–momentum tensor will behave like that of a perfect fluid with $\varpi = -1 (p_o = -\rho_o = constant)$. In such a case Eq. (9.57) indicates that the universe expands in an exponentially accelerated way. According to Eq. (9.87), if $\varpi = -1$ the particle horizon can become so large as required whenever the inflationary phase continues for sufficient time. In this way the different parts of the universe nowadays visible can turn to be causally connected. The inflationary era ended because the state of the scalar field that gives rise to inflation is unstable. After some time the scalar field underwent a non-adiabatic transition to a state of lower energy, thus creating a large quantity of entropy and transforming its energy into particles and radiation. After this process took place, the universe began to be dominated by the radiation.

inflationary models

Inflationary models also have an answer for another standard Big-Bang model dilemma. In a universe dominated by radiation or matter (in general, in universes whose constituents have $\varpi > -1/3$), $\Omega = 1$ is an unstable fixed point for the equations governing the dynamics of the universe. This means that if at a given time it is $\Omega = 1$ then this condition will remain at any time (flat universe is forever flat). But if $\Omega \neq 1$ then the expansion of the universe will make that Ω goes away from 1 even more. This characteristic of the expansion can be tested by means of Eq. (9.63) applied to a universe whose constituents have $\varpi_i > -1/3$. It can be easily noticed that $\Omega = \sum \Omega_i$ goes to 1 if z goes to infinity (i.e., when the scale factor goes to zero), but it goes monotonically away from 1 as z decreases (i.e., when a increases):

$$\sum_i \Omega_i(z) = \frac{\sum_i \Omega_{\text{today}_i}\, (1+z)^{1+3\varpi_i}}{\sum_j \Omega_{\text{today}\ j}\, (1+z)^{1+3\varpi_j} + (1 - \sum_j \Omega_{\text{today}_j})} \xrightarrow[z\to\infty]{} 1$$

It is then surprising that the value of Ω for the present-day universe is so close to 1, in spite of the fact that the universe has expanded for a long time. This means that the value of Ω in the early universe had to be notably close to 1. For instance, let us compute Ω^{rad} at the time of nucleosynthesis ($1+z = a_{today}/a_{nucleosynthesis} = T^{phot}{}_{nucleosynthesis}/T^{phot}{}_{today} \sim 10^9$); using Eq. (9.63) we obtain that $1-\Omega^{rad}\ (t_{nucleosynthesis}) \sim 10^{-6}$ (see Figure 9.18). Going back to the time when protons, neutrons, and the rest of hadrons did not exist but were decomposed in their constituents—the quarks—we shall reach temperatures higher than 10^{13} K; in such a case it results that $1-\Omega^{rad}$ is smaller than 10^{-10}. It is at least intriguing that the universe requires a so extreme fine tuning of its initial conditions in order to explain its present-day state. The inflationary models can solve this puzzle, because inflationary era is governed by a constituent with $\varpi = -1$. Being $\varpi < -1/3$, then Eq. (9.63) indicates that $\Omega = 1$ is a stable fixed point: the universe goes toward $\Omega = 1$ when it expands. If the inflationary era stands enough time, the value of Ω will become very close to 1. After the transition ending the inflationary era, this value of Ω will be inherited by the resulting radiation-dominated universe.

To the above-mentioned achievements, we must add that inflationary models allow exploiting the quantum fluctuations of the scalar field to build a causal theory of the spectrum of density fluctuations that gave rise to the faint anisotropies of the cosmic background radiation, and were the seeds from which the matter gravitationally collapsed to form the structures of galaxies and galaxy clusters of the present-day universe.

9.9. EXPERIMENTAL TESTS

The lack of a full accomplishment of Mach's Principle in General Relativity is not an obstacle to the complete success of Einstein's theory (Newtonian mechanics is also successful in its range of application, although it uses the notion of absolute space). General Relativity theory has been confirmed each time it was subjected to a test. Other metric theories of gravity, like Brans and Dicke's (1961) that attempts a fuller realization of Mach's Principle, have not shown a better aptitude to describe the gravitational phenomena.

The identity between inertial and gravitational masses, on which Principle of equivalence is based, has been verified through torsion balances up to an accuracy of one part in 10^{12} (Braginsky and Panov, 1971). A similar precision has been reached through the comparison of Earth and Moon accelerations in the solar gravitational field by means of Lunar laser ranging. Instead the atomic-scale tests are very far from such degree of precision (Fray et al., 2004). The accuracy will be increased up to 10^{-17} by STEP project (NASA-ESA), which will study freely falling bodies in an Earth-orbiting spacecraft. The application of the Principle of equivalence to electromagnetic wave propagation leads to the gravitational frequency shift (i.e. the dependence of a clock rate on the gravitational potential), which has been verified by Pound and Rebka in 1960. In 1979 the frequency of a hydrogen maser fixed to the Earth was compared with the frequency of a similar maser placed in a rocket launched vertically upward to 10,000 km; the gravitational frequency shift was verified with an accuracy of 10^{-4} (Vessot and Levine, 1979; Vessot and Levine et al., 1980). In 1993 the *Global Positioning System* (GPS) began working, which uses a satellite network to determine the coordinates of a point on the Earth surface with a margin of error of a few meters. To reach such performance, the system must take into account the effect of the terrestrial gravitational potential on the rate of the atomic clocks transported by the satellites.

Schwarzschild geometry within the Solar System was tested in several ways. The deflection of electromagnetic waves in the solar gravitational field was verified with an accuracy of 10^{-3} (Robertson, Carter, and Dillinger, 1991; Lebach et al., 1995). The gravitational field not only deflects the electromagnetic waves but also causes a delay in its travel time (Shapiro, 1964). The delay for radar signals reflected at Venus and Mercury (around 200 μs for pulses going near the solar limb) was measured by Shapiro in 1968. Viking mission to Mars confirmed the relativistic delay with an accuracy of 10^{-3} (Reasenberg et al., 1979). Cassini probe, in a journey toward Saturn, increased the precision up to 10^{-5} (Bertotti, Iess, and Tortora, 2003). The experimental verification of the relativistic contribution to Mercury's perihelion shift (Shapiro, 1972) did not reach a similar margin of confidence, because it was hindered by the lack of a reliable value for the solar quadrupole—the solar mass does not display a perfectly spherical distribution—which makes a Newtonian contribution to the effect.

The geometry associated with the gravitational field can also be examined by means of gyroscopes. In 1916, de Sitter calculated the change of direction for a gyroscope orbiting in a Schwarzschild geometry. The *geodesic precession* or *de Sitter effect* has been measured to 0.7% error (Williams, 1996) for the gyroscope composed by the Earth–Moon system in solar orbit (19 milliarcseconds per year), using data coming from the laser ranging of lunar motion. The *Gravity Probe-B*, launched in 2004, carries a set of superconducting gyroscopes, which will make possible to establish the de Sitter effect with an accuracy of 10^{-4}. This experiment will detect not only the geodesic precession due to the static terrestrial gravitational field (6.6 arcseconds per year for an orbit at a height of 650 km), but also a much more subtle contribution (42 milliarcseconds per year) produced by the Earth's rotation action on the geometrical structure of the surrounding space-time. Differing from Newtonian gravity, General Relativity takes as sources of gravity not only matter distribution but also its state of motion (i.e., all the components of the energy–momentum tensor). In this sense, it resembles electromagnetism, where not only the charge distribution but also the electric current acts as a source of the electromagnetic field; this similarity is evident in the linear approximation (8.55) for Einstein equations. Thus, in General Relativity the Earth's rotation causes an effect on its gravitational field, which resembles the magnetic field due to a rotating charged sphere.[22] Before General Relativity, Friedlaender brothers (1896) and Föppl (1904) searched in the laboratory some evidence about the existence of *gravitomagnetism.* In General Relativity, H. Thirring and J. Lense derived the gravitomagnetism in 1918. The Gravity Probe B will make possible to discern the contribution of terrestrial gravitomagnetism to the precession of the orbiting gyroscopes with a margin of error of 1%. Meanwhile, there are already two types of evidence about gravitomagnetism. In 1998, Ciufolini announced the observation of gravitomagnetic effects on the orbits of satellites LAGEOS I and II (see also Ciufolini and Pavlis, 2004). According to General Relativity, the gravitomagnetism due to the terrestrial rotation should produce a drag of the satellite orbital plane in the sense of the Earth's rotation. The effect is extremely weak: for a polar orbit at a height of 5800 km, the nodal points displace 2 m per year; however it can be revealed by LAGEOS satellites since their orbits are tracked with a great accuracy by means of laser ranging (these satellites belong to a system aimed at detecting changes in the Earth's crust through the monitoring of its orbits, whose points are taken as reference points in the space). Also in 1998, Cui, Zhang, and Chen announced that the quasiperiodic modulation observed in the X-ray emissions from binary systems containing black hole candidates could come from the precession of the black hole matter accretion disk produced by gravitomagnetic effects associated with the black hole rotation.

The set of experimental data confirming General Relativity was greatly enriched with the observation of the binary pulsar PSR 1913+16 located in our galaxy at a distance of 5 kpc, discovered by R.A. Hulse and J.H. Taylor (1975), who won the Nobel Prize in Physics 1993. The pulsar is a neutron star that rotates at high velocity (17 turns per second, in this case) emitting a concentrated beam of electromagnetic radiation (like a lighthouse); each time the pulsar turns, a pulse of the emitted radiation reaches the Earth. The detailed study of the systematic variations of the arriving times of pulses made possible to infer that the pulsar was not alone but it took part in a binary system—where both bodies orbit around the common center of mass. Moreover, the set of parameters characterizing their orbits have been obtained. The nature of the orbits converted the binary pulsar in an ideal system to measure meaningful relativistic effects: they are two similar bodies of 1.4 solar masses, which orbit with a period of 7.75 hours, being their maximum approach and separation of 1.1 and 4.8 solar radius respectively. Such a system should have a periastron shift 35,000 times larger than Mercury's, beside exhibiting other effects well studied within the Solar System like the Shapiro delay and the effect of the gravitational potential on the clock rates (in this case measured in terms of the succession of pulses). The binary

---

[22] The vacuum solution with axial symmetry is Kerr geometry (1963) which depends on two integration constants M and J, the source mass and intrinsic angular momentum (spin). This solution genuinely differs from Schwarzschild's, because it is not possible to pass from one of them to the other through a coordinate change, what constitutes another evidence of the absolute character of the rotation in General Relativity.

pulsar showed an excellent agreement between these features and the theoretical predictions of General Relativity. But the more important contribution of the binary pulsar to General Relativity was that it exhibited, although indirectly, the first evidence of the existence of gravitational waves. According to General Relativity, the described binary system is a variable quadrupole that should lose energy by emission of gravitational waves. The theory says that the loss of energy has to produce a decrease in the 7.75 hours orbital period at a rate of 7.6×10^{-5} s per year. The decrease in the period shifts the periastron times, so leading to an accumulative effect that was observed during 18 years; an excellent agreement with the theoretical prediction was obtained, within a margin of error of 0.35% (Taylor, 1992). Other binary pulsars discovered in the 1990s will increase the flux of experimental data to test General Relativity.

Noteworthy gravitational waves can be produced in events where major mass distribution changes happen, like supernovae explosions, gravitational collapses of massive objects,[23] collisions between black holes or galaxies, and mass transfer between close objects. These gravitational waves could be detected in terrestrial laboratories. "Bar" detectors use as antenna a metallic bar that should resonate with the gravitational wave. Typically the bars weigh some tons, and have a length of around 3 m and a diameter of 0.5 m; they have resonance frequencies of about 1 kHz. There are several bar detectors working in different places on the planet—ALLEGRO, AURIGA, EXPLORER, NAUTILUS, and NIOBE. Until now they did not succeed in registering events that could be ascribed to the passing of a gravitational wave; so the experimenters are trying to improve the sensitivity and diminish the background noise. Other gravitational wave detectors use as antenna a device based on Michelson interferometer; the idea is to detect the phase shifts that would be produced under the variations of the interferometer dimensions caused by the passing of a gravitational wave. One of these interferometric antennas called LIGO (*Laser Interferometer Gravitational wave Observatory*) began working at the end of 1999. Others are close to start operations (VIRGO, GEO 600, TAMA 300). Besides, there is a project to build an interferometer in the space, whose arms will have several millions of kilometers, by combining satellites in solar orbit (LISA).

The deflection of light rays by gravity causes a *gravitational lens* effect that distorts the images of astronomical objects or produces multiple images of them. The effect appears when the light coming from far objects is deviated by some massive astronomical object, which stands in its journey toward the observation place. The consequence of this deviation is similar to what happens

---

[23] Nevertheless, if the mass distribution keeps a spherical symmetry there will not be emitted gravitational waves. As Birkhoff (1923) proved, Schwarzschild solution is the only spherically symmetric vacuum solution. This implies that it is not possible to emit waves by using as a source a variable spherically symmetric mass distribution. As in electromagnetism, there is not monopolar radiation of gravitational waves. Gravitational dipolar radiation does not exist either (identical conclusion would be valid in electromagnetism for a system of charges having the same charge–mass ratio and provided with motions conserving the total linear and angular momenta). The multipolar expansion for the gravitational radiation begins at the quadrupolar term.

when one looks through an irregular glass. The first observation of this effect was due to Walsh, Carswell, and Weymann (1979), who analyzed two twin images of a same quasar (QSO 0957+561 A, B) separated by 5.7 arcseconds, at $z = 1.405$. Nowadays, a large number of images produced by gravitational lenses are known, some of them obtained by the Hubble space telescope. Also *microlensing* effects have been reported, which would prove the existence of MACHOs (*Massive Compact Halo Objects*) in our galactic halo, with masses a few smaller than the solar mass (Alcock et al., 1997). The microlensing effect manifests itself through the temporary increasing of the apparent brightness of a star when one of these objects passes in front of the star.

General Relativity has also been successful in affording an appropriate framework to explain the cosmological redshift and the cosmic background radiation. On the other hand, the black holes predicted by the theory could form from the gravitational collapse of a quantity of matter large enough. The fate of a star able to retain more than 2 or 3 solar masses is a black hole; the mass of the star will compact behind its event horizon once the star exhausted its capacity of producing the thermonuclear reactions that keep it in equilibrium (Oppenheimer and Snyder, 1939).[24] The presence of a black hole is inferred from the suction that it exerts on the neighboring matter. Matter spirals toward the black hole, like it would make near a big drain, forming an accretion disk of matter around the black hole. While falling, the matter accelerates, compresses, and increases its temperature. Under such conditions, the matter in the accretion disk emits large quantities of radiation before going through the event horizon; the frequency of this radiation reaches the range of X-rays. These features make possible to recognize black holes in binary systems where one of the stars had evolved to a black hole and sucked matter from the companion star. The velocities of the matter around active galactic nuclei (AGN) indicates that these nuclei host black holes of $10^6 \sim 10^9$ solar masses. The large amount of radiation and the matter jets emitted by AGNs are manifestations of the dynamics of the matter surrounding the black hole. The radio-source Sagitarius A* at the center of our galaxy—the Milky Way—coincides with the location of a black hole of 2 or 3 millions of solar masses, as it has been inferred from the motion of neighboring stars (Ghez et al., 1998).

[24] Once the thermonuclear reactions are exhausted, stars of masses smaller than 1.4 solar masses (Chandrasekhar limit, 1931) reaches the equilibrium when the gravitational collapse is stopped by the pressure of the degenerated electron gas; in that way, they become *white dwarfs* (the temperature at the surface is around 10^4 K and the radius is about 10^4 km). In neutron stars the density is similar to that of an atomic nucleus, and gravity is balanced by the combination of the pressure of the degenerated neutron gas and nuclear forces. Masses of neutron stars vary between 0.1 and 2-3 solar masses, and their radii are about 10 km (Oppenheimer and Volkoff, 1939).

Appendix

A.1. EULER–LAGRANGE EQUATION

As differential calculus teaches, a given function $y(x)$ is *stationary* (it has a maximum, a minimum, or an inflection point) at those values of its variable x where the derivative $y'(x)$ vanishes. An analogous issue can be studied for a "function of a function" or *functional* $I[y(x)]$ of the type

$$I\,[y(x)] = \int_{x_1}^{x_2} L\,(y'(x), y(x), x)\mathrm{d}x \tag{A.1}$$

If the interval of integration (x_1, x_2) is fixed, then the value of $I[y(x)]$ will exclusively depend on function $y(x)$. We could wonder how function $y(x)$ should behave in the interval (x_1, x_2) in order that the functional $I[y(x)]$ reaches an extreme value—maximum or minimum—or, in general, be stationary. This question appears in different problems of mathematical physics. The simpler example is the curve joining two given points through the minimal possible length. In Euclidean plane, the length of a curve joining two points A and B having coordinates $(x_1\ ,\ y_1)$ and $(x_2\ ,\ y_2)$, is the integral:

$$\int_A^B \mathrm{d}l = \int_A^B \sqrt{\mathrm{d}x^2 + \mathrm{d}y^2} = \int_{x_1}^{x_2} \sqrt{1 + y'\,(x)^2}\ \mathrm{d}x \tag{A.2}$$

where the function $y(x)$ designs each curve joining A with B; therefore

$$y(x_1) = y_1 \qquad\qquad y(x_2) = y_2 \tag{A.3}$$

We know that functional (A.2) is minimum when points A and B are joined through a straight line; then the solution to the problem is the linear function $y(x) = y_1 + \frac{y_2 - y_1}{x_2 - x_1}(x - x_1)$. In most of the cases the solution to this kind of problems does not appear in such obvious way. Therefore, we need a method to find out the function $y(x)$ making the functional $I\,[y(x)]$ stationary in the interval $(x_1,\ x_2)$, under certain *boundary conditions* of type (A.3). This is the subject of the *calculus* of *variations*.

The central point of the calculus of variations is that the functional $I[y(x)]$ is stationary if its value does not change, at the lower order, under infinitesimal variations of its argument $y(x)$ leaving unaltered the boundary conditions (A.3):

$$y(x) \to y(x) + \delta y(x) \qquad \delta y(x_1) = 0 = \delta y(x_2) \tag{A.4}$$

The variation (A.4) indicates the substitution of function $y(x)$ with a different function that differs from the original one in $\delta y(x)$ at each value of x. The change of functional $I[y(x)]$ under the variation $\delta y(x)$ can then be written as

$$\begin{aligned} \delta I &= \int_{x_1}^{x_2} \left[\frac{\partial L}{\partial y'} (\delta y)' + \frac{\partial L}{\partial y} \delta y \right] \mathrm{d}x \\ &= \left[\frac{\partial L}{\partial y'} \delta y \right]_{x_1}^{x_2} - \int_{x_1}^{x_2} \left[\left(\frac{\partial L}{\partial y'} \right)' \delta y - \frac{\partial L}{\partial y} \delta y \right] \mathrm{d}x \end{aligned} \tag{A.5}$$

where an integration by parts has been performed. According to Eq. (A.4) the variations δy must vanish at the ends of the interval of integration. Therefore the boundary term in the integration by parts is null. Apart from that, the variation $\delta y(x)$ is an arbitrary function (which should not be mistaken for the differential $\mathrm{d}y = y'(x)\mathrm{d}x$). Then the necessary and sufficient condition for the functional $I[y(x)]$ being stationary is that function $y(x)$ fulfils the equation

$$\left(\frac{\partial L}{\partial y'} \right)' - \frac{\partial L}{\partial y} = 0 \tag{A.6}$$

Equation (A.6) is called *Euler–Lagrange equation.* It is a second order differential equation for $y(x)$ having the boundary conditions (A.3).

The variational method here displayed can be extended to a functional depending on several functions of one or more variables; in such a case, it will result an equation like Eq. (A.6) for each function of its argument, each of them having so many partial derivatives as variables have the functions. The application of the method to functionals depending on derivatives of larger order would lead to differential equations of higher order and a consequent enlargement of the set of boundary conditions.

A.2. THE ACTION FUNCTIONAL

The fundamental dynamical laws of Physics have the form of Euler–Lagrange equations. This means that physical systems evolve in such a way that certain functional becomes stationary. This functional is called *action* S, and its integrand L is called *Lagrangian.* In classical mechanics, for instance, the equation governing the behavior of a particle with mass m and potential energy $V(r)$ is

$$m\ddot{\mathbf{r}} = -\nabla V \qquad \text{or} \qquad m\,\frac{\mathrm{d}^2 x^\alpha}{\mathrm{d}t^2} = -\frac{\partial V}{\partial x^\alpha} \tag{A.7}$$

together with given boundary conditions $\mathbf{r}(t_1) = \mathbf{r}_1, \mathbf{r}(t_2) = \mathbf{r}_2$. Since $u^2 = \dot{\mathbf{r}}^2 = \Sigma(\dot{x}^\alpha)^2$, then the acceleration can be written

$$\frac{d^2 x^\alpha}{dt^2} = \frac{1}{2} \frac{d}{dt}\left(\frac{\partial \dot{\mathbf{r}}^2}{\partial \dot{x}^\alpha}\right)$$

which shows that Eq. (A.7) is an Euler–Lagrange equation for the Lagrangian

$$L(\dot{x}^\alpha, x^\alpha) = \frac{1}{2} m \dot{\mathbf{r}}^2 - V(\mathbf{r}) \tag{A.8}$$

In this case the action functional $S[\mathbf{r}(t)] = \int_{t_1}^{t_2} L\, dt$ depends on three functions of one variable (the three components of vector $\mathbf{r}(t)$). In particular, in a gravitational potential $\Phi(\mathbf{r})$ the Lagrangian is $L = m\, u^2/2 - m\, \Phi$.

In §8.5 it was proved that the motion of a freely gravitating relativistic particle, i.e., the timelike world lines satisfying the geodesic equation (8.19), results from the variation of the action (8.16) whose Lagrangian is Eq. (8.17). It is easy to verify that the action for an electrically charged relativistic particle that gravitates in the presence of an electromagnetic potential A_i is the magnitude

$$S = -mc \int ds - q \int A_i\, dx^i \tag{A.9}$$

The first term is the action of the freely gravitating particle, and the second one describes the interaction between the charge q and the electromagnetic field. The respective Lagrangian is then

$$L\left(\frac{dx^i}{d\lambda}, x^i(\lambda)\right) = -mc\sqrt{g_{ij}\frac{dx^i}{d\lambda}\frac{dx^j}{d\lambda}} - qA_i(x^k)\frac{dx^i}{d\lambda} \tag{A.10}$$

In fact, the Euler–Lagrange equations deriving from this Lagrangian become $(Dp/D\tau)_i = q\, F_{ij}\, U^j$ (see the Lorentz four-force in Eq. (7.90)), and are subjected to the boundary conditions $x^i(\lambda_1) = x^i{}_1,\ x^i(\lambda_2) = x^i{}_2$.

Maxwell's laws (7.84) for the electromagnetic field can be regarded as the result of making stationary the action[1]

$$S[A_i(x^j)] = -\frac{1}{4\, c\, \mu_o} \int g^{kl} g^{rs} F_{kr} F_{ls} \sqrt{|g|}\, d^4x - \frac{1}{c} \int A_i j^i \sqrt{|g|}\, d^4x \tag{A.11}$$

The second term in Eq. (A.11) describes the interaction between the electromagnetic field and a distribution of electric current j^i ; if this current comes from the motion of a charged particle, then the form of the interaction in action (A.9)

[1] The expressions in Gaussian System of units are obtained through the replacements: $\mu_o \to 4\pi c^{-2}$, $A_i \to A_i c^{-1}$, $F_{ij} \to F_{ij} c^{-1}$.

will be recovered. Differing from the former cases, action (A.11) is a functional of a *field* $A_i(x^j)$. $S[A_i(x^j)]$ is an integral in the four-volume contained between two spacelike hypersurfaces having the role played by the boundaries t_1 and t_2, or λ_1 and λ_2 in the previous mechanical examples. Besides the condition that the field does not vary on these hypersurfaces, the cancellation of the boundary term resulting from the integration by parts now requires that the field vanishes on the lateral boundary of the four-volume (the spatial infinity). Since the Lagrangian is a function of $\partial_j A_i$ and A_i, the Euler–Lagrange equations will be differential equations in partial derivatives ∂_j. The only dependence on A_i comes from the interaction term because F_{ij} depends just on the potential derivatives (incidentally, notice that the symmetry of Christoffel symbols implies $F_{ij} = A_{j;i} - A_{i;j} = A_{j,i} - A_{i,j}$). Therefore the Euler–Lagrange equations yield

$$-\frac{1}{4\mu_o}\partial_j\left(\sqrt{|g|}\, g^{kl}\, g^{rs}\frac{\partial(F_{kr}\, F_{ls})}{\partial(\partial_j A_i)}\right) + \sqrt{|g|}\, j^i = 0 \tag{A.12}$$

where

$$g^{kl}\, g^{rs}\frac{\partial(F_{kr}\, F_{ls})}{\partial(\partial_j A_i)} = (g^{jl} g^{is} - g^{il} g^{js})F_{ls} + (g^{kj} g^{ri} - g^{ki} g^{rj})F_{kr} = -4F^{ij}$$

Taking into account Eq. (8.36), Eq. (A.12) becomes

$$F^{ij}{}_{;j} = -\mu_o\, j^i \tag{A.13}$$

As was remarked in §8.6, Eq. (A.13) is the law corresponding to Eq. (7.84) when the Principle of equivalence is implemented by means of the minimal coupling prescription. By using expressions (8.36) for the divergence of vectors and antisymmetric tensors, it can be proved that $F^{ij}{}_{;ji} \equiv 0$ for any antisymmetric tensor F^{ij}. The *automatic conservation* of the left side in Eq. (A.13) forces the source conservation: $j^i{}_{;i} = 0$. In other words, Maxwell's laws have to be formulated with sources conserving the charge; otherwise they would be inconsistent.

Also Einstein equations accept a variational formulation. As David Hilbert (1915) proved, the variation of the action

$$S[g_{ij}(x^k)] \propto \int R\sqrt{|g|}\, \mathrm{d}^4x \tag{A.14}$$

provides the geometric terms entering in Einstein equations. In fact

$$\delta\left(R\sqrt{|g|}\right) = \delta\left(g^{ij}\, R_{ij}\sqrt{|g|}\right) = \sqrt{|g|}\, R_{ij}\, \delta g^{ij} + R\, \delta\sqrt{|g|} + \sqrt{|g|}\, g^{ij}\, \delta R_{ij}$$

where the variation of $|g|$ under metric tensor changes is (see Note 17 in Chap. 8)

$$\delta\sqrt{|g|} = \frac{1}{2}\sqrt{|g|}\, g^{ij}\, \delta g_{ij} = -\frac{1}{2}\sqrt{|g|}\, g_{ij}\, \delta g^{ij} \tag{A.15}$$

Thus

$$\delta\int R\sqrt{|g|}\, \mathrm{d}^4x = \int (R_{ij} - \tfrac{1}{2} g_{ij}\, R)\delta g^{ij}\sqrt{|g|}\, \mathrm{d}^4x + \int g^{ij}\, \delta\, R_{ij}\sqrt{|g|}\, \mathrm{d}^4x \tag{A.16}$$

The last term in this expression does not contribute to the variation of Hilbert action because it becomes a boundary term. To see this, let us consider Riemann tensor definition (8.40) in a locally inertial frame with origin at O ($\Gamma^c{}_{a\ b}(\mathrm{O}) = 0 = g^{a\ b}{}_{,c}(\mathrm{O})$); then the terms of the form $\Gamma\ \delta\Gamma$ do not contribute to the variation at O, so yielding

$$g^{ab}\delta R_{ab} = g^{ab}\ \delta R^c{}_{acb} = g^{ab}(\delta\Gamma^c_{ba,c} - \delta\Gamma^c_{ca,b})$$
$$= (g^{ac}\delta\Gamma^b_{ca} - g^{ab}\delta\Gamma^c_{ca})_{,b} \qquad \text{(A.17)}$$

Here it is important to notice that, although the connection elements do not transform as tensor components, the difference of two different connections does tensorially transform (see Eq. (8.23)). This means that $\delta\,\Gamma^j_{ik}$ has tensor character, thus the expression (A.17) is the divergence of a four-vector whose form in an arbitrary coordinate system is

$$g^{ij}\,\delta R_{ij} = (g^{ik}\,\delta\Gamma^j_{ki} - g^{ij}\,\delta\Gamma^k_{ki})_{;j} \qquad \text{(A.18)}$$

Therefore the last term in Eq. (A.16) is the integral of the divergence of a four-vector in a four-volume. This term can be written as an integral on the boundary by applying the divergence theorem (see Section A.5). Equation (A.16) then shows that Hilbert action is stationary when metric satisfies the vacuum ($T^{ij} = 0$) Einstein equations. The cosmological constant can be included by replacing R with $R + 2\Lambda$ in the integrand of Hilbert action.

A.3. METRIC ENERGY–MOMENTUM TENSOR

To obtain Einstein equations with sources ($T^{ij} \neq 0$) we must add Hilbert action with the actions belonging to the rest of the present fields and matter. The variation of the total action with respect to metric tensor will then lead to Eq. (8.46). This statement implies that the energy–momentum tensor of the rest of fields and matter comes from varying the action with respect to the metric tensor. For instance, the energy–momentum tensor (7.96) emerges from varying the action of the electromagnetic field without sources ($j^i = 0$) with respect to the metric tensor:

$$-\frac{1}{4c\mu_o}\delta(g^{kl}\ g^{rs}\ F_{kr}\ F_{ls}\sqrt{|g|})$$
$$= -\frac{1}{4c\mu_o}(g^{rs}\ F_{kr}\ F_{ls}\ \delta\ g^{kl} + g^{kl}\ F_{kr}\ F_{ls}\ \delta\ g^{rs})\sqrt{|g|}$$
$$+\frac{1}{4c\mu_o}g^{kl}g^{rs}\ F_{kr}\ F_{ls}\tfrac{1}{2}\sqrt{|g|}\ g_{ij}\ \delta\ g^{ij}$$
$$= \frac{1}{2c\mu_o}(-F_{ir}\ F_j{}^r + \frac{1}{4}g_{ij}\ F^{kr}\ F_{kr})\sqrt{|g|}\ \delta\,g^{ij} = \frac{1}{2c}T_{ij}\sqrt{|g|}\ \delta\ g^{ij} \qquad \text{(A.19)}$$

Actually the energy–momentum tensor of the electromagnetic field (7.96) was not univocally defined in §7.7, due to the possibility of adding a term being divergenceless or having a divergence that becomes null as a consequence of Maxwell's laws. This ambiguity obscured the meaning of the localization of the electromagnetic energy. Since the geometry is sensitive to the energy location, no ambiguity in this point could be accepted in General Relativity. Thus, the metric energy–momentum tensor gives a unique and satisfactory answer to the question.[2,3]

With Eqs. (A.16) and (A.19) we can determine the factor needed in the Hilbert action in order that its variation should lead to Einstein equations (8.46):

$$S[g_{ij}(x^k)] = -\frac{c^3}{16\pi G}\int (R+2\Lambda)\sqrt{|g|}\ \mathrm{d}^4x \tag{A.20}$$

A.4. GAUGE TRANSFORMATIONS AND SOURCE CONSERVATION

As seen in §7.7, field tensor F_{ij} is not sensitive to a gauge transformation $A_i \to A_i + \xi_{,i}$. For Maxwell's laws, two A_i field configurations differing in a gauge transformation are entirely equivalent. Then the action should not discern among these equivalent field configurations. Therefore the action should be automatically stationary if the field variation had the form $\delta A_i = \xi_{,i}$. This action symmetry should be verified for any configuration; even those not solving Euler–Lagrange equations. In this way, the null variation of the action under a gauge transformation would not add any information about the field dynamics. The existence of field variations that are not perceived by the action indicates that the action is not able to dictate the dynamics of a sector of the field. The degrees of freedom whose dynamics does not result to be determined by the action constitute the field gauge freedom, which can be "frozen" by means of a "gauge fixing" (see §7.7 and §8.7).

The change of the electromagnetic action (A.11) under a variation δA_i is

$$\delta S[A_i(x^j)] = -\frac{1}{c}\int (\mu_0^{-1}F^{ij}{}_{;j} + j^i)\delta A_i\sqrt{|g|}\ \mathrm{d}^4x + \text{boundary terms}$$

---

[2] Notice that action (A.11) does not contain derivatives of the metric tensor only because the tensor F_{ij} does not require covariant derivatives of A_i . If the action of a field included covariant derivatives then the metric energy–momentum tensor would incorporate higher field derivatives coming from the integration by parts of those terms containing derivatives of the metric tensor.

[3] The action describing the interaction between the electromagnetic field and charged particles depends only on the dynamical variables associated with the field and the particles; there is not contributions due to the metric tensor (in Eq. (A.9) the interaction is described by the integral $-q\int A_i\ \mathrm{d}x^i$ along the particle world line). Therefore there is not interaction term in the energy–momentum tensor of the particles-field system: $T^{ij} = T^{ij}{}_{\text{particles}} + T^{ij}{}_{\text{field}}$.

as it is inferred from Eqs. (A.12) and (A.13). In a gauge transformation it results

$$\delta_{\text{gauge}}\ S[A_i(x^j)] = -\frac{1}{c}\int (\mu_o^{-1}\ F^{ij}{}_{;j} + j^i)\xi_{,i}\ \sqrt{|g|}\ \mathrm{d}^4x + \text{boundary terms}$$

Taking into account that

$$(F^{ij}{}_{;j}\ \xi)_{;i} = F^{ij}{}_{;\,ji}\ \xi + F^{ij}{}_{;j}\ \xi_{,i} = F^{ij}{}_{;j}\ \xi_{,i}$$

$$(j^i\xi)_{;i} = j^i{}_{;i}\ \xi + j^i\xi_{,i}$$

where we exploited that $F^{ij}{}_{;ji} \equiv 0$, and applying the divergence theorem (see Section A.5) it results [4]

$$\delta_{\text{gauge}}\ S\,[A_i(x^j)] = \frac{1}{c}\int j^i{}_{;i}\ \xi\ \sqrt{|g|}\ \mathrm{d}^4x + \text{boundary terms} \qquad \text{(A.21)}$$

The gauge invariance of the action then requires that sources conserve the charge ($j^i{}_{;i} = 0$). Naturally, this conclusion is not new since Maxwell's laws require the charge conservation as well. To illustrate this issue let us come back to the example (A.9) for the charged particle. Obviously, a moving charge q conserves the charge, and clearly its action is gauge invariant:

$$\begin{aligned}\delta_{\text{gauge}}\int A_i\ \mathrm{d}x^i &= \int (\delta_{\text{gauge}}\ A_i)\ \mathrm{d}x^i = \int_{\lambda 1}^{\lambda_2} \xi_{,i}\frac{\mathrm{d}x^i}{\mathrm{d}\lambda}\mathrm{d}\lambda \\ &= \int_{\lambda_1}^{\lambda 2}\frac{\mathrm{d}\xi}{\mathrm{d}\lambda}\mathrm{d}\lambda = \text{boundary terms}\end{aligned} \qquad \text{(A.22)}$$

While the gauge invariance marks the existence of spurious degrees of freedom among the dynamical variables in the action, on the other hand it imposes restrictions to the behavior of sources.

The treatment given to the gauge invariance of the electromagnetic field action can also be applied to the gravitational field action. Hilbert action is not sensitive to variations of the metric tensor having the form

$$\delta\ g_{ij} = \xi^k\ g_{ij,k} + g_{ik}\ \xi^k{}_{,j} + g_{kj}\ \xi^k{}_{,i} = \xi_{i;\,j} + \xi_{j;i} \qquad \text{(A.23)}$$

Owing to the automatic conservation of Einstein tensor, this type of variations only produces boundary terms for any metric in which the action is evaluated. When variations (A.23) are regarded in a coordinate system where the components of vector ξ^k are constant in a vicinity of $x^i(\xi^k{}_{,j}(x^i) = 0)$, they correspond to the substitution of the metric tensor at x^i with the metric tensor at the close event

[4] The evident gauge invariance of the action first term thus reappears as the result of the automatic conservation $F^{ij}{}_{;ij} \equiv 0$.

of coordinates $x^i + \xi^i$. The "gauge transformation" then consists in "dragging" the components of the metric tensor along the direction of a vector ξ^k. Apart from boundary terms, this operation does not modify Hilbert action. In order that the source terms—i.e., the action of the rest of present fields and matter—will not be sensitive to these gauge transformations either, the sources have to be conserved. In other words, Einstein equations do not admit sources having arbitrary evolutions. The energy–momentum distribution of sources, characterized by the tensor T^{ij}, is compelled to satisfy the condition $T^{ij}{}_{;j} = 0$. Since this condition emerges from the fulfillment of the dynamical equations for the sources, which in turn depends on the space-time geometry, this means that Einstein equations obligate to solve the evolution of geometry and sources in a joint way.

A.5. KILLING VECTORS AND ENERGY–MOMENTUM CONSERVATION

According to the expression (8.36), the divergence of any four-vector j^i verifies

$$\int_{\text{four-volume}} j^i{}_{;i}\sqrt{|g|}\;\mathrm{d}^4x = \int_{\text{four-volume}} \left(\sqrt{|g|}\;j^i\right)_{,i}\,\mathrm{d}^4x$$

Then, by using the divergence theorem for four variables (see §7.5), one obtains

$$= \int_{\substack{\text{boundary of the}\\ \text{four-volume}}} \sqrt{|g|}\;j^i\,\varepsilon_{i|jkl|}\frac{\partial(x^j, x^k, x^l)}{\partial(a, b, c)}\mathrm{d}a\;\mathrm{d}b\;\mathrm{d}c$$

where ε_{ijkl} is the Levi–Civita symbol and $(a,\ b,\ c)$ are three coordinates in the boundary of the four-volume. The Levi–Civita symbol multiplied by $\sqrt{|g|}$ behaves as a pseudotensor (see §8.6 and Note 12 in Chapter 7). The factors accompanying j^i in the last expression constitute the hypersurface element, which is a four-vector normal to the boundary of the four-volume. Joining this result with the starting expression, it is obtained

$$\int_{\text{four-volume}} j^i{}_{;i}\;\sqrt{|g|}\;\mathrm{d}^4x = \int_{\substack{\text{boundary of the}\\ \text{four-volume}}} j^i\;\mathrm{d}\Sigma_i \tag{A.24}$$

If j^i has zero divergence, then

$$\int_{\substack{\text{boundary of the}\\ \text{four-volume}}} j^i\;\mathrm{d}\;\Sigma_i = 0 \tag{A.25}$$

that, as seen in §7.6, expresses the local conservation of the *charge* defined as

$$\text{charge} = \int_{\substack{\text{spacelike}\\ \text{hypersurface}}} c^{-1}\, j^i\, d\Sigma_i \tag{A.26}$$

Instead, there is not an expression like (8.36) for a symmetric tensor. Thus, the zero divergence of the energy–momentum tensor T^{ij} does not allow to conclude the conservation of a magnitude like (A.26). Nevertheless, in order to reach the result (A.26) we could build a four-vector by means of a contraction $T^{ij}K_i$, where K_i is a four-vector that should provide with zero divergence to $T^{ij}K_i$. So, K_i should verify

$$0 = (T^{ij}\, K_i)_{;j} = T^{ij}{}_{;j}\, K_i + T^{ij}\, K_{i;j} = T^{ij}\, K_{i;j}$$

Since T^{ij} is symmetric, K_i should then satisfy the condition

$$K_{i;j} + K_{j;i} = 0 \tag{A.27}$$

which is equivalent to

$$K^k\, g_{ij,k} + g_{ki}\, K^k{}_{,j} + g_{kj}\, K^k{}_{,i} = 0 \tag{A.28}$$

In such a case the conserved magnitude is

$$P = \int_{\substack{\text{spacelike}\\ \text{hypersurface}}} c^{-1}\, T^{ij}\, K_i\, d\Sigma_j \tag{A.29}$$

i.e., an energy or a momentum according to K_i is timelike or spacelike (remember that $d\Sigma_j$ is timelike because it is normal to a spacelike hypersurface).

A vector fulfilling Eqs. (A.27–A.28) is called *Killing vector*. The left side in Eq. (A.28) is a derivation of metric in K direction, so Killing vectors signal invariances of the geometry. The conservation of the energy or a momentum component of a freely gravitating system is then a consequence of an invariance of the geometry under temporal or spatial displacements. This result was already obtained for a freely gravitating particle: in Eq. (8.18) the conservation of the component p_k happens if and only if $g_{ij,k} = 0\ \forall i, j$; in this case there exists a Killing vector whose kth component is equal to 1 and the rest zero. In general, if K is a Killing vector for the geometry and p is the energy–momentum four-vector for a freely gravitating particle, then $p \cdot K$ will be conserved.

In Minkowski space-time, any vector whose Cartesian components are constant is a Killing vector. If the Cartesian components of K in a given frame are (1, 0, 0, 0) then the magnitude (A.29) will be the energy (multiplied by c^{-1}); if K is (0, 1, 0, 0) then Eq. (A.29) will become the x momentum component. Other Killing vectors of Minkowski space-time relate to the angular momentum conservation: the Killing vector whose Cartesian components are $(0, y, -x, 0)$ makes (A.29) to become the z angular momentum component of the physical system.

Bibliography

Alexander, H.G. (ed.) (1998), *The Leibniz-Clarke Correspondence Together with Extracts from Newton's Principia and Opticks*, Manchester: Manchester University Press.

Anderson, J.L. (1967), *Principles of Relativity Physics*, N.Y.: Academic Press.

Barbour, J. and Pfister, H. (eds.) (1995), *Mach's Principle: From Newton's Bucket to Quantum Gravity*, Boston: Birkhäuser.

Cheng, T.-P. (2005), *Relativity, Gravitation and Cosmology: A Basic Introduction*, Oxford: Oxford University Press.

Clotfelter, B.E. (1970), *Reference Systems and Inertia (the nature of space)*, Ames: The Iowa State University Press.

Coles, P. and Lucchin, F. (2002), *Cosmology: The Origin and Evolution of Cosmic Structures*, Chichester: John Wiley & Sons.

Einstein, A., Lorentz, H.A., Minkowski, H., and Weyl, H. (1958), *The Principle of Relativity: A Collection of Original Memoirs*, N.Y.: Dover.

Friedman, M. (1986), *Foundations of Space-Time Theories: Relativistic Physics and Philosophy of Science*, Princeton: Princeton University Press.

Hecht, E. and Zajac, A. (1974), *Optics*, Reading (Mass.): Addison-Wesley Pub. Co.

Jammer, M. (1994), *Concepts of Space: The History of Theories of Space in Physics*, N.Y.: Dover.

Landau, L.D. and Lifshitz, E.M. (1971), *The Classical Theory of Fields*, London: Pergamon.

Misner, C.W., Thorne, K.S., and Wheeler, J.A. (1973), *Gravitation*, San Francisco: Freeman.

Møller, C. (1952), *The Theory of Relativity*, Oxford: Clarendon Press.

Pais, A. (1982), *"Subtle is the Lord": The science and the Life of Albert Einstein*, Oxford/New York: Oxford University Press.

Peebles, P.J.E. (1993), *Principles of Physical Cosmology*, Princeton: Princeton University Press.

Rindler, W. (2001), *Relativity: Special, General and Cosmological*, N.Y.: Oxford University Press.

Schutz, B.F. (1985), *A First Course in General Relativity*, Cambridge: Cambridge University Press.

Shankland, R.S. (1963), "Conversations with Albert Einstein", American Journal of Physics **31**, 47–57.

Stewart, A.B. (1964), "The discovery of stellar aberration", Scientific American **210** (3), 100–108.

Swenson, L.S., "The Michelson-Morley-Miller experiments before and after 1905", Journal for the History of Astronomy **1** (1970), 56–78.
Taylor, E.F. and Wheeler, J.A. (1992), *Space-Time Physics*, San Francisco: Freeman.
Will, C.M. (1993), *Theory and Experiment in Gravitational Physics*, Cambridge: Cambridge University Press.

Cited Papers

Airy, G.B. (1871), Proc. Roy. Soc. (London) **20**, 35.
Alcock, C. et al. (1997), Astrophys. J. **486**, 697.
Arago, D.F.J. (1810), C. R. Acad. Sci. Paris **34** (1853), 38, and *Œuvres completes* (1858) Vol. 7, Paris: Gide.
Babacock, G.C. and Bergman, T.G. (1964), J. Opt. Soc. Am. **54**, 147.
Balbi, A. et al. (2000), Astrophys. J. Lett. **545**, 1.
Bartholin, E., (1669), *Experimenta crystalli Islandici disdiaclastici quibus mira & insolita refractio detegitur*, Hafniæ.
Bennett, C.L. et al. (2003), Astrophys. J. Suppl. **148**, 175.
Bertotti, B., Iess, L., and Tortora, P. (2003), Nature **425**, 374.
Bessel, F.W. (1838), Astron. Nachr. **16**, 65.
Birkhoff, G.D. (1923), *Relativity and Modern Physics*, Cambridge (Mass.): Harvard University Press.
Bondi, H., Pirani, F.A.E., and Robinson, I. (1959), Proc. Roy. Soc. London A **251**, 519.
Brace, D.B. (1904), Phil. Mag. (6) **7**, 317.
Bradley, J. (1728), Phil. Trans. R. Soc. London **35**, 637.
Braginsky, V.B. and Panov, V.I. (1971), Zh. Eksp. Teor. Fiz. **61**, 873; Sov. Phys. JETP **34** (1972), 463.
Brans, C. and Dicke, R.H. (1961), Phys. Rev. **124**, 925.
Brillouin, L. (1960), *Wave Propagation and Group Velocity*, N.Y.: Academic Press.
Bucherer, A.H. (1908), Verh. dtsch. phys. Ges. **6**, 688; Phys. Z. **9**, 755.
Cannon, R. et al. (2006), Mont. Not. Roy. Astron. Soc. **372**, 425.
Chandrasekhar, S. (1931), Phil. Mag. **11**, 592; Astrophys. J. **74**, 81.
Ciufolini, I. and Pavlis, E.C. (2004), Nature **431**, 958.
Ciufolini, I., Pavlis, E., Chieppa, F., Fernandes-Vieira, E., and Pérez-Mercader, J. (1998), Science **279**, 2100.
Colless, M. et al. (2001), Mont. Not. Roy. Astron. Soc. **328**, 1039.
Compton, A.H. (1921), Phys. Rev. **21**, 483.
Cui, W., Zhang, S.N., and Chen, W. (1998), Astrophys. J. Lett. **492**, L53.
de Bernardis, P. et al. (2000), Nature **404**, 955.
de Broglie, L. (1924), *Thèse*, Paris; Annales de Physique **3** (1925), 22.
de Sitter, W. (1916), Mont. Not. Roy. Astron. Soc. **77**, 155; 481.
de Sitter, W. (1917), Proc. Sect. Sci. K. Ned. Akad. **19**, 1217; **20**, 229.
Dicke, R.H., Peebles, P.J.E., Roll, P.G., and Wilkinson, D.T. (1965), Astrophys. J. **142**, 41.
Dyson, F.W., Eddington, A.S.E., and Davidson, C.R. (1920), Phil. Trans. Roy. Soc. A **220**, 291.
Einstein, A. (1905a), Annalen der Physik **17**, 132.

Einstein, A. (1905b), Annalen der Physik **17**, 891.
Einstein, A. (1905c), Annalen der Physik **18**, 639.
Einstein, A. (1907), Jb. Radioakt. Elektronik **4**, 411.
Einstein, A. (1910), Arch. Sci. Phys. et Nat. **29**, 125.
Einstein, A. (1911), Annalen der Physik **35**, 898.
Einstein, A. (1912), Vierteljahrsschrift für gerichtliche Medizin und öffentliches Sanitätswesen **44**, 37.
Einstein, A. (1915a), Sber. Preuss. Akad. Wiss. Berlin **47**, 831.
Einstein, A. (1915b), Sber. Preuss. Akad. Wiss. Berlin **47**, 844.
Einstein, A. (1916), Annalen der Physik **49**, 769.
Einstein, A. (1917a), Phys. Z. **18**, 121.
Einstein, A. (1917b), Sber. Preuss. Akad. Wiss. Berlin, 142.
Einstein, A. (1918), Annalen der Physik **55**, 241.
Einstein, A. (1922), *The Meaning of Relativity*, Princeton: Princeton University Press.
Einstein, A. (1924), Berliner Tageblatt, April 20.
Einstein, A. (1934), *The World as I See It*, New York: Covici Friede.
Eötvös, R.V. (1889), Math. Naturw. Ber. aus Ungarn **8**, 65.
Eötvös, R.V. , Pékar, D., and Fekete, E. (1922), Annalen der Physik **68**, 11.
FitzGerald, G.F. (1889), Science **13**, 390.
Fizeau, H.L. (1851), C. R. Acad. Sci. Paris **33**, 349; Ann. Chim. Phys.(3[a] série) **57** (1859), 385.
Föppl, A. (1904), Sitzung. Bayer. Akad. Wiss., Math.-Phys. Kl. **34**, 5; 383 (partially translated in Barbour and Pfister (1995)).
Frank, P. (1909), Sitz. Ber. Akad. Wiss. Wien. IIa, No. 118, 373.
Fray, S., Álvarez Diez, C., Hänsch, T.W., and Weitz, M. (2004), Phys. Rev. Lett. **93**, 240404.
Fresnel, A.J. (1818), Ann. Chim. Phys. **9**, 57.
Friedlaender, B. and Friedlaender I. (1896), *Absolute oder Relative Bewegung?* Berlin: Leonhard Simion (partially translated in Barbour and Pfister (1995)).
Friedmann, A. (1922), Z. Phys. **10**, 377.
Frisch, D.H. and Smith, J.H. (1963), Am. J. Phys. **31**, 342.
Galilei, G. (1638), *Discorsi e dimostrazioni matematiche intorno a due nuove scienze*, Leiden; *Dialogues Concerning Two New Sciences*, N.Y.: Dover (1954).
Gamow, G. (1948), Nature **162**, 680.
Ghez, A. et al. (1998), Astrophys. J. **509**, 678.
Gödel, K. (1949), Rev. Mod. Phys. **21**, 447.
Grimaldi, F.M. (1665), *Physico-mathesis de lumine coloribus et iride, libri 2 in quorum 1. afferuntur noua experimenta, Bologna*: Vittorio Benati.
Gurvits, L.I., Kellermann, K.I. and Frey, S. (1999), Astron. Astrophys. **342**, 378.
Hafele, J.C. and Keating, R.E. (1972), Science **177**, 166, 168.
Hilbert, D. (1915), Konigl. Gesell. D. Wiss. Göttingen, Nachr., Math.-Phys. Kl., 395.
Hoek, M. (1868), Arch. Néerl. Sci. **3**, 180.
Hofmann, W. (1904), *Kritische Beleuchtung der beiden Grundbegriffe der Mechanik: Bewegung und Trägheit und daraus gezogene Folgerungen betreffs der Achsendrehung der Erde und des Foucault'schen Pendelversuchs*, Wien and Leipzig: M. Kuppitsch Wwe. (partially translated in Barbour and Pfister (1995)).
Hubble, E.P. (1929), Proc. Natl. Acad. Sci. U.S. **15**, 169.
Hughes, V.H., Robinson, H.G., and Beltrán-López, V. (1960), Phys. Rev. Lett. **4**, 342.
Hulse, R.A. and Taylor, J.H. (1975), Astrophys. J. Lett. **195**, L51.
Huygens, C. (1690), *Traité de la lumière*, Leiden; *Treatise on light*, http://www.gutenberg.org/etext/14725 (2005).

Illingworth, K.K. (1927), Phys. Rev. **30**, 692.
Ives, H.E. and Stillwell, G.R. (1938), J. Opt. Soc. Am. **28**, 215.
Joos, G. (1930), Annalen der Physik (5) **7**, 385.
Kennedy, R.J. (1926), Proc. Natl. Acad. Sci. U.S. **12**, 621.
Kennedy, R.J. and Thorndike, E.M. (1932), Phys. Rev. **42**, 400.
Kerr, R.P. (1963), Phys. Rev. Lett. **11**, 237.
Kruskal, M.D. (1960), Phys. Rev. **119**, 1743.
Larmor, J. (1900), *Æther and Matter*, Cambridge: Cambridge University Press.
Lebach, D.E. et al. (1995), Phys. Rev. Lett. **75**, 1439.
Lemaître, G.E. (1925), J. Math. Phys. **4**, 188.
Lewis, G.N. (1926), Nature **118**, 874.
Lodge, O.J. (1893), Phil. Trans. **184**, 727.
Lovelock, D. (1971), J. Math. Phys. **12**, 498.
Lorentz, H.A. (1886), Koninklijke Akademie van Wetenschappen te Amsterdam. Afdeeling Natuurkunde. Verslagen en Mededeelingen **2**, 297.
Lorentz, H.A. (1892), Verh. K. Akad. Wet. (Amsterdam) **1**, 74.
Lorentz, H.A. (1895), *Versuch einer Theorie der electrischen und optischen Erscheiningen in bewegten Körpen*, Leiden: Brill.
Lorentz, H.A. (1899), Verh. K. Akad. Wet. (Amsterdam) **7**, 507 (translated in Proceedings of the section of sciences, Koninklijke Akademie van Wetenschappen te Amsterdam **1**, 427).
Lorentz, H.A. (1904), Verh. K. Akad. Wet. (Amsterdam) **12**, 986 (translated in Proceedings of the section of sciences, Koninklijke Akademie van Wetenschappen te Amsterdam **6**, 809).
Lorentz, H.A. (1909), *Theory of Electrons: And Its Applications to the Phenomena of Light and Radiant Heat*, Leipzig: B.G. Teubner; *The Theory of Electrons*, N.Y.: Dover (2004).
Mach, E. (1872), *Die Geschichte und die Wurzel des Satzes von der Erhaltung der Arbeit*, Praha: Calve'sche Buchhandlung (*History and Root of the Principle of the Conservation of Energy*, Chicago: Open Court).
Mach, E. (1883), *Die Mechanik in ihrer Entwicklung. Historisch-kritisch dargestellt*, Leipzig: F.A. Brockhaus (*The Science of Mechanics: A Critical and Historical Account of Its Development*, La Salle (Illinois): Open Court.
Mather, J.C. et al. (1994), Astrophys. J. **420**, 439.
Matzner, R.A., Shepley, L.C., and Warren, J.B. (1970), Ann. Phys. **57**, 401.
Maxwell, J.C. (1862), Phil Mag. (4) **23**, 12.
Maxwell, J.C. (1879), *Letter to D.P. Todd, March 19*, published in Nature **21** (1880), 314. See also the article *Ether* in the 9th edition of the Encyclopædia Britannica (1878).
Michelson, A.A. (1881), Am. J. Sci. (3) **22**, 120.
Michelson, A.A. and Morley, E.W. (1886), Am. J. Sci. **31**, 377.
Michelson, A.A. and Morley, E.W. (1887), Am. J. Sci. (3) **34**, 333; Phil. Mag. (5) **24**, 449.
Michelson, A.A. (1897), Am. J. Sci. **3,** 475.
Miller, D.C. (1922), Phys. Rev. **19**, 407; Science **55**, 496.
Miller, D.C. (1925), Proc. Natl. Acad. Sci. U.S. **11**, 311.
Minkowski, H. (1909), Phys. Zeitschr. **10**, 104.
Morley, E.W. and Miller, D.C. (1898), Phys. Rev. **7**, 283.
Morley, E.W. and Miller, D.C. (1905), Phil. Mag. (6) **9**, 680.
Morley, E.W. and Miller, D.C. (1906), Science (2) **23**, 417; **25** (1907), 525.
Mössbauer, R.L. (1958), Z. Phys. **15**, 124.
Newton, I. (1687), *Philosophiae Naturalis Principia Mathematica*, London; *The Principia: Mathematical Principles of Natural Philosophy and His System of the World*, Whitefish: Kessinger Publishing (2003).

Newton, I. (1704), *Opticks: or, a Treatise of the Reflexions, Refractions, Inflexions and Colours of Light*, London; *Opticks*, Amherst (N.Y.): Prometheus Books (2003).
Oppenheimer, J.R. and Snyder, H. (1939), Phys. Rev. **56**, 455.
Oppenheimer, J.R. and Volkoff, G.M. (1939), Phys. Rev. **55**, 374.
Ozsváth, I. and Schücking, E.L. (1962), Nature **193** 1168; (1969), Ann. Phys. **55**, 166.
Penzias, A.A. and Wilson, R.W. (1965), Astrophys. J. **142**, 419.
Perlmutter, S. et al. (1999), Astrophys. J. **517**, 565.
Picard, A. and Stahel, E. (1926), C. R. Acad. Sci. Paris **183**, 420.
Picard, A. and Stahel, E. (1927), C. R. Acad. Sci. Paris **185**, 1198.
Poincaré, J.H. (1899), *Lecture in La Sorbonne*; see also in *Rapports du Congrés International de Physique*, C. Guillaume and L. Poincaré (eds.), Paris: Gauthier-Villars (1900), Vol.1, 1.
Poincaré, J.H. (1909), *Sechs Vorträge aus der Reinen Mathematik und Mathematischen Physik*, Leipzig: Teubner, 1910.
Pound, R.V. and Rebka, G.A. (1960), Phys. Rev. Lett. **4**, 337.
Rayleigh, Lord (1902), Phil. Mag. (6) **4**, 678.
Reasenberg, R.D., Shapiro, I.I. et al. (1979), Astrophys. J. Lett. **234**, L219.
Reissner, H. (1914), Phys. Z. **15**, 371.
Riess, A.G. et al. (1998), Astron. J. **116**, 1009.
Ritz, W. (1908), Ann. Chem. Phys. **13**, 145.
Robertson, H.P. (1935), Astrophys. J. **82**, 248; **83** (1936), 187, 257.
Robertson, D.S., Carter, W.E., and Dillinger, W.H. (1991), Nature **349**, 768.
Roll, P.G., Krotkov, R., and Dicke, R.H. (1964), Ann. Phys. (N.Y.) **26**, 442–517.
Rossi, B. and Hall, D.B. (1941), Phys. Rev. **59**, 223.
Rubin, V., Ford, W.K., and Thonnard, N. (1980), Astrophys. J. **238**, 471.
Schrödinger, E. (1925), Annalen der Physik **77**, 325.
Schwarzschild, K. (1916), Sitzber. Deut. Akad. Wiss. Berlin, Kl. Math.-Phys. Tech., 189.
Shapiro, I.I. (1964), Phys. Rev. Lett. **13**, 789.
Shapiro, I.I. (1968), Phys. Rev. Lett. **20**, 1265.
Shapiro, I.I. (1972), Gen. Rel. Grav. **3**, 175.
Smoot, G.F. et al. (1992), Astrophys. J. **396**, L 1.
Stark, J. (1909), Phys. Z. **10**, 902.
Stokes, G.G. (1845), Phil. Mag. (3) **27**, 9.
Szekeres, G. (1960), Publ. Mat. Debrecen **7**, 285.
Taylor, J. (1992), in Proceedings of the 13th International Conference on General Relativity and Gravitation, Córdoba (Argentina) June–July 1992, eds. R.J. Gleiser, C.N. Kozameh, and O.M. Moreschi, Bristol: IOP Publishing, 1993.
Thirring, H. and Lense, J. (1918), Phys. Z. **19**, 156 (translated in Gen. Rel. Grav. **16** (1984), 711).
Thomas, L.H. (1927), Phil. Mag. **3**, 1.
Tomaschek, R. (1924), Ann. Physik Z. **10**, 185.
Trouton, F.T. and Noble, H.R. (1903), Phil. Trans. **202**, 165.
Vessot, R.F.C. and Levine, M.W. (1979), Gen. Rel. Grav. **10**, 181.
Vessot, R.F.C., Levine, M.W. et al. (1980), Phys. Rev. Lett. **45**, 2081.
Voigt, W. (1887), Goett. Nachr., 41.
Walker, A.G. (1936), Proc. London Math. Soc. **42**, 90.
Walsh, D., Carswell, R.F. and Weymann, R.J. (1979), Nature **279**, 381.
Williams, J.G., Newhall, X.X. and Dickey, J.O. (1996), Phys. Rev. D **53**, 6730.

Index

F